Operating and Testing of Electrical Power Apparatus

Operating and Testing of Electrical Power Apparatus

Contributors :
Julio Posada-Roman,
Jose A. Garcia-Souto, *et al.*

AURIS REFERENCE LTD.
London, UK

Operating and Testing of Electrical Power Apparatus
Contributors : Julio Posada-Roman *and* Jose A. Garcia-Souto, *et al.*

Auris Reference Ltd., UK

www.aurisreference.com

United Kingdom

Copyright 2016

Printed in 2017 for Sale in the Indian Subcontinent

Operating and Testing of Electrical Power Apparatus

ISBN: 978-1-78154-514-0

British Library Cataloguing in Publication Data
A CIP record for this book is available from the British Library

Exclusively distributed by CBS Publishers & Distributors Pvt. Ltd.

Sales & Distribution Rights only for India, Pakistan, Bangladesh, Sri Lanka, Nepal and Bhutan.This book is not to be sold outside these territories.

PREFACE

Electrical perspective it is well known that equipment such as light fittings, switches, motors or virtually any type of electrical apparatus may cause sparking or could generate In addition, static electricity built up on a person, clothing or item may discharge with a spark. In either case a spark could ignite an explosive atmosphere with potentially devastating results. Therefore if an explosive environment is unavoidable in a workplace, suitably rated and protected electrical equipment must be used to prevent the possibility of ignition from a spark, or other source of heat. In addition special precautions such as anti-static clothing and shoes may be necessary to prevent the possibility of a static electrical discharge potentially igniting a flammable gas or combustible dust.

This comprehensive text gives students a strong foundation for an understanding of the behaviour, operation, and testing of electric power apparatus under normal, overload, and fault conditions. It provides up-to-date methods for preventive maintenance, presents logical methods by which the more common troubles may be identified and localized, and recommends emergency repairs that will keep the equipment in operation until it can be scheduled out for service. Also included are outlines of inspection programs that will help ensure safe, efficient, economical, and dependable operation.

This page left intentionally blank.

CONTENTS

LIST OF CONTRIBUTORS

Julio Posada-Roman

Department of Electronics Technology, Optoelectronics and Laser Technology Group, Universidad Carlos III de Madrid, Av. Universidad 30, E-28911 Leganés, Madrid, Spain; E-Mails: jsouto@ing.uc3m.es (J.A.G.-S.); jrserran@ing.uc3m.es (J.R.-S.)

Jose A. Garcia-Souto

Department of Electronics Technology, Optoelectronics and Laser Technology Group, Universidad Carlos III de Madrid, Av. Universidad 30, E-28911 Leganés, Madrid, Spain; E-Mails: jsouto@ing.uc3m.es (J.A.G.-S.); jrserran@ing.uc3m.es (J.R.-S.)

Jesus Rubio-Serrano

Department of Electronics Technology, Optoelectronics and Laser Technology Group, Universidad Carlos III de Madrid, Av. Universidad 30, E-28911 Leganés, Madrid, Spain; E-Mails: jsouto@ing.uc3m.es (J.A.G.-S.); jrserran@ing.uc3m.es (J.R.-S.)

This page left intentionally blank.

Chapter 1

ELECTRICAL SYSTEMS

Electrical systems differ around the world — both in voltage and less critically, frequency. The physical interface (plugs and sockets) are also different and often incompatible. However, travellers with electrical appliances can take a few steps to ensure that they can be safely used at their destination.

UNDERSTAND

Voltage and Frequency

Start by taking a look at the back of the device you want to use. If it says something like "100-240V, 50/60 Hz", it will work anywhere in the world with the right plugs.

Dealing with electricity differences can be daunting, but it actually isn't too hard. There are only two main types of **electric systems** used around the world, with varying **physical connections** :

- 100-127 volt, at 60 hertz frequency (*in general : North and Central Americas, Western Japan*)
- 220-240 volt, at 50 hertz frequency (*in general : the rest of the world, with some exceptions*)

Occasionally, you will find 100-127 volts at 50 Hz, such as in Tokyo, Madagascar, and some Caribbean islands. On the other hand, there's 220-240 volts at 60 Hz, such as in South Korea, Peru, some states of Brazil and Guyana. A few other countries using 60 Hz are internally divided, with 100-127 volts in some locations, and 220-240 volts in others, such as in Brazil, the Philippines, and Saudi Arabia. Be extra careful each time you travel to a new destination within these countries, and ask about the voltage. Be aware of multiphase electrical systems.

If the voltage and frequency for your device is the same as where you are travelling, then you need to worry only about the physical plug. (The small difference between 110V and 120V is within the tolerances of most electrical devices. Likewise for 220V and 240V.)

If the voltage provided by the local supply is not within the range accepted by your device, then you will need a **transformer** or **converter** to convert the voltage. Most travel accessory sources offer them and come with several plug adapters to solve all but the most exotic needs.

Plugs and Adapters

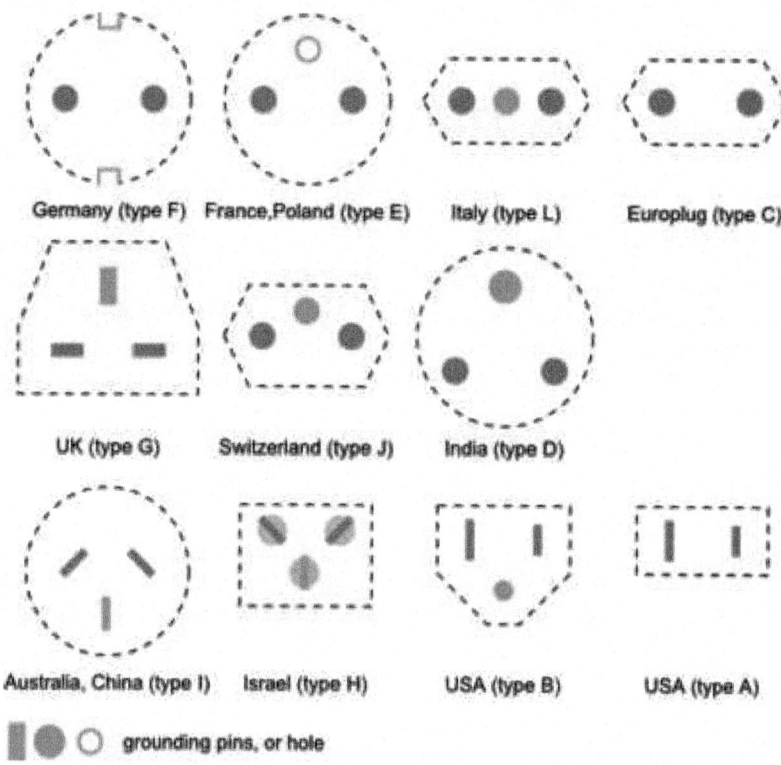

Fig. : Plug types.

A device that lets you insert a plug into a different socket is an **adapter** : these are small, cheap and safe. For example, between Britain and Germany, you need only an adapter. You stick your British plug in the adapter, which connects the rectangular phase/live and neutral prongs to the round German ones and puts the ground where the German outlet expects it. Then, you're good to go.

Unfortunately, there are many different plugs in the world. The three most widespread standards are the following :

- The "American" (Type A or B) plug, with two vertical pins
- The "European" (Type C or F) plug, with two round pins
- The "British" (Type G) plug, with three rectangular pins.

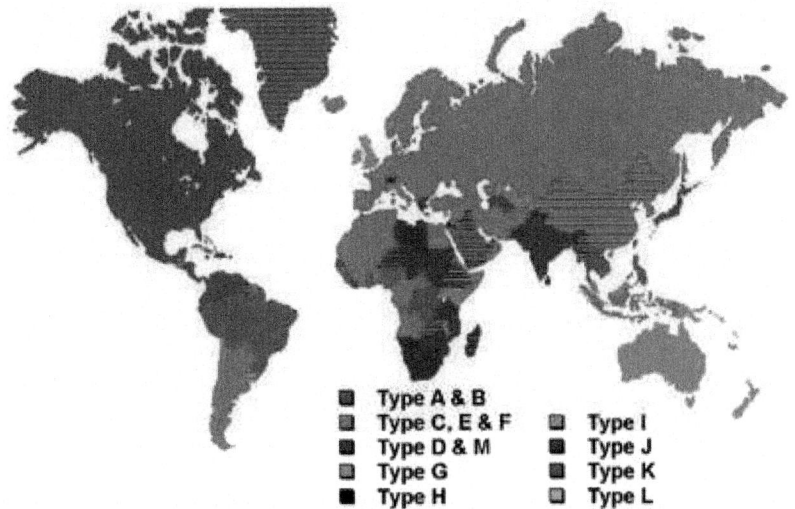

Type A & B
Type C, E & F
Type D & M
Type G
Type H
Type I
Type J
Type K
Type L

Fig. : Map of the world coloured by type of plug used.

If your device has one of these plugs and you can adapt it to the others, you have 90% of the world covered. Adapters between Type A and Type C and *from C to G* are tiny and cheap; converting Type A into G or Type G into anything else, on the other hand, needs a bulkier model.

For hobbyists : if you can't find an adapter, and you're staying for a longer time, just buy a separate plug at your destination, remove the existing plug, and attach the new one. Unlike adapters, plugs are always available, and they're generally cheaper too. Caution : try this only if you know what you're doing! (Fire and/or electrocution are possible if you are inexperienced.)

As a last resort, a Type C plug can be forced into a Type G socket without any converter at all if you ignore what your mother told you and stick a pen or similar pointy object into the center (ground) hole, which fools the socket into thinking a ground pin has been inserted and opens up the other holes. (This is, in fact, exactly what cheap C-to-G adapters do.) Disable the power to the socket and try to use something non-conductive (a dry non-metallic object) to do this! This procedure will damage the socket and could be unlawful in some countries, so expect the owner to be displeased.

There's one more complication to consider : any two-pin socket is **ungrounded**, but all three-pin plugs are **grounded**. Trying to get grounding to work makes life more difficult, as any of sockets C, D, E, F, H, J, K or L will happily accept the ungrounded plug C but will not work with any grounded variant other than their own. Do **not** use an adapter to turn a three-pin into a two-pin : this will disable grounding, potentially leaving you vulnerable to electrocution and other electrical nastiness.

A last word of warning : many developing countries use **multi-plug** sockets that accept (say) both Type A and Type C. *Don't assume the voltage is correct just* *because the plug fits,* since a Thai Type A+C socket still carries 220V and may destroy American (110V) Type A devices.

Transformer or Converter?

Technical details. The difference between a transformer and converter is the way that they deal with the wave-form of the electricity. Converters simply chop the wave in half. This is relatively simple and can be done in a small amount of space, so converters are comparatively light-weight and inexpensive. Transformers alter the length of the wave. This is more complicated and takes up more space : transformers are basically chunks of iron specially-wrapped in wires so they are larger, heavier, and more expensive. Electric appliances can function with either a full or half-sine wave, whereas electronic devices must have a full sine wave.

If you are using a 220V-240V appliance at 110V, you will need a transformer.

If you are using a 110V appliance at 220V-240V, you can also use a transformer but may be able to get away with a (cheaper) converter.

If your device is an electric appliance with a heating element or mechanical motor such as an iron or hair-dryer, then you can probably just use a converter, but *make sure your transformer or converter is fully rated to deliver the amps and watts your device needs.* If your device is *electronic* and uses electronic chips or circuits, such as a computer, printer, TV, microwave, VCR or even a battery charger, you will need a transformer.

Transformers

- They have two different types : **"step-up"** and **"step-down"**. Step-up transformers allow you to *plug a higher-voltage device into a lower-voltage power socket* (such as using a UK device in the US). Step-down transformers allow you to plug *a lower-voltage device into a higher-voltage socket* (such as using a US device in the UK). Some transformers offer both. Take care to use the right type : if you plugged a 110-to-220 V step-up transformer into a 220 V socket, you would get 440 V and a fried device.

- You must also make sure that the **power rating** *(wattage)* of your transformer is at least 10% greater than that of the device; otherwise, the transformer can overheat and even catch fire. Before buying a transformer, look for the "input" figure : usually on the device's plug or in the manual. Some don't display wattage, but you can work it out simply by multiplying the voltage (V) and the current (amps (A); if it is miliamps (mA), divide by 1,000). The resulting figure is the same as the wattage.

- Transformers can be used with both electronic devices (such as those with chips and circuits) or electrical appliances (such as those with heating elements and motors). They can usually operate for a much longer time than converters.

Converters

These lighter-weight, less expensive devices can handle large wattage loads of up to 1600 watts but only step-down voltage, not raise it. They are suitable for those in 110-120V countries travelling to where the voltage is 220-240V. Converters are designed to operate for **only an hour or two at a time**, *not* continuously. As stated above, **they cannot be used with electronic devices** : devices that use chips or circuits, such as a computers, printers, VCRs ,or even battery chargers.

Nowadays, many electronic devices actually come with a converter which plugs into the power mains and converts the current to DC. However, if this won't accept a foreign voltage (check the plug) do **not** place a second converter behind it. You must use a heavier transformer instead. Fortunately, in the past few years, more and more devices come with a universal AC/DC converter already included, and the most you would need is a plug adapter.

Frequency (Hz)

Frequency is generally not a problem-most travel items will work on either 50 or 60 Hz. If all the electrical appliance does is produce heat or light (except fluorescent lighting), then the frequency is unlikely to matter.

Frequency is most likely to affect clocks and devices with motors. They may run faster or slower than they should and may be damaged in the long-run as a result. Again, though, some motorised devices may function correctly on either 50 or 60 Hz-especially if they also operate on batteries. Just look on the label or plug.

However, you still may need to be careful if you have a sensitive or expensive device that converts AC (power from the wall) into DC (battery-like current)-especially if you also need to convert the voltage. A device will convert AC to DC either to save battery power by allowing you to plug into the mains or charge a battery in the device. The design of power supplies where AC is converted into DC does take frequency into account.

Even though 60 Hz converts a little more easily to DC than 50 Hz does, there's enough tolerance in *most* small appliances and electronic gadgets that you can ignore frequency. However, if you also need to change the voltage (because the voltage of your device is different from the mains power voltage), you *cannot* use a switching-type converter. You must use the heavier iron-core transformer. If in doubt, consult a reputable electrical goods dealer.

If your device won't operate with a different frequency (powerful motors and non-quartz clocks), there is really nothing you can do to change it. Unlike voltage, frequency cannot easily be converted. Foreign embassies may have to use huge generators to provide current compatible with equipment from home.

If you desperately need to have power at your home country's frequency, you might try using a 12V DC to AC converter intended for vehicle use. However, most of these (especially those commonly found in stores) output a "sawtooth" wave instead of a sine wave. (Check the manufacturer's website if you need a sine wave output. It may be special order.) Make sure the wattage of the converter is sufficient for whatever device you need to operate, and the 12V battery has enough amps for the job. For example, 12V times 15 amps gives 180 watts (or less after losses are included).

Japan is a special case. East Japan (*e.g.* Tokyo) uses 50 Hz and west Japan (eg Osaka) uses 60 Hz. Equipment made for the Japanese market may have a switch to select 50 Hz or 60 Hz.

Unstable Supply

In many developing countries, electrical supply is highly erratic and you need to take precautions to protect your equipment.

The main danger is **power spikes**, where the voltage supplied temporarily surges to dangerous levels, with potentially catastrophic consequences. In developed countries, the main source of spikes is lightning strikes, but, in developing countries, they're most often associated with **power outages** since when the power comes back on, it rarely does so smoothly. The cheapest method of protection is thus simply to *disconnect electronic devices* as soon as the power goes out and wait a few minutes after the power comes back on until plugging them back in.

Surge protectors are devices designed specially to protect against spikes and surges, and some are available in portable travel-sized versions. Some surge protectors can also be fitted to a telephone line to protect your phone or laptop modem. The most common variety use a **metal oxide varistor** (MOV), which shorts to ground if a given voltage is exceeded. These are easily destroyed by larger spikes, and better models will have a light indicating when the MOV has broken down, but you still need to keep an eye on them as the device will still continue to give power even if the protection is gone. There are also surge protectors with **fuses**, which are fail-safe (a blown fuse will stop power) and replaceable, but there is still a risk of a short, sharp spike which can pass through and damage your device before the fuse blows.

In some (mostly poor) regions, you may experience electricity **voltage drops**. Instead of 240V for example, you may get only 200V or even less (50% of the nominal supply voltage is not unknown). This happens especially if you're at "the end of the line" (far from the source or transformer) and is caused by the resistance of the electric lines themselves. Some appliances, such as light bulbs and heating equipment just keep working under a lower voltage, although a 20% voltage drop will cause a 36% power drop. Most electronic devices also keep working, but voltage drops are critical for fluorescent lamps, refrigerators, and air conditioners which may stop working altogether (usually without being damaged : when the voltage returns to normal, they will start working again).

Voltage drops can be solved with a special device called a **voltage stabiliser** or **AVR** (Automatic Voltage Regulator). A stabiliser will raise the voltage again to its normal level. The principle is the same as for switching converters, except that stabilisers will produce a stable output, even with an unstable input. Stabilisers come in different power ranges, but they're all large, bulky and not practical to carry around. Be aware that some appliances, such as refrigerators, briefly consume 2 or 3 times more power at start-up; the stabiliser should be able to provide this power. Voltage stabilisers can introduce surges if there is a power outage. The cheaper and most common relay type can also damage electronic equipment.

APPLIANCES

If you are buying new appliances, get in the habit of checking the voltage. A dual-voltage hair straightener will cost you no more than a single voltage one, and save considerable hassle when travelling.

Laptop Computers

Virtually all laptop computers (including those with internal power supplies) will handle well a range of 100 to 240 volts and a frequency of 50 to 60Hz. In other words, you won't need a converter/transformer; most power supplies have supported ranges printed directly on them so have a look. You will still need to check that you have the plug that matches the outlet for the country you are going to to see if you need to buy an adaptor.

Laptop computer power supplies are generally very good at accepting a poor or varying supply. Many manufacturers use the same type of supply, so getting spares is not too hard. The type used by HP/Compaq is very common. It is very easy and cheap to get a spare supply from sites such as Ebay. However, make sure it is a genuine manufacturer replacement and not a cheap copy. With a spare, you can take a risk with an unknown supply. Of course, do not take any risks if your laptop is one of the few with an internal supply.

If you are taking a laptop, you can use it to charge other items using a USB port on the laptop, even if they are normally not connected to it - this can save you a bundle of transformers in your luggage. Just make sure you have the correct USB cable - there are many different types.

Radios

Radios also tend to be interchangeable from country-to-country. The exact FM range being used in a few countries is different, so you may not be able to access all stations. In the US, only odd channels (88.1,88.3, 100.1 etc.) are used. A radio intended exclusively for the US market will not work well in most other countries. Japan, in particular, has an FM band from 76 MHz to 90 MHz rather than the more common 87.5 MHz to 108 MHz. The countries of the former Soviet Union have also used a similar band. For the medium wave band, channel spacings (the difference between each valid frequency) can be 9kHz or 10kHz (for USA). Some digital radios will have a switch or setting to choose which channel spacing is used. Without this, they will not work correctly outside their intended market. Old-fashioned analogue dial tuners don't have this limitation.

If you need a new radio for international travel, consider one that includes the **shortwave band** (SW). This way, you can receive news and information from all over the world (BBC, CBC, Voice of America, Radio Australia, etc.) Shortwave is above the medium wave band (in frequency terms), but travels a lot further, especially after dark. In the past few years, the size and price have come down considerably for AM/SW/FM radios, and they are much easier to use. A handful of stations now require the single sideband (SSB) function (normally used by hobbyists for voice communication), but for most people it's not essential.

Digital Radio

Digital radio is in use in some countries, but has generally not attracted large audiences. So radio listening remains a predominantly analogue world, unlike television. The most common systems are DAB (Europe), DAB+, DRM and HD Radio (US). For travellers, an analogue radio (especially those with digital displays) are the best choice.

Mobile Phones and Digital Cameras

Chargers for these may work with both 110V and 240V systems, though you may still need an adaptor plug or have to use the shaver socket. You may be able to get a second charger for the other voltage system, or even a dual voltage charger designed for both systems. However, your mobile phone handset may not be compatible with the country's network, or you may be limited to certain cellular providers.

Equipment Using Standard Batteries

Battery sizes and voltages tend to be standard from place to place, and equipment that uses off-the-shelf batteries tends to be interchangeable. It may be difficult to get good quality batteries in some countries, especially alkaline batteries which are needed by most electronic equipment. If a cheaper battery is used, make sure to remove it as soon as it is exhausted or if the equipment will not be used for a while due to the risk of leakage.

Dual voltage battery chargers for NiCad and NiMH generally cost no more than single voltage ones, but you need to look for this feature before you buy. If an existing single-voltage charger uses a 12 volt DC adapter, find a quality dual-voltage adapter (110V - 240V) at 12 Volts DC with its DC current rating (in milli Amps) equal or higher, and the same size plug on the charger end. (This is not possible if the charger plugs directly into the power mains without any cord.)

BE CAUTIOUS

Large Appliance Power Mains (Single, Split Phase or Three Phase Supply)

In most countries, electrical power is distributed using a three phase system. This means that there are 3 different live/phase wires and optionally a single neutral wire. Domestic outlets are invariably single phase. A domestic outlet will receive just one of these phase wires and a neutral wire. Depending on the country, a mains supply into a residential building may be a single phase supply or three phase supply. Most larger buildings will receive a three phase supply.

There are good reasons why electricity networks use 3 phase supplies. They are also most appropriate for running large machines with motors (large air conditioners, industrial/commercial ovens and other power hungry appliances). The voltage between any of the three phases and neutral is the same as the domestic single phase outlet voltage (110V, 230V, 240V etc). However, the voltage between two phase wires is typically 380-415V or 208V in a 110V single phase system.

In North America there is a further variation, the split phase system. In the split phase system, two phase wires are used, but the voltage between these wires is double the single phase voltage - 240V. These systems are used to run large appliances such as cookers. However, it is also not uncommon to use two of the phase wires from the three phase system to give 208V.

In some places (mostly industrial buildings) three phase outlets may be available (often coloured red). Adapter cables or plugs can be used which take one of the phase connections and neutral for standard single phase appliances. Regardless of how many phases are in use, the frequency (50Hz or 60Hz) of the supply remains unchanged.

In countries with a poor distribution network, it is not unknown for single phase sockets to be connected across 2 phases to boost the voltage. This is dangerous and can damage electronic equipment. As a general rule, do not attempt to connect your personal electrical items to a socket or wiring system that is a three phase system.

Generators

In many countries without fully developed electrical power distribution systems, the use of generators is common. Generator supplies can be very good; however, in many places they are bad and can cause damage to sensitive equipment connected. The voltage, frequency, and waveform shape (it should be a smooth sine-wave) can vary. In some places, people modify generators to run faster. This gives more voltage and power but increases the frequency too. The part of a generator that keeps it running at a constant speed is called the governor. If this is tampered with, the output voltage could rise sufficiently to cause damage. The best advice is not to connect valuable equipment to the supply, or at the very least disconnect it as soon as you're finished.

If you are unsure about the quality of the generator in use, there are a few simple rules. If it runs from petrol/gasoline, it is bad : anyone serious about using generator power uses a diesel oil powered system. A good quality generator will have a low engine speed. 1500 RPM for 50Hz or 1800 RPM for 60 Hz. If the engine speed is 3000 RPM or more, it is not a good machine.

Lamps

Lamps and their light bulbs are very sensitive to voltage. If you shift between voltage systems, you will need to change the light bulbs to match the voltage, unless the lamp is designed to operate on both systems, say through a low voltage adaptor. If you buy a lamp abroad, you may need to have an electrician completely rewire a lamp when you get home to comply with your country's electrical safety standards. This may not be a problem for a one-off special item, but if you are going into the importing business it could be a showstopper.

Also watch out for the light bulb connection. In 100-127V systems this is often a screw connector while in 220-240V systems it is often a bayonet connector. These connectors also come in at least two different sizes. Be sure you can obtain light bulbs of the right voltage, size, and connector shape in the country you intend to use the lamp, and at a reasonable price, or the lamp may become little more than junk when the bulb fails.

Note that fluorescent and LED lighting contains electronics and must use a heavy iron-core transformer to convert voltage. Converters are **not** acceptable. Some fluorescent units might be sensitive to changes in frequency (50 or 60 Hz) if it's not the same as what is specified. This type of lighting has its own "flickering" frequency, which is suppose to be too rapid to notice. However, old and defective units often produce an annoying, visible flicker, and the wrong electrical frequency might have the same effect as well.

Electric Motors

The electric motors in things like refrigerators, vacuum cleaners, washing machines and other whiteware are often sensitive to frequency. Older hairdryers and electric shavers might be also. Even if you use a step-up or step-down transformer, the different supply frequencies mean that motors will run at the wrong speed and quickly burn out. The larger and more powerful the motor is, the more this is true. Don't, for example, bring a vacuum cleaner from the US to Europe (or *vice versa*). It's almost guaranteed to fail-even if you have a voltage converter.

Electric Shavers

Hotels often provide a special electrical outlet specifically for electric shavers. They allow any voltage shaver to be plugged into them and be used safely in front of the bathroom mirror. They may also accept your cellphone adaptor or similar low power battery charging unit. Many-**but not all**-electric shavers sold today are dual voltage 50/60Hz and some will even recharge the battery at 12V DC (such as in an automobile). Check the label and instructions for compatibility.

Hairdryers

Hairdryers are a particular risk; if you accidentally plug your 100-120V hairdryer into a 240V outlet. you may find it catching fire in your hands! Newer models should have a thermal switch, though. Allow 15-20 mins for it to cool down, then use a voltage converter (if the dryer is 50 Hz compatible). Similarly, a 220-240V hairdryer in a 120V outlet may run slowly and not heat up enough. Most good hotels and motels will be able to supply a hairdryer, and it may even be a room fitting. However, it may be worthwhile buying or borrowing a hairdryer suited for the electrical system of countries you'll be traveling in.

Many new hairdryers sold in 100-120V countries are dual voltage with settings for 100-120V and 220-240V. Even though it's motorized, it will work on either 50 or 60 Hz. Don't forget to lockout the high setting with a flat screwdriver or something similar. At 220-240V, the low setting becomes as powerful as the high setting was at home (with 'low' unavailable).

Clocks

An electric clock of any sort is sensitive to voltage. If the voltage is doubled or halved, it will not function and may burn out. Furthermore, the electric frequency (50 or 60 Hz) is used in cheap clocks (such as many clock-radio style clocks) to keep the time. Thus, if a clock made for North America were used in Europe – even with a voltage adapter – it would lose 10 min/h! Obviously, not a great idea if you have a train to catch. On the other hand, if the clock has a quartz crystal, this is used for the timekeeping, and it operates independently of the line frequency. Inexpensive, battery-operated, digital LCD travel clocks (with a push button back light) are also available. These are recommended for destinations with frequent blackouts.

Video Equipment

Televisions, many radios, video and DVD players, as well as videotapes, are often specific to the broadcast system used in the country that they are sold in, usually associated with the frequency of the country's electric current. For example, North America is 60 Hz and its television is 30 frames per second, while Europe is 50 Hz and its television is 25 frames per second. The main three analogue television broadcast systems are **PAL**, the closest to a worldwide standard, **NTSC**, used mostly in the Americas and some East Asian countries and **SECAM**, originally from France and adopted by much of Eastern Europe and the Middle East, but there are various incompatibilities even within these supposed standards. There is no difference between PAL and SECAM for unconverted *DIGITAL* video including DVDs. However, any analog output to a television set would be in the native format of the country of location. Brazil uses a hybrid PAL/NTSC standard called

"PAL-M". In Brazil, DVDs and video tape are the same as NTSC, but all players and TV sets are useless outside the country unless they have a separate NTSC setting.

Before purchasing any video equipment, read the manual and warranty carefully. For TVs and VCRs, don't forget about cable television frequencies; they may not be the same, even if everything else is. Television sets often won't work correctly in another country from where they were sold, even if the voltage and video standard are the same. For example, a television set made for the USA will skip a few channels in Japan. Furthermore, many countries have or are in the process of switching to digital over-the-air broadcasting, (dates by country). Unless you have an internationally compatible device, you may find your expensive looking system is little more than worthless junk in another country because it won't work with your country's broadcast system. Your warranty is probably only valid in the country of purchase, and you may need to return the goods to the place you purchased them from.

The final problem with transporting TVs is that many European countries, notoriously the UK, require a license to watch any live TV (over-the-air, cable, satellite, and even live-streams on the internet). Fines can be hefty (in addition to being charged for the license).

DVD and Blu-Ray, infuriatingly, have completely artificial limitations introduced in the form of **region coding**, which attempts to limit the region where the discs can be used, as a technique to keep the various regions as separate markets. For example, a Region 1 player for North America will not play a Region 3 DVD for Hong Kong. The workarounds are to obtain either a regionless DVD player which ignores the code, purchase multi-region discs (Regions 1 and 3 in this case), or better yet, Region 0 discs, which can be played on any device. Blu-Ray discs **cannot** be played at all in a standard DVD player - not even at a lower resolution. However, Blu-Ray discs **played on a Blu-Ray player** can be displayed on a standard def. television, provided you have the correct cables and connections. (HDMI cables are *not* compatible.)

Technically, there is no such thing as an NTSC or PAL DVD disc, as all colour information is the same for both. When discs are labelled as such, what they're referring to is the picture size and frame rate (*i.e.* number of frames per second) that are used in **most** (but not all !) countries that have TV broadcasts on this same system. Many NTSC players cannot play PAL DVDs, unless that's a specific feature included (many Philips and JVC models include this). PAL DVD players are generally much better at playing NTSC, but it's not a certainty. If all else fails, a computer DVD-ROM can play any DVD movie, though there's a limit on how many times you can change the region code. Unlike analogue television sets, computer monitors can automatically handle both 25 (PAL and SECAM) and 30 (NTSC) frames per second, as well as various picture sizes. This also applies to LCD and plasma "flat panel" television sets, but don't expect their tuner to be compatible outside the country in which they were sold.

Video cameras can usually be charged with both electrical systems so you can record during travels and view it back home. Digital cameras and video cameras can usually output to both **PAL**, **NTSC**, and **SECAM**, so you can view your recording while travelling. Bring an RCA (yellow plug) to SCART adaptor if you plan to view video from a camcorder on a European television set.

If you have something on VHS video tape, it's best to convert to DVD before travelling. (Conversions between PAL and NTSC can be done before burning.) Use a video capture card for recording the VHS into a digital file on your computer. Then with DVD-making software, burn the file to a blank DVD.

Note that to be playable on a television set using a connected DVD player, a burned disc **must** be in the native DVD format (same as Hollywood movies) with the "AUDIO_TS" and "VIDEO_TS" folders. If you burn a Windows media, Powerpoint, Quicktime, Adobe flash, etc., file to a DVD, it can **only** be played back on another computer. This may be totally inadequate for a presentation. Unless your company or organization is already equipped, locating a computer video projector in a foreign country can be a challenge. Travelling with one is not recommended either, as they are expensive, fragile, and somewhat bulky and heavy. Exception : many newer DVD players can play "JPG" still picture files as a slideshow. Some even have an SD card slot, so you can view your photos taken from a digital camera. Caution : **NEVER** computer-edit anything directly on a photo-media card (SD, CF, Sony memory stick, etc.) Copy it to the hard drive or a USB jump drive first, then edit.

If required, converting DVDs from one format to another (PAL, NTSC), can be done on a computer with a fast CPU, or you can get it done professionally. Allow plenty of time, as this can take many hours. Regular blank discs work fine for making **copies** of a foreign format, as it's all just a bunch of ones and zeros and no different than copying anything else. Copies can be made quickly, while conversions cannot.

Stay Safe

The electrical engineer's maxim : The smoke that escapes from a device or a component is its spirit without which it cannot work. In other words : if smoke rises from the device, then it's destroyed.

The first time you use electrical equipment on a voltage system you haven't used before, watch for excessive heat, strange smells, and smoke. This is especially true for those residing in countries with 120V (USA, Canada, Japan, etc.) visiting places with the higher voltage. Smoke is a sure sign your equipment cannot cope with the voltage system.

If your electrical equipment gets very hot, smells of burning (there is a distinct smell of electrically fried circuit boards) or starts to smoke, turn it off at the wall or the main switch immediately, then carefully unplug the equipment. Do not disconnect or unplug by just grabbing the smoking device, its plug or cord, and then unplugging it, as these parts are probably very hot, and the insulation could be melted or unsafe, which could result in electrocution.

You may find your expensive equipment has been fried and needs to be replaced because the wrong voltage was used. However, if the equipment only got hot and did not smoke or produce strange burning smells you may be lucky. Some older devices have fuses that you may be able to replace. New devices, such as gaming consoles, will trip a circuit breaker. Disconnect them from all power and leave them for 60 minutes or so, and the circuit breaker will normally reset.

Do not rely on fuses to protect your equipment. If a fuse does blow, you should have things checked by an electrician before using the suspect equipment again.

In Third World countries with frequent blackouts, it's not at all uncommon for a visitor to plug something in and have the power go out coincidentally. Always check the neighbourhood first, before blaming the appliance or looking at the fuse/circuit breaker.

Chapter 2

INSULATOR (ELECTRICITY)

An **electrical insulator** is a material whose internal electric charges do not flow freely, and therefore make it very hard to conduct an electric current under the influence of an electric field. A perfect insulator does not exist, but some materials such as glass, paper and Teflon, which have high resistivity, are very good electrical insulators. A much larger class of materials, even though they may have lower bulk resistivity, are still good enough to insulate electrical wiring and cables. Examples include rubber-like polymers and most plastics. Such materials can serve as practical and safe insulators for low to moderate voltages (hundreds, or even thousands, of volts).

Insulators are used in electrical equipment to support and separate electrical conductors without allowing current through themselves. An insulating material used in bulk to wrap electrical cables or other equipment is called *insulation*. The term *insulator* is also used more specifically to refer to insulating supports used to attach electric power distribution or transmission lines to utility poles and transmission towers. They support the weight of the suspended wires without allowing the current to flow through the tower to ground.

PHYSICS OF CONDUCTION IN SOLIDS

Electrical insulation is the absence of electrical conduction. Electronic band theory (a branch of physics) says that a charge flows if states are available into which electrons can be excited. This allows electrons to gain energy and thereby move through a conductor such as a metal. If no such states are available, the material is an insulator.

Most insulators have a large band gap. This occurs because the "valence" band containing the highest energy electrons is full, and a large energy gap separates this band from the next band above it. There is always some voltage (called the breakdown voltage) that gives electrons enough energy to be excited into this band. Once this voltage is exceeded the material ceases being an insulator, and charge begins to pass through it. However, it is usually accompanied by physical or chemical changes that permanently degrade the material's insulating properties.

Materials that lack electron conduction are insulators if they lack other mobile charges as well. For example, if a liquid or gas contains ions, then the ions can be made to flow as an electric current, and the material is a conductor. Electrolytes and plasmas contain ions and act as conductors whether or not electron flow is involved.

BREAKDOWN

When subjected to a high enough voltage, insulators suffer from the phenomenon of electrical breakdown. When the electric field applied across an insulating substance exceeds in any location the threshold breakdown field for that substance, the insulator suddenly becomes a conductor, causing a large increase in current, an electric arc through the substance. Electrical breakdown occurs when the electric field in the material is strong enough to accelerate free charge carriers (electrons and ions, which are always present at low concentrations) to a high enough velocity to knock electrons from atoms when they strike them, ionizing the atoms. These freed electrons and ions are in turn accelerated and strike other atoms, creating more charge carriers, in a chain reaction. Rapidly the insulator becomes filled with mobile charge carriers, and its resistance drops to a low level. In a solid, the breakdown voltage is proportional to the band gap energy. The air in a region around a high-voltage conductor can break down and ionise without a catastrophic increase in current; this is called "corona discharge". However if the region of air breakdown extends to another conductor at a different voltage it creates a conductive path between them, and a large current flows through the air, creating an *electric arc*. Even a vacuum can suffer a sort of breakdown, but in this case the breakdown or vacuum arc involves charges ejected from the surface of metal electrodes rather than produced by the vacuum itself. In case of some insulators, the conduction may take place at a very high temperature as then the energy acquired by the valence electrons is sufficient to take them into conduction band.

USES

A flexible coating of an insulator is often applied to electric wire and cable, this is called *insulated wire*. Since air is an insulator, in principle no other substance is needed to keep power where it should be. High-voltage power lines commonly use just air, since a solid (*e.g.*, plastic) coating is impractical. However, wires that touch each other produce cross-connections, short circuits, and fire hazards. In coaxial cable the center conductor must be supported exactly in the middle of the hollow shield in order to prevent EM wave reflections. Finally, wires that expose voltages higher than 60V can cause human shock and electrocution hazards. Insulating coatings help to prevent all of these problems.

Some wires have a mechanical covering with no voltage rating — *e.g.* : service-drop, welding, doorbell, thermostat wire. An insulated wire or cable has a voltage rating and a maximum conductor temperature rating. It may not have an ampacity (current-carrying capacity) rating, since this is dependent upon the surrounding environment (*e.g.* ambient temperature).

In electronic systems, printed circuit boards are made from epoxy plastic and fibreglass. The non-conductive boards support layers of copper foil conductors. In electronic devices, the tiny and delicate active components are embedded within non-conductive epoxy or phenolic plastics, or within baked glass or ceramic coatings.

In microelectronic components such as transistors and ICs, the silicon material is normally a conductor because of doping, but it can easily be selectively transformed into a good insulator by the application of heat and oxygen. Oxidised silicon is quartz, *i.e.* silicon dioxide, the primary component of glass.

In high voltage systems containing transformers and capacitors, liquid insulator oil is the typical method used for preventing arcs. The oil replaces air in spaces that must support significant voltage without electrical breakdown. Other high voltage system insulation materials include ceramic or glass wire holders, gas, vacuum, and simply placing wires enough far apart to use air as insulation.

TELEGRAPH AND POWER TRANSMISSION INSULATORS

Overhead conductors for high-voltage electric power transmission are bare, and are insulated by the surrounding air. Conductors for lower voltages in distribution may have some insulation but are often bare as well. Insulating supports called *insulators* are required at the points where they are supported by utility poles or transmission towers. Insulators are also required where the wire enters buildings or electrical devices, such as transformers or circuit breakers, to insulate the wire from the case. These hollow insulators with a conductor inside them are called bushings.

MATERIAL

Insulators used for high-voltage power transmission are made from glass, porcelain or composite polymer materials. Porcelain insulators are made from clay, quartz or alumina and feldspar, and are covered with a smooth glaze to shed water. Insulators made from porcelain rich in alumina are used where high mechanical strength is a criterion. Porcelain has a dielectric strength of about 4–10 kV/mm. Glass has a higher dielectric strength, but it attracts condensation and the thick irregular shapes needed for insulators are difficult to cast without internal strains. Some insulator manufacturers stopped making glass insulators in the late 1960s, switching to ceramic materials.

Recently, some electric utilities have begun converting to polymer composite materials for some types of insulators. These are typically composed of a central rod made of fibre reinforced plastic and an outer weathershed made of silicone rubber or ethylene propylene diene monomer rubber (EPDM). Composite insulators are less costly, lighter in weight, and have excellent hydrophobic capability. This combination makes them ideal for service in polluted areas. However, these materials do not yet have the long-term proven service life of glass and porcelain.

DESIGN

The electrical breakdown of an insulator due to excessive voltage can occur in one of two ways :

- A *puncture arc* is a breakdown and conduction of the material of the insulator, causing an electric arc through the interior of the insulator. The heat resulting from the arc usually damages the insulator irreparably. *Puncture voltage* is the voltage across the insulator (when installed in its normal manner) that causes a puncture arc.

- A *flashover arc* is a breakdown and conduction of the air around or along the surface of the insulator, causing an arc along the outside of the insulator. They are usually designed to withstand this without damage. *Flashover voltage* is the voltage that causes a flash-over arc.

Most high voltage insulators are designed with a lower flashover voltage than puncture voltage, so they flash over before they puncture, to avoid damage.

Dirt, pollution, salt, and particularly water on the surface of a high voltage insulator can create a conductive path across it, causing leakage currents and flashovers. The flashover voltage can be more than 50% lower when the insulator is wet. High voltage insulators for outdoor use are shaped to maximise the length of the leakage path along the surface from one end to the other, called the creepage length, to minimise these leakage currents. To accomplish this the surface is moulded into a series of corrugations or concentric disc shapes. These usually include one or more *sheds*; downward facing cup-shaped surfaces that act as umbrellas to ensure that the part of the surface leakage path under the 'cup' stays dry in wet weather. Minimum creepage distances are 20–25 mm/kV, but must be increased in high pollution or airborne sea-salt areas.

TYPES OF INSULATORS

These are the common classes of insulator :

- *Pin type insulator* - As the name suggests, the pin type insulator is mounted on a pin on the cross-arm on the pole. There is a groove on the upper end of the insulator. The conductor passes through this groove and is tied to the insulator with annealed wire of the same material as the conductor. Pin type insulators are used for transmission and distribution of electric power at voltages up to 33 kV. Beyond operating voltage of 33 kV, the pin type insulators become too bulky and hence uneconomical.

- *Suspension insulator* - For voltages greater than 33 kV, it is a usual practice to use suspension type insulators shown in Figure. Consist of a number of porcelain discs connected in series by metal links in the form of a string. The conductor is suspended at the bottom end of this string while the other end of the string is secured to the cross-arm of the tower. The number of disc units used depends on the voltage.

- *Strain insulator* - A *dead end* or *anchor* pole or tower is used where a straight section of line ends, or angles off in another direction. These poles must withstand the lateral (horizontal) tension of the long straight section of wire. In order to support this lateral load, strain insulators are used. For low voltage lines (less than 11 kV), shackle insulators are used as strain insulators. However, for high voltage transmission lines, strings of cap-and-pin (disc) insulators are used, attached to the cross-arm in a horizontal direction. When the tension load in lines is exceedingly high, such as at long river spans, two or more strings are used in parallel.

- *Shackle insulator* - In early days, the shackle insulators were used as strain insulators. But now a day, they are frequently used for low voltage distribution lines. Such insulators can be used either in a horizontal position or in a vertical position. They can be directly fixed to the pole with a bolt or to the cross arm.

- *Line post insulator*

- *Station post insulator*

- *Cut-out.*

CAP AND PIN INSULATORS

Higher voltage transmission lines usually use modular *cap and pin* insulator designs *(pictures, left)*. The wires are suspended from a 'string' of identical disc-shaped insulators that attach to each other with metal clevis pin or ball and socket links. The advantage of this design is that insulator strings with different breakdown voltages, for use with different line voltages, can be constructed by using different numbers of the basic units. Also, if one of the insulator units in the string breaks, it can be replaced without discarding the entire string.

Each unit is constructed of a ceramic or glass disc with a metal cap and pin cemented to opposite sides. In order to make defective units obvious, glass units are designed with Class B construction, so that an overvoltage causes a puncture arc through the glass instead of a flashover. The glass is heat-treated so it shatters, making the damaged unit visible. However the mechanical strength of the unit is unchanged, so the insulator string stays together.

Standard disc insulator units are 25 centimetres (9.8 in) in diameter and 15 cm (6 in) long, can support a load of 80-120 kN (18-27 klbf), have a dry flashover voltage of about 72 kV, and are rated at an operating voltage of 10-12 kV. However, the flashover voltage of a string is less than the sum of its component discs, because the electric field is not distributed evenly across the string but is strongest at the disc nearest to the conductor, which flashes over first. Metal *grading rings* are sometimes added around the disc at the high voltage end, to reduce the electric field across that disc and improve flashover voltage.

In very high voltage lines the insulator may be surrounded by corona rings. These typically consist of toruses of aluminum (most commonly) or copper tubing attached to the line. They are designed to reduce the electric field at the point where the insulator is attached to the line, to prevent corona discharge, which results in power losses.

Typical number of disc insulator units for standard line voltages	
Line voltage (kV)	Discs
34.5	3
46	4
69	5
92	7
115	8
138	9
161	11
196	13
230	15
287	19
345	22
360	23

Fig. : A recent photo of an open wire telegraph pole route with porcelain insulators. Quidenham, Norfolk, United Kingdom.

HISTORY

The first electrical systems to make use of insulators were telegraph lines; direct attachment of wires to wooden poles was found to give very poor results, especially during damp weather.

The first glass insulators used in large quantities had an unthreaded pinhole. These pieces of glass were positioned on a tapered wooden pin, vertically extending upwards from the pole's cross-arm (commonly only two insulators to a pole and maybe one on top of the pole itself). Natural contraction and expansion of the wires tied to these "threadless insulators" resulted in insulators unseating from their pins, requiring manual reseating.

Amongst the first to produce ceramic insulators were companies in the United Kingdom, with Stiff and Doulton using stoneware from the mid-1840s, Joseph Bourne (later renamed Denby) producing them from around 1860 and Bullers from 1868. Utility patent number 48,906 was granted to Louis A. Cauvet on July 25, 1865 for a process to produce insulators with a threaded pinhole. To this day, pin-type insulators still have threaded pinholes.

The invention of suspension-type insulators made high-voltage power transmission possible. Pin-type insulators were unsatisfactory over about 60,000 volts.

A large variety of telephone, telegraph and power insulators have been made; some people collect them, both for their historic interest and for the aesthetic quality of many insulator designs and finishes. The National Insulator Association (NIA) is currently the largest insulator collecting community (with over 9000 members as of 2013). National insulator shows have been held by the NIA since its formation in 1973.

INSULATION OF ANTENNAS

Often a broadcasting radio antenna is built as a mast radiator, which means that the entire mast structure is energised with high voltage and must be insulated from the ground. Steatite mountings are used. They have to withstand not only the voltage of the mast radiator to ground, which can reach values up to 400 kV at some antennas, but also the weight of the mast construction and dynamic forces. Arcing horns and lightning arresters are necessary because lightning strikes to the mast are common.

Guy wires supporting antenna masts usually have strain insulators inserted in the cable run, to keep the high voltages on the antenna from short circuiting to ground or creating a shock hazard. Often guy cables have several insulators, placed to break up the cable into lengths that are not sub-multiples of the transmitting wavelength to avoid unwanted electrical resonances in the guy. These insulators are usually ceramic and cylindrical or egg-shaped. This construction has the advantage that the ceramic is under compression rather than tension, so it can withstand greater load, and that if the insulator breaks, the cable ends are still linked.

These insulators also have to be equipped with overvoltage protection equipment. For the dimensions of the guy insulation, static charges on guys have to be considered. At high masts these can be much higher than the voltage caused by the transmitter, requiring guys divided by insulators in multiple sections on the highest masts. In this case, guys which are grounded at the anchor basements via a coil - or if possible, directly - are the better choice.

Feedlines attaching antennas to radio equipment, particularly twin lead type, often must be kept at a distance from metal structures. The insulated supports used for this purpose are called *standoff insulators*.

INSULATION IN ELECTRICAL APPARATUS

The most important insulation material is air. A variety of solid, liquid, and gaseous insulators are also used in electrical apparatus. In smaller transformers, generators, and electric motors, insulation on the wire coils consists of up to four thin layers of polymer varnish film. Film insulated **magnet wire** permits a manufacturer to obtain the maximum number of turns within the available space. Windings that use thicker conductors are often wrapped with supplemental fiber glass insulating tape. Windings may also be impregnated with insulating varnishes to prevent electrical corona and reduce magnetically induced wire vibration. Large power transformer windings are still mostly insulated with paper, wood, varnish, and mineral oil; although these materials have been used for more than 100 years, they still provide a good balance of economy and adequate performance. Busbars and circuit breakers in switchgear may be insulated with glass-reinforced plastic insulation, treated to have low flame spread and to prevent tracking of current across the material.

In older apparatus made up to the early 1970s, boards made of compressed asbestos may be found; while this is an adequate insulator at power frequencies, handling or repairs to asbestos material can release dangerous fibers into the air and must be carried cautiously. Wire insulated with felted asbestos was used in high-temperature and rugged applications from the 1920s. Wire of this type was sold by General Electric under the trade name "Deltabeston."

Live-front switchboards up to the early part of the 20th century were made of slate or marble. Some high voltage equipment is designed to operate within a high pressure insulating gas such as sulfur hexafluoride. Insulation materials that perform well at power and low frequencies may be unsatisfactory at radio frequency, due to heating from excessive dielectric dissipation.

Electrical wires may be insulated with polyethylene, cross-linked polyethylene (either through electron beam processing or chemical cross-linking), PVC, Kapton, rubber-like polymers, oil impregnated paper, Teflon, silicone, or modified ethylene tetrafluoroethylene (ETFE). Larger power cables may use compressed inorganic powder, depending on the application.

Flexible insulating materials such as PVC (polyvinyl chloride) are used to insulate the circuit and prevent human contact with a 'live' wire – one having voltage of 600 volts or less. Alternative materials are likely to become increasingly used due to EU safety and environmental legislation making PVC less economic.

APPLIANCE CLASSES

In the electrical appliance manufacturing industry, the following **IEC protection classes** are used to differentiate between the protective-earth connection requirements of devices.

Class 0

These appliances have no protective-earth connection and feature only a single level of insulation and were intended for use in dry areas. A single fault could cause an electric shock or other dangerous occurrence. Sales of these items have been banned in the UK since 1975.

Class I

Fig. : green/yellow ground.

Fig. : Class I symbol.

These appliances must have their chassis connected to electrical earth (US : ground) by a separate earth conductor (coloured green/yellow in most countries, green in the US, Canada and Japan). The earth connection is achieved with a 3-conductor mains cable, typically ending with 3-prong AC connector which plugs into a corresponding AC outlet. The basic requirement is that no single failure can result in dangerous voltage becoming exposed so that it might cause an electric shock and that if a fault occurs the supply will be removed automatically (this is sometimes referred to as ADS = Automatic Disconnection of Supply).

A fault in the appliance which causes a live conductor to contact the casing will cause a current to flow in the earth conductor. If large enough, this current will trip an over-current device (fuse or circuit breaker (CB)) and disconnect the supply. The disconnection time has to be fast enough not to allow fibrillation to start if a person is in contact with the casing at the time. This time and the current rating in turn sets a maximum earth resistance permissible. To provide supplementary protection against high-impedance faults it is common to recommend a residual-current device (RCD) also known as a residual current circuit breaker (RCCB), ground fault circuit interrupter (GFCI), or residual current operated circuit-breaker with integral over-current protection (RCBO), which will cut off the supply of electricity to the appliance if the currents in the two poles of the supply are not equal and opposite.

Class 0I

Electrical installations where the chassis is connected to earth with a separate terminal.

Class II

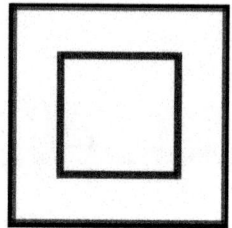

Fig. : Class II symbol.

A Class II or **double insulated** electrical appliance is one which has been designed in such a way that it does not require a safety connection to electrical earth (ground).

The basic requirement is that no single failure can result in dangerous voltage becoming exposed so that it might cause an electric shock and that this is achieved without relying on an earthed metal casing. This is usually achieved at least in part by having two layers of insulating material surrounding live parts or by using reinforced insulation.

In Europe, a double insulated appliance must be labelled *Class II, double insulated,* or bear the double insulation symbol (a square inside another square).

Insulated AC/DC power supplies (such as cell-phone chargers) are typically designated as Class II, meaning that the DC output wires are isolated from the AC input. The designation "Class II" should not be confused with the designation "Class 2", as the latter is unrelated to insulation (it originates from standard UL 1310, setting limits on maximum output voltage/current/power).

Class III

Fig. : Class III symbol.

A Class III appliance is designed to be supplied from a separated/safety extra-low voltage (SELV) power source. The voltage from a SELV supply is low enough that under normal conditions a person can safely come into contact with it without risk of electrical shock. The extra safety features built into Class I and Class II appliances are therefore not required. For medical devices, compliance with Class III is *not* considered sufficient protection, and further more-stringent regulations apply to such equipment.

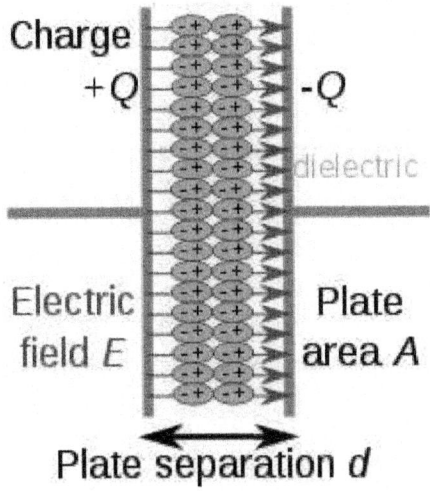

Fig. : A polarized dielectric material.

DIELECTRIC

A **dielectric material** (**dielectric** for short) is an electrical insulator that can be polarized by an applied electric field. When a dielectric is placed in an electric field, electric charges do not flow through the material as they do in a conductor, but only slightly shift from their average equilibrium positions causing **dielectric polarization**. Because of dielectric polarization, positive charges are displaced toward the field and negative charges shift in the opposite direction. This creates an internal electric field that reduces the overall field within the dielectric itself. If a dielectric is composed of weakly bonded molecules, those molecules not only become polarized, but also reorient so that their symmetry axis aligns to the field.

The study of dielectric properties concerns storage and dissipation of electric and magnetic energy in materials. It is important to explain various phenomena in electronics, optics, and solid-state physics.

Terminology

While the term *insulator* implies low electrical conduction, *dielectric* typically means materials with a high polarizability. The latter is expressed by a number called the relative permittivity (also known in older texts as dielectric constant). The term insulator is generally used to indicate electrical obstruction while the term dielectric is used to indicate the energy storing capacity of the material (by means of polarization). A common example of a dielectric is the electrically insulating material between the metallic plates of a capacitor. The polarization of the dielectric by the applied electric field increases the capacitor's surface charge.

The term "dielectric" was coined by William Whewell (from "dia-electric") in response to a request from Michael Faraday. A *perfect dielectric* is a material with zero electrical conductivity. (cf. perfect conductor), thus exhibiting only a displacement current; therefore it stores and returns electrical energy as if it were an ideal capacitor.

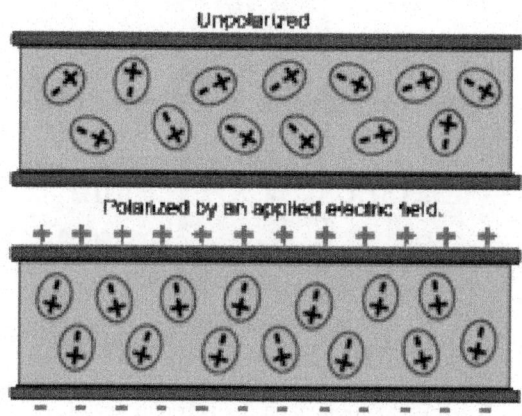

Fig. : A dielectric medium showing orientation of charged particles creating polarization effects. Such a medium can have a higher ratio of electric flux to charge (permittivity) than empty space.

Permittivity

In electromagnetism, **absolute permittivity** is the measure of the resistance that is encountered when forming an electric field in a medium. In other words, permittivity is a measure of how an electric field affects, and is affected by, a dielectric medium. The permittivity of a medium describes how much electric field (more correctly, flux) is 'generated' per unit charge in that medium. More electric flux exists in a medium with a high permittivity (per unit charge) because of polarization effects. Permittivity is directly related to electric susceptibility, which is a measure of how easily a dielectric polarizes in response to an electric field. Thus, permittivity relates to a material's ability to transmit (or "permit") an electric field.

In SI units, permittivity ε is measured in farads per meter (F/m); electric susceptibility χ is dimensionless. They are related to each other through

$$\varepsilon = \varepsilon_r \varepsilon_0 = (1 + x)\, \varepsilon_0$$

where ε_r is the relative permittivity of the material, and $\varepsilon_0 = 8.8541878176.. \times 10^{-12}$ F/m is the vacuum permittivity.

Explanation

In electromagnetism, the electric displacement field **D** represents how an electric field **E** influences the organization of electric charges in a given medium, including charge migration and electric dipole reorientation. Its relation to permittivity in the very simple case of *linear, homogeneous, isotropic* materials with *"instantaneous" response* to changes in electric field is

$$\mathbf{D} = \varepsilon \, \mathbf{E}$$

where the permittivity ε is a scalar. If the medium is anisotropic, the permittivity is a second rank tensor.

In general, permittivity is not a constant, as it can vary with the position in the medium, the frequency of the field applied, humidity, temperature, and other parameters. In a non-linear medium, the permittivity can depend on the strength of the electric field. Permittivity as a function of frequency can take on real or complex values.

In SI units, permittivity is measured in farads per meter (F/m or $A^2 \, s^4 \, kg^{-1} \, m^{-3}$). The displacement field **D** is measured in units of coulombs per square meter (C/m^2), while the electric field **E** is measured in volts per meter (V/m). **D** and **E** describe the interaction between charged objects. **D** is related to the *charge densities* associated with this interaction, while **E** is related to the *forces* and *potential differences*.

Vacuum Permittivity

The vacuum permittivity ε_0 (also called **permittivity of free space** or the **electric constant**) is the ratio **D/E** in free space. It also appears in the Coulomb force constant, $k_e = 1/(4\pi\varepsilon_0)$.

Its value is

$$\varepsilon_0 \overset{\text{def}}{=} \frac{1}{c_0^2 \mu_0} = \frac{1}{35950207149.4727056\pi} \frac{F}{m} \approx 8.8541878176... \times 10^{-12} \; F/m$$

where :

  c_0 is the speed of light in free space,

  μ_0 is the vacuum permeability.

Constants c_0 and μ_0 are defined in SI units to have exact numerical values, shifting responsibility of experiment to the determination of the meter and the ampere. (The approximation in the second value of ε_0 above stems from π being an irrational number.)

Relative Permittivity

The linear permittivity of a homogeneous material is usually given relative to that of free space, as a relative permittivity ε_r (also called dielectric constant, although this sometimes only refers to the static, zero-frequency relative permittivity). In an anisotropic material, the relative permittivity may be a tensor, causing bire-fringence. The actual permittivity is then calculated by multiplying the relative permittivity by ε_0 :

$$\varepsilon = \varepsilon_r\varepsilon_0 = (1 + x) \varepsilon_0,$$

where χ (frequently written χ_e) is the electric susceptibility of the material.

The susceptibility is defined as the constant of proportionality (which may be a tensor) relating an electric field $\mathbf{E}$ to the induced dielectric polarization density $\mathbf{P}$ such that

$$\mathbf{P} = \varepsilon_0 \mathbf{X} \mathbf{E},$$

where ε_0 is the electric permittivity of free space.

The susceptibility of a medium is related to its relative permittivity ε_r by

$$X = \varepsilon_r - 1.$$

So in the case of a vacuum,

$$X = 0.$$

The susceptibility is also related to the polarizability of individual particles in the medium by the Clausius-Mossotti relation.

The electric displacement $\mathbf{D}$ is related to the polarization density $\mathbf{P}$ by

$$\mathbf{D} = \varepsilon_0 \mathbf{E} + \mathbf{P} = \varepsilon_0 (1 + x) \mathbf{E} = \varepsilon_r \varepsilon_0 \mathbf{E}.$$

The permittivity ε and permeability μ of a medium together determine the phase velocity $v = c/n$ of electromagnetic radiation through that medium :

$$\varepsilon\mu = \frac{1}{v^2}.$$

Dispersion and Causality

In general, a material cannot polarize instantaneously in response to an applied field, and so the more general formulation as a function of time is

$$P(t) = \varepsilon_0 \int_{-\infty}^{t} x(t - t') \mathbf{E}(t') \, dt'.$$

That is, the polarization is a convolution of the electric field at previous times with time-dependent susceptibility given by $\chi(\Delta t)$. The upper limit of this integral can be extended to infinity as well if one defines $\chi(\Delta t) = 0$ for $\Delta t < 0$. An instanta-neous response corresponds to Dirac delta function susceptibility $\chi(\Delta t) = \chi \, \delta(\Delta t)$.

It is more convenient in a linear system to take the Fourier transform and write this relationship as a function of frequency. Because of the convolution theorem, the integral becomes a simple product,

$$P(\omega) = \varepsilon_0 \chi(\omega) E(\omega).$$

This frequency dependence of the susceptibility leads to frequency depend-ence of the permittivity. The shape of the susceptibility with respect to frequency characterizes the dispersion properties of the material.

Moreover, the fact that the polarization can only depend on the electric field at previous times (*i.e.* $\chi(\Delta t) = 0$ for $\Delta t < 0$), a consequence of causality, imposes Kramers–Kronig constraints on the susceptibility $\chi(0)$.

Complex Permittivity

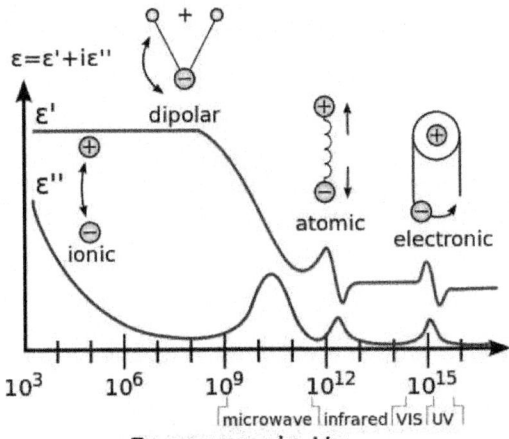

Fig. : A dielectric permittivity spectrum over a wide range of frequencies. ε' and ε'' denote the real and the imaginary part of the permittivity, respectively. Various processes are labeled on the image : ionic and dipolar relaxation, and atomic and electronic resonances at higher energies.

As opposed to the response of a vacuum, the response of normal materials to external fields generally depends on the frequency of the field. This frequency dependence reflects the fact that a material's polarization does not respond instantaneously to an applied field. The response must always be *causal* (arising after the applied field) which can be represented by a phase difference. For this reason, permittivity is often treated as a complex function of the (angular) frequency of the applied field ω : $\varepsilon \to \hat{\varepsilon}(\omega)$ (since complex numbers allow specification of magnitude and phase). The definition of permittivity therefore becomes

$$D_0 e^{-i\omega t} = \hat{\varepsilon}(\omega) E_0 e^{-i\omega t}$$

where :

D_0 and E_0 are the amplitudes of the displacement and electric fields, respectively,

i is the imaginary unit, $i^2 = -1$.

The response of a medium to static electric fields is described by the low-frequency limit of permittivity, also called the static permittivity ε_s (also ε_{DC}) :

$$\varepsilon_s = \lim_{\omega \to 0} \hat{\varepsilon}(\omega).$$

At the high-frequency limit, the complex permittivity is commonly referred to as ε_∞. At the plasma frequency and above, dielectrics behave as ideal metals, with electron gas behaviour. The static permittivity is a good approximation for alternating fields of low frequencies, and as the frequency increases a measurable phase difference δ emerges between $\mathbf{D}$ and $\mathbf{E}$. The frequency at which the phase shift becomes noticeable depends on temperature and the details of the medium. For moderate fields strength (E_0), $\mathbf{D}$ and $\mathbf{E}$ remain proportional, and

$$\hat{\varepsilon} = \frac{D_0}{E_0} = |\varepsilon| e^{i\delta}.$$

Since the response of materials to alternating fields is characterized by a complex permittivity, it is natural to separate its real and imaginary parts, which is done by convention in the following way :

$$\hat{\varepsilon}(\omega) = \varepsilon'(\omega) + i\varepsilon''(\omega) = (\cos \delta + i \sin \delta).$$

where

> ε' is the real part of the permittivity, which is related to the stored energy within the medium;
>
> ε'' is the imaginary part of the permittivity, which is related to the dissipation (or loss) of energy within the medium;
>
> δ is the loss angle.

It is important to realize that the choice of sign for time-dependence, exp(-$i\omega t$), dictates the sign convention for the imaginary part of permittivity. The signs used here correspond to those commonly used in physics, whereas for the engineering convention one should reverse all imaginary quantities.

The complex permittivity is usually a complicated function of frequency ω, since it is a superimposed description of dispersion phenomena occurring at multiple frequencies. The dielectric function $\varepsilon(\omega)$ must have poles only for frequencies with positive imaginary parts, and therefore satisfies the Kramers–Kronig relations. However, in the narrow frequency ranges that are often studied in practice, the permittivity can be approximated as frequency-independent or by model functions.

At a given frequency, the imaginary part of $\hat{\varepsilon}$ leads to absorption loss if it is positive (in the above sign convention) and gain if it is negative. More generally, the imaginary parts of the eigen values of the anisotropic dielectric tensor should be considered.

In the case of solids, the complex dielectric function is intimately connected to band structure. The primary quantity that characterizes the electronic structure of any crystalline material is the probability of photon absorption, which is directly related to the imaginary part of the optical dielectric function $\varepsilon(\omega)$. The optical dielectric function is given by the fundamental expression :

$$\varepsilon(\omega) = 1 + \frac{8\pi^2 e^2}{m^2} \sum_{c,v} \int W_{c,v}(E) \left[\varphi(\hbar\omega - E) - \varphi(\hbar\omega - E)\right] dx.$$

In this expression, $W_{c,v}(E)$ represents the product of the Brillouin zone-averaged transition probability at the energy E with the joint density of states, $J_{c,v}(E)$; φ is a broadening function, representing the role of scattering in smearing out the energy levels. In general, the broadening is intermediate between Lorentzian and Gaussian; for an alloy it is somewhat closer to Gaussian because of strong scattering from statistical fluctuations in the local composition on a nanometer scale.

Tensorial Permittivity

According to the Drude model of magnetized plasma, a more general expression which takes into account the interaction of the carriers with an alternating electric field at millimeter and microwave frequencies in an axially magnetized semiconductor requires the expression of the permittivity as a non-diagonal tensor.

$$\mathbf{D}(\omega) = \begin{vmatrix} \varepsilon_1 & -i\varepsilon_2 & 0 \\ i\varepsilon_2 & \varepsilon_1 & 0 \\ 0 & 0 & \varepsilon_z \end{vmatrix} \mathbf{E}(\omega)$$

If ε_2 vanishes, then the tensor is diagonal but not proportional to the identity and the medium is said an uniaxial medium.

Classification of Materials

$\sigma/(\omega\varepsilon')$	Current conduction	Field propagation
0		perfect dielectric lossless medium
$\ll 1$	low-conductivity material poor conductor	low-loss medium good dielectric
≈ 1	lossy conducting material	lossy propagation medium
$\gg 1$	high-conductivity material good conductor	high-loss medium poor dielectric
∞	perfect conductor	

Materials can be classified according to their complex-valued permittivity ε, upon comparison of its real ε' and imaginary ε'' components (or, equivalently, conductivity, σ, when it›s accounted for in the latter). A *perfect conductor* has infinite conductivity, $\sigma = \infty$, while a *perfect dielectric* is a material that has no conductivity at all, $\sigma = 0$; this latter case, of real-valued permittivity (or complex-valued permittivity with zero imaginary component) is also associated with the name *lossless media*. Generally, when $\sigma/(\omega\varepsilon') \ll 1$ we consider the material to be a *low-loss dielectric* (nearly though not exactly lossless), whereas $\sigma/(\omega\varepsilon') \gg 1$ is associated with a *good conductor*; such materials with non-negligible conductivity yield a large amount of loss that inhibit the propagation of electromagnetic waves, thus are also said to be *lossy media*. Those materials that do not fall under either limit are considered to be general media.

Lossy Medium

In the case of lossy medium, *i.e.* when the conduction current is not negligible, the total current density flowing is :

$$J_{tot} = J_c + J_d = \sigma E - i\omega\varepsilon'E = -i\omega\,\hat{\varepsilon}\,E$$

where :

σ is the conductivity of the medium;

ε' is the real part of the permittivity.

$\hat{\varepsilon}$ is the complex permittivity.

The size of the displacement current is dependent on the frequency ω of the applied field E; there is no displacement current in a constant field.

In this formalism, the complex permittivity is defined as :

$$\hat{\varepsilon} = \varepsilon' + i\frac{\sigma}{\omega}$$

In general, the absorption of electromagnetic energy by dielectrics is covered by a few different mechanisms that influence the shape of the permittivity as a function of frequency :

- First, are the relaxation effects associated with permanent and induced molecular dipoles. At low frequencies the field changes slowly enough to allow dipoles to reach equilibrium before the field has measurably changed. For frequencies at which dipole orientations cannot follow the applied field because of the viscosity of the medium, absorption of the field's energy leads to energy dissipation. The mechanism of dipoles relaxing is called dielectric relaxation and for ideal dipoles is described by classic Debye relaxation.

- Second are the resonance effects, which arise from the rotations or vibrations of atoms, ions, or electrons. These processes are observed in the neighbourhood of their characteristic absorption frequencies.

The above effects often combine to cause non-linear effects within capacitors. For example, dielectric absorption refers to the inability of a capacitor that has been charged for a long time to completely discharge when briefly discharged. Although an ideal capacitor would remain at zero volts after being discharged, real capacitors will develop a small voltage, a phenomenon that is also called *soakage* or *battery action*. For some dielectrics, such as many polymer films, the resulting voltage may be less than 1-2% of the original voltage. However, it can be as much as 15 - 25% in the case of electrolytic capacitors or supercapacitors.

Quantum-mechanical Interpretation

In terms of quantum mechanics, permittivity is explained by atomic and molecular interactions.

At low frequencies, molecules in polar dielectrics are polarized by an applied electric field, which induces periodic rotations. For example, at the microwave frequency, the microwave field causes the periodic rotation of water molecules, sufficient to break hydrogen bonds. The field does work against the bonds and the energy is absorbed by the material as heat. This is why microwave ovens work very well for materials containing water. There are two maxima of the imaginary component (the absorptive index) of water, one at the microwave frequency, and the other at far ultra-violet (UV) frequency. Both of these resonances are at higher frequencies than the operating frequency of microwave ovens.

At moderate frequencies, the energy is too high to cause rotation, yet too low to affect electrons directly, and is absorbed in the form of resonant molecular vibrations. In water, this is where the absorptive index starts to drop sharply, and the minimum of the imaginary permittivity is at the frequency of blue light (optical regime).

At high frequencies (such as UV and above), molecules cannot relax, and the energy is purely absorbed by atoms, exciting electron energy levels. Thus, these frequencies are classified as ionizing radiation.

While carrying out a complete *ab initio* (that is, first-principles) modelling is now computationally possible, it has not been widely applied yet. Thus, a phenomenological model is accepted as being an adequate method of capturing experimental behaviours. The Debye model and the Lorentz model use a 1st-order and 2nd-order (respectively) lumped system parameter linear representation (such as an RC and an LRC resonant circuit).

Measurement

The dielectric constant of a material can be found by a variety of static electrical measurements. The complex permittivity is evaluated over a wide range of frequencies by using different variants of dielectric spectroscopy, covering nearly 21 orders of magnitude from 10^{-6} to 10^{15} Hz. Also, by using cryostats and ovens, the dielectric properties of a medium can be characterized over an array of temperatures. In order to study systems for such diverse excitation fields, a number of measurement setups are used, each adequate for a special frequency range.

Various microwave measurement techniques are outlined in Chen *et. al.*. Typical errors for the Hakki-Coleman method employing a puck of material between conducting planes are about 0.3%.

- Low-frequency time domain measurements (10^{-6}-10^3 Hz)
- Low-frequency frequency domain measurements (10^{-5}-10^6 Hz)
- Reflective coaxial methods (10^6-10^{10} Hz)
- Transmission coaxial method (10^8-10^{11} Hz)

- Quasi-optical methods (10^9-10^{10} Hz)
- Terahertz time-domain spectroscopy (10^{11}-10^{13} Hz)
- Fourier-transform methods (10^{11}-10^{15} Hz)

At infrared and optical frequencies, a common technique is ellipsometry. Dual polarisation interferometry is also used to measure the complex refractive index for very thin films at optical frequencies.

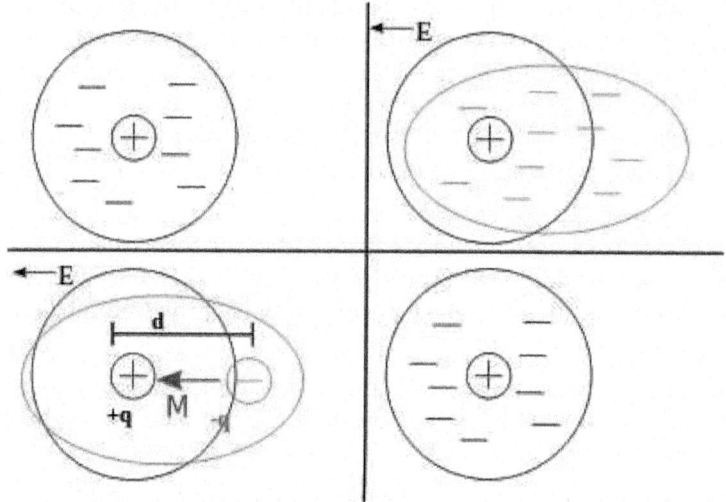

Fig. : Electric field interaction with an atom under the classical dielectric model.

DIELECTRIC POLARIZATION

Basic Atomic Model

In the classical approach to the dielectric model, a material is made up of atoms. Each atom consists of a cloud of negative charge (Electrons) bound to and surrounding a positive point charge at its center. In the presence of an electric field the charge cloud is distorted.

This can be reduced to a simple dipole using the superposition principle. A dipole is characterized by its dipole moment. It is the relationship between the electric field and the dipole moment that gives rise to the behaviour of the dielectric. (Note that the dipole moment points in the same direction as the electric field. This isn't always correct, and is a major simplification, but is suitable for many materials.)

When the electric field is removed the atom returns to its original state. The time required to do so is the so-called relaxation time; an exponential decay.

This is the essence of the model in physics. The behaviour of the dielectric now depends on the situation. The more complicated the situation, the richer the model must be to accurately describe the behaviour. Important questions are :

- Is the electric field constant or does it vary with time?
  - If the electric field does vary, at what rate?
- What are the characteristics of the material?
  - Is the direction of the field important (isotropy)?
  - Is the material the same all the way through (homogeneous)?
  - Do any boundaries/interfaces have to be taken into account?
- Is the system linear, or do non-linearities have to be taken into account?

The relationship between the electric field **E** and the dipole moment **M** gives rise to the behaviour of the dielectric, which, for a given material, can be characterized by the function **F** defined by the equation :

$$M = F(E).$$

When both the type of electric field and the type of material have been defined, one then chooses the simplest function F that correctly predicts the phenomena of interest. Examples of phenomena that can be so modeled include :

- Refractive index
- Group velocity dispersion
- Birefringence
- Self-focusing
- Harmonic generation

Dipolar Polarization

Dipolar polarization is a polarization that is either inherent to polar molecules (**orientation polarization**), or can be induced in any molecule in which the asymmetric distortion of the nuclei is possible (**distortion polarization**). Orientation polarization results from a permanent dipole, *e.g.*, that arising from the 104.45° angle between the asymmetric bonds between oxygen and hydrogen atoms in the water molecule, which retains polarization in the absence of an external electric field. The assembly of these dipoles forms a macroscopic polarization.

When an external electric field is applied, the distance between charges, which is related to chemical bonding, remains constant in orientation polarization; however, the polarization itself rotates. This rotation occurs on a timescale that depends on the torque and surrounding local viscosity of the molecules. Because the rotation is not instantaneous, dipolar polarizations lose the response to electric fields at the lowest frequency in polarizations. A molecule rotates about 1ps per radian in a fluid, thus this loss occurs at about 10^{11} Hz (in the microwave region). The delay of the response to the change of the electric field causes friction and heat.

When an external electric field is applied in the infrared, a molecule is bent and stretched by the field and the molecular moment changes in response. The molecular vibration frequency is approximately the inverse of the time taken for the molecule to bend, and the **distortion polarization** disappears above the infrared.

Ionic Polarization

Ionic polarization is polarization caused by relative displacements between positive and negative ions in ionic crystals (for example, NaCl).

If crystals or molecules do not consist of only atoms of the same kind, the distribution of charges around an atom in the crystals or molecules leans to positive or negative. As a result, when lattice vibrations or molecular vibrations induce relative displacements of the atoms, the centers of positive and negative charges might be in different locations. These center positions are affected by the symmetry of the displacements. When the centers don't correspond, polarizations arise in molecules or crystals. This polarization is called **ionic polarization**.

Ionic polarization causes ferroelectric transition as well as dipolar polarization. The transition, which is caused by the order of the directional orientations of permanent dipoles along a particular direction, is called **order-disorder phase transition**. Transition caused by ionic polarizations in crystals is called **displacive phase transition**.

Dielectric Dispersion

In physics, **dielectric dispersion** is the dependence of the permittivity of a dielectric material on the frequency of an applied electric field. Because there is always a lag between changes in polarization and changes in an electric field, the permittivity of the dielectric is a complicated, complex-valued function of frequency of the electric field. It is very important for the application of dielectric materials and the analysis of polarization systems.

This is one instance of a general phenomenon known as material dispersion : a frequency-dependent response of a medium for wave propagation.

When the frequency becomes higher :

1. it becomes impossible for dipolar polarization to follow the electric field in the microwave region around 10^{10} Hz;
2. in the infrared or far-infrared region around 10^{13} Hz, ionic polarization and molecular distortion polarization lose the response to the electric field;
3. electronic polarization loses its response in the ultraviolet region around 10^{15} Hz.

In the frequency region above ultra-violet, permittivity approaches the constant ε_0 in every substance, where ε_0 is the permittivity of the free space. Because permittivity indicates the strength of the relation between an electric field and polarization, if a polarization process loses its response, permittivity decreases.

Dielectric Relaxation

Dielectric relaxation is the momentary delay (or lag) in the dielectric constant of a material. This is usually caused by the delay in molecular polarization with respect to a changing electric field in a dielectric medium (*e.g.*, inside capacitors or between two large conducting surfaces). Dielectric relaxation in changing electric fields could be considered analogous to hysteresis in changing magnetic fields (for inductors or transformers). Relaxation in general is a delay or lag in the response of a linear system, and therefore dielectric relaxation is measured relative to the expected linear steady state (equilibrium) dielectric values. The time lag between electrical field and polarization implies an irreversible degradation of free energy (G).

In physics, **dielectric relaxation** refers to the relaxation response of a dielectric medium to an external electric field of microwave frequencies. This relaxation is often described in terms of permittivity as a function of frequency, which can, for ideal systems, be described by the Debye equation. On the other hand, the distortion related to ionic and electronic polarization shows behaviour of the resonance or oscillator type. The character of the distortion process depends on the structure, composition, and surroundings of the sample.

The number of possible wavelengths of emitted radiation due to dielectric relaxation can be equated using Hemmings' first law (named after Mark Hemmings)

$$n = \frac{l^2 - l}{2}$$

where

n is the number of different possible wavelengths of emitted radiation

l is the number of energy levels (including ground level).

Debye Relaxation

Debye relaxation is the dielectric relaxation response of an ideal, non-interacting population of dipoles to an alternating external electric field. It is usually expressed in the complex permittivity ε of a medium as a function of the field's frequency ω :

$$\hat{\varepsilon}(\omega) = \varepsilon_\infty + \frac{\Delta\varepsilon}{1 + i\omega t},$$

where ε_∞ is the permittivity at the high frequency limit, $\Delta\varepsilon = \varepsilon_s - \varepsilon_\infty$ where ε_s is the static, low frequency permittivity, and τ is the characteristic relaxation time of the medium.

This relaxation model was introduced by and named after the physicist Peter Debye (1913).

Variants of the Debye Equation

- Cole–Cole equation
- Cole–Davidson equation
- Havriliak–Negami relaxation
- Kohlrausch–Williams–Watts function (Fourier transform of stretched exponential function).

Para-electricity

Para-electricity is the ability of many materials (specifically ceramic crystals) to become polarized under an applied electric field. Unlike ferroelectricity, this can happen even if there is no permanent electric dipole that exists in the material, and removal of the fields results in the polarization in the material returning to zero. The mechanisms that cause **paraelectric** behaviour are the distortion of individual ions (displacement of the electron cloud from the nucleus) and polarization of molecules or combinations of ions or defects.

Para-electricity occurs in crystal phases where electric dipoles are unaligned (*i.e.*, unordered domains that are electrically charged) and thus have the potential to align in an external electric field and strengthen it. In comparison to the ferroelectric phase, the domains are unordered and the internal field is weak.

The $LiNbO_3$ crystal is ferroelectric below 1430 K, and above this temperature it transforms into a disordered paraelectric phase. Similarly, other perovskites also exhibit Para-electricity at high temperatures.

Para-electricity has been explored as a possible refrigeration mechanism; polarizing a paraelectric by applying an electric field under adiabatic process conditions raises the temperature, while removing the field lowers the temperature. A heat pump that polarizes the paraelectric, allows it to return to ambient temperature, then brings it into contact with the object to be cooled, and depolarizes it, would result in refrigeration.

Tunability

Tunable dielectrics are insulators whose ability to store electrical charge changes when a voltage is applied.

Generally, strontium titanate ($SrTiO_3$) is used for devices operating at low temperatures, while barium strontium titanate ($Ba1-xSrxTiO_3$) substitutes for room temperature devices. Other potential materials include microwave dielectrics and carbon nanotube (CNT) composites.

In 2013 multi-sheet layers of strontium titanate interleaved with single layers of strontium oxide produced a dielectric capable of operating at up to 125GHz. The material was created via molecular beam epitaxy. The two have mismatched crystal spacing that produces strain within the strontium titanate layer that makes it less stable and tunable.

Systems such as Ba1-xSrxTiO₃ have a paraelectric–ferroelectric transition just below ambient temperature, providing high tunability. Such films suffer significant losses arising from defects.Here we report the experimental realization of a highly tunable ground state arising from the emergence of a local ferroelectric instability in biaxially strained Srn+1TinO3n+1 phases with n ≥ 3 at frequencies up to 125 GHz. In contrast to traditional methods of modifying ferroelectrics — doping or strain — in this unique system an increase in the separation between the (SrO)² planes, which can be achieved by changing n, bolsters the local ferroelectric instability. This new control parameter, n, can be exploited to achieve a figure of merit at room temperature that rivals all known tunable microwave dielectrics.

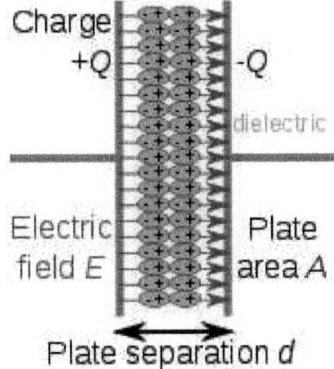

Fig. : Charge separation in a parallel-plate capacitor causes an internal electric field. A dielectric (orange) reduces the field and increases the capacitance.

Applications

Capacitors

Commercially manufactured capacitors typically use a solid dielectric material with high permittivity as the intervening medium between the stored positive and negative charges. This material is often referred to in technical contexts as the *capacitor dielectric*.

The most obvious advantage to using such a dielectric material is that it prevents the conducting plates the charges are stored on from coming into direct electrical contact. More significantly, however, a high permittivity allows a greater stored charge at a given voltage. This can be seen by treating the case of a linear dielectric with permittivity ε and thickness d between two conducting plates with uniform charge density σ_ε. In this case the charge density is given by

$$\sigma_\varepsilon = \varepsilon \frac{V}{d}$$

and the capacitance per unit area by

$$c = \frac{\sigma_\varepsilon}{V} = \frac{\varepsilon}{d}$$

From this, it can easily be seen that a larger ε leads to greater charge stored and thus greater capacitance.

Dielectric materials used for capacitors are also chosen such that they are resistant to ionization. This allows the capacitor to operate at higher voltages before the insulating dielectric ionizes and begins to allow undesirable current.

Dielectric Resonator

A *dielectric resonator oscillator* (DRO) is an electronic component that exhibits resonance for a narrow range of frequencies, generally in the microwave band. It consists of a "puck" of ceramic that has a large dielectric constant and a low dissipation factor. Such resonators are often used to provide a frequency reference in an oscillator circuit. An unshielded dielectric resonator can be used as a Dielectric Resonator Antenna (DRA).

Some Practical Dielectrics

Dielectric materials can be solids, liquids, or gases. In addition, a high vacuum can also be a useful, nearly lossless dielectric even though its relative dielectric constant is only unity.

Solid dielectrics are perhaps the most commonly used dielectrics in electrical engineering, and many solids are very good insulators. Some examples include porcelain, glass, and most plastics. Air, nitrogen and sulfur hexafluoride are the three most commonly used gaseous dielectrics.

- Industrial coatings such as parylene provide a dielectric barrier between the substrate and its environment.
- Mineral oil is used extensively inside electrical transformers as a fluid dielectric and to assist in cooling. Dielectric fluids with higher dielectric constants, such as electrical grade castor oil, are often used in high voltage capacitors to help prevent corona discharge and increase capacitance.
- Because dielectrics resist the flow of electricity, the surface of a dielectric may retain *stranded* excess electrical charges. This may occur accidentally when the dielectric is rubbed (the triboelectric effect). This can be useful, as in a Van de Graaff generator or electrophorus, or it can be potentially destructive as in the case of electrostatic discharge.
- Specially processed dielectrics, called electrets (which should not be confused with ferroelectrics), may retain excess internal charge or "frozen in" polarization. Electrets have a semipermanent external electric field, and are the electrostatic equivalent to magnets. Electrets have numerous practical applications in the home and industry.
- Some dielectrics can generate a potential difference when subjected to mechanical stress, or change physical shape if an external voltage is applied across the material. This property is called piezoelectricity. Piezoelectric materials are another class of very useful dielectrics.

- Some ionic crystals and polymer dielectrics exhibit a spontaneous dipole moment, which can be reversed by an externally applied electric field. This behaviour is called the ferroelectric effect. These materials are analogous to the way ferromagnetic materials behave within an externally applied magnetic field. Ferroelectric materials often have very high dielectric constants, making them quite useful for capacitors.

Chapter 3

ELECTRICAL BREAKDOWN

Electrical breakdown or **dielectric breakdown** refers to a rapid reduction in the resistance of an electrical insulator when the voltage applied across it exceeds the breakdown voltage. This results in a portion of the insulator becoming electrically conductive. Electrical breakdown may be a momentary event (as in an electrostatic discharge), or may lead to a continuous arc discharge if protective devices fail to interrupt the current in a high power circuit.

Under sufficient electrical stress, electrical breakdown can occur within solids, liquids, gases or vacuum. However, the specific breakdown mechanisms are significantly different for each, particularly in different kinds of dielectric medium.

FAILURE OF ELECTRICAL INSULATION

Electrical breakdown is often associated with the failure of solid or liquid insulating materials used inside high voltage transformers or capacitors in the electricity distribution grid, usually resulting in a short circuit or a blown fuse. Electrical breakdown can also occur across the insulators that suspend overhead power lines, within underground power cables, or lines arcing to nearby branches of trees.

MECHANISM

Electrical breakdown occurs within a gas (or mixture of gases, such as air) when the dielectric strength of the gas is exceeded. Regions of high electrical stress can cause nearby gas to partially ionize and begin conducting. This is done deliberately in low pressure discharges such as in fluorescent lights.

Partial electrical breakdown of the air causes the "fresh air" smell of ozone during thunderstorms or around high-voltage equipment. Although air is normally an excellent insulator, when stressed by a sufficiently high voltage (an electric field strength of about $3 \times 10^{6V}/m$), air can begin to break down, becoming partially conductive. If the voltage is sufficiently high, complete electrical breakdown of the air will culminate in an electrical spark or an electric arc that bridges the entire gap. While the small sparks generated by static electricity may barely be audible, larger sparks are often accompanied by a loud snap or bang. Lightning is an example of an immense spark that can be many miles long. The colour of the spark depends upon the gases that make up the gaseous media.

If a fuse or circuit breaker fails to interrupt the current through a spark in a power circuit, current may continue, forming a very hot electric arc. The colour of an arc depends primarily upon the conductor materials (as they are vapourized and mix within the hot plasma in the arc). The free ions in and around the arc recombine to create new chemical compounds (ozone, carbon monoxide, nitrous oxide and other compounds). Ozone is most easily noticed due to its distinct odour. Although sparks and arcs are usually undesirable, they can be useful in everyday applications such as spark plugs for gasoline engines, electrical welding of metals, or for metal melting in an electric arc furnace.

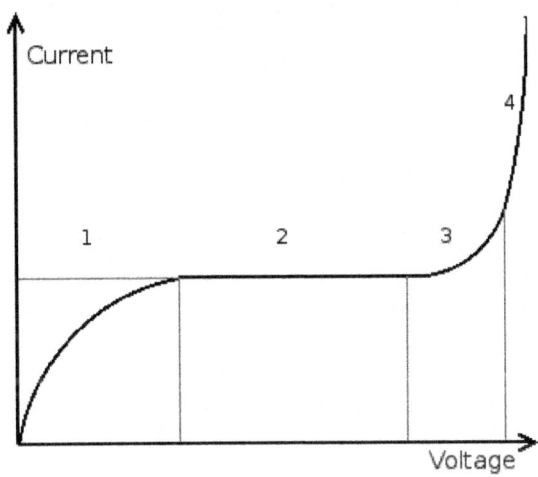

Fig. : Voltage-current relation before breakdown.

The vacuum itself is expected to undergo electrical breakdown at or near the Schwinger limit.

Voltage-current Relation

Before gas breakdown, there is a non-linear relation between voltage and current. In region 1, there are free ions that can be accelerated by the field and induce a current. These will be saturated after a certain voltage and give a constant current, region 2. Region 3 and 4 are.

Corona Breakdown

Partial breakdown of the air occurs as a corona discharge on high voltage conductors at points with the highest electrical stress. As the dielectric strength of the material surrounding the conductor determines the maximum strength of the electric field the surrounding material can tolerate before becoming conductive, conductors that consist of sharp points, or balls with small radii, are more prone to causing dielectric breakdown. Corona is sometimes seen as a bluish glow around high voltage wires and heard as a sizzling sound along high voltage power lines. Corona also generates radio frequency noise that can also be heard as "static" or buzzing on radio receivers. Corona can also occur naturally at high points (such as church spires, treetops, or ship masts) during thunderstorms as St. Elmo's Fire. Although corona discharge is usually undesirable, until recently it was essential in the operation of photocopiers (xerography) and laser printers. Many modern copiers and laser printers now charge the photoconductor drum with an electrically conductive roller, reducing undesirable indoor ozone pollution. Additionally, lightning rods use corona discharge to create conductive paths in the air that point towards the rod, deflecting potentially-damaging lightning away from buildings and other structures.

Corona discharge ozone generators have been used for more than 30 years in the water purification process. Ozone is a toxic gas, even more potent than chlorine. In a typical drinking water treatment plant, the ozone gas is dissolved into the filtered water to kill bacteria and viruses. Ozone also removes the bad odours and taste from the water. The main advantage of ozone is that the overdose (residual) decomposes to gaseous oxygen well before the water reaches the consumer. This is in contrast with chlorine which stays in the water and can be tasted by the consumer.

Corona discharges are also used to modify the surface properties of many polymers. An example is the corona treatment of plastic materials which allows paint or ink to adhere properly.

DISRUPTIVE DEVICES

A disruptive device is designed to electrically overstress a dielectric beyond its dielectric strength so as to intentionally cause electrical breakdown of the device. This results in the sudden transition of a portion of the dielectric from an insulating state to a highly conductive state. This transition is characterized by the formation of an electric spark (plasma channel), possibly followed by an electric arc through part of the dielectric material. If the dielectric happens to be a solid, permanent physical and chemical changes along the path of the discharge will significantly reduce the material's dielectric strength, and the device can only be used one time. However, if the dielectric material is a fluid or gas, the dielectric can fully recover its insulating properties once current through the plasma channel has been externally interrupted. Commercial Spark gaps use this property to abruptly switch high voltages in pulsed power systems, to provide surge protection for telecommunication and electrical power systems, and ignite fuel via spark plugs in many internal combustion engines.

ARC FLASH

An arc flash (also called a flashover), which is distinctly different from the arc blast, is part of an arc fault, a type of electrical explosion that results from a low-impedance connection to ground or another voltage phase in an electrical system.

Definition

An arc flash is the light and heat produced from an electric arc supplied with sufficient electrical energy to cause substantial damage, harm, fire, or injury. Electrical arcs experience negative resistance, which causes the electrical resistance to decrease as the arc temperature increases. Therefore, as the arc develops and gets hotter the resistance drops, drawing more and more current (runaway) until some part of the system melts, trips, or evapourates, providing enough distance to break the circuit and extinguish the arc. Electrical arcs, when well controlled and fed by limited energy, produce very bright light, and are used in arc lamps (enclosed, or with open electrodes), for welding, plasma cutting, and other industrial applications. Welding arcs can easily turn steel into a liquid with an average of only 24 DC volts. When an uncontrolled arc forms at high voltages, arc flashes can produce deafening noises, supersonic concussive-forces, super-heated shrapnel, temperatures far greater than the Sun's, and intense, high-energy radiation capable of vapourizing nearby materials.

Arc flash temperatures can reach or exceed 35,000 °F (19,400 °C) at the arc terminals. The massive energy released in the fault rapidly vapourizes the metal conductors involved, blasting molten metal and expanding plasma outward with extraordinary force. A typical arc flash incident can be inconsequential but could conceivably easily produce a more severe explosion. The result of the violent event can cause destruction of equipment involved, fire, and injury not only to an electrical worker but also to bystanders. During the arc flash, electrical energy vapourizes the metal, which changes from solid state to gas vapour, expanding it with explosive force. For example, when copper vapourizes it suddenly expands by a factor of 67,000 times in volume.

In addition to the explosive blast, called the **arc blast** of such a fault, destruction also arises from the intense radiant heat produced by the arc. The metal plasma arc produces tremendous amounts of light energy from far infrared to ultra-violet. Surfaces of nearby objects, including people absorb this energy and are instantly heated to vapourizing temperatures. The effects of this can be seen on adjacent walls and equipment - they are often ablated and eroded from the radiant effects.

Examples

One of the most common examples of an arc flash occurs when an incandescent light bulb burns out. When the filament breaks, an arc is sustained across the filament, enveloping it in plasma with a bright, blue flash. Most household lightbulbs have a built-in fuse, to prevent a sustained arc-flash from forming and blowing fuses in the circuit panel. Most 480 V electrical services have sufficient capacity to cause an arc flash hazard. Medium-voltage equipment (above 600 V) is higher potential and therefore a higher risk for an arc flash hazard. Higher voltages can cause a spark to jump, initiating an arc flash without the need for physical contact, and can sustain an arc across longer gaps. Most power-lines use voltages exceeding 1000 volts, and can be an arc-flash hazard to birds, squirrels, people, or equipment such as vehicles or ladders. Arc flashes are often witnessed from lines or transformers just before a power outage, creating bright flashes like lightning that can be seen for long distances.

High-tension power-lines often operate in the range of tens to hundreds of kilovolts. Care must usually be taken to ensure that the lines are insulated with a proper "flashover rating" and sufficiently spaced from each other, or an arc flash can spontaneously develop. If the high-tension lines become too close, either to each other or ground, a corona discharge may form between the conductors. This is typically a blue or reddish light caused by ionization of the air, accompanied by a hissing or frying sound. The corona discharge can easily lead to an arc flash, by creating a conductive pathway between the lines. This ionization can be enhanced during electrical storms, causing spontaneous arc-flashes and leading to power outages.

One of the most common causes of arc flash injuries happens when switching-on electrical circuits and, especially, tripped circuit-breakers. A tripped circuit-breaker often indicates a fault has occurred somewhere down the line from the panel. The fault must usually be isolated before switching the power on, or an arc flash can easily be generated. Small arcs usually form in switches when the contacts first touch, and can provide a place for an arc flash to develop. If the voltage is high enough, and the wires leading to the fault are large enough to allow a substantial amount of current, an arc flash can form within the panel when the switch is turned on. Generally, either an electric motor with shorted windings or a shorted power-transformer are the culprits, being capable of drawing the energy needed to sustain a dangerous arc-flash. Circuit breakers are often the primary defense against current runaway, especially if there are no secondary fuses, so if an arc flash develops in a breaker there may be nothing to stop a flash from going out of control. Once an arc flash begins in a breaker, it can quickly migrate from a single circuit to the phases of the panel itself, allowing very high energies to flow. Precautions must usually be used when switching circuit breakers, such as standing off to the side while switching to keep the body out of the way, wearing protective clothing, or turning-off equipment, circuits and panels downline prior to switching. Very large switchgear is often able to handle very high energies, and, thus, many places require the use of full protective equipment before turning it on.

As an example of the energy released in an arc flash incident, a single phase-to-phase fault on a 480 V system with 20,000 amps of fault current. The resulting power is 9.6 MW. If the fault lasts for 10 cycles at 60 Hz, the resulting energy would be 1600 kilojoules. For comparison, TNT releases 2175 J/g or more when detonated (a conventional value of 4,184 J/g is used for TNT equivalent). Thus, this fault energy is equivalent to 380 grams (approximately 0.8 pounds) of TNT. The character of an arc flash blast is quite different from a chemical explosion (more heat and light, less mechanical shock), but the resulting devastation is comparable. The rapidly expanding superheated vapour produced by the arc can cause serious injury or damage, and the intense UV, visible, and IR light produced by the arc can temporarily and sometimes even permanently blind or cause eye damage to people.

There are four different arc flash type events to be assessed when designing safety programs :

- Open Air Arc Flashes
- Ejected Arc Flashes
- Equipment Focused Arc Flashes (Arc-in-a-box)
- Tracking Arc Flashes.

Protecting Personnel

There are many methods of protecting personnel from arc flash hazards. This can include personnel wearing arc flash personal protective equipment (PPE) or modifying the design and configuration of electrical equipment. The best way to remove the hazards of an arc flash is to de-energize electrical equipment when interacting with it, however de-energizing electrical equipment is in and of itself an arc flash hazard. In this case, one of the newest solutions is to allow the operator to stand far back from the electrical equipment by operating equipment remotely, this is called remote racking.

Arc Flash Protection Equipment

With recent increased awareness of the dangers of arc flash, there have been many companies that offer arc flash personal protective equipment (PPE). The materials are tested for their arc rating. The arc rating is the maximum incident energy resistance demonstrated by a material prior to breakopen (a hole in the material) or necessary to pass through and cause with 50% probability a second or third degree burn. Arc rating is normally expressed in cal/cm² (or small calories of heat energy per square centimeter). The tests for determining arc rating are defined in ASTM F1506 *Standard Performance Specification for Flame Resistant Textile Materials for Wearing Apparel for Use by Electrical Workers Exposed to Momentary Electric Arc and Related Thermal Hazards*. Among the best fabrics for protection against electric arc flash are the Modacrylic-cotton blends.

Selection of appropriate PPE, given a certain task to be performed, is normally handled in one of two possible ways. For example when working on 600 V switchgear and performing a removal of bolted covers to expose bare, energized parts. The minimum rating of PPE necessary for any category is the maximum available energy for that category. For example, a Category 3 arc-flash hazard requires PPE rated for no less than 25 cal/cm² (1.05 MJ/m²).

The second method of selecting PPE is to perform an arc flash hazard calculation to determine the available incident arc energy. Once the incident energy is calculated the appropriate ensemble of PPE that offers protection greater than the energy available can be selected.

PPE provides protection after an arc flash incident has occurred and should be viewed as the last line of protection. Reducing the frequency and severity of incidents should be the first option and this can be achieved through a complete arc flash hazard assessment and through the application of technology such as high-resistance grounding which has been proven to reduce the frequency and severity of incidents.

Reducing Hazard by Design

Three key factors determine the intensity of an arc flash on personnel. These factors are the quantity of fault current available in a system, the time until an arc flash fault is cleared, and the distance an individual is from a fault arc. Various design and equipment configuration choices can be made to affect these factors and in turn reduce the arc flash hazard.

Fault Current

Fault current can be limited by using current limiting devices such as grounding resistors or fuses. If the fault current is limited to 5 amperes or less, then many ground faults self-extinguish and do not propagate into phase-to-phase faults.

Arcing Time

Arcing time can be reduced by temporarily setting upstream protective devices to lower setpoints during maintenance periods, or by employing zone-selective interlocking protection (ZSIP). With zone-selective interlocking, a downstream breaker that detects a fault communicates with an upstream breaker to delay its instantaneous tripping function. In this way "selectivity" will be preserved, in other words faults in the circuit are cleared by the breaker nearest to the fault, minimizing the effect on the entire system. A fault on a branch circuit will be detected by all breakers upstream of the fault (closer to the source of power). The circuit breaker closest to the downstream fault will send a restraining signal to prevent upstream breakers from tripping instantaneously. The presence of the fault will nevertheless activate the preset trip delay timer(s) of the upstream circuit breaker(s); this will allow an upstream circuit breaker to interrupt the fault, if still necessary after the preset time has elapsed. The ZSIP system allows faster instantaneous trip settings to be used, without loss of selectivity. The faster trip times reduce the total energy in an arc fault discharge.

Arcing time can significantly be reduced by protection based on detection of arc-flash light. Optical detection is often combined with overcurrent information. Light and current based protection can be set up with dedicated arc-flash protective relays, or by using normal protective relays equipped with an add-on arc-flash option.

The most efficient means to reduce arcing time is to use an arc eliminator that will extinguish the arc within a few milliseconds.

Distance

The distance from an arc flash source within which an unprotected person has a 50% chance of receiving a second degree burn is referred to as the "flash protection boundary". Those conducting flash hazard analyses must consider this boundary, and then must determine what PPE should be worn within the flash protection boundary. Remote operators or robots can be used to perform activities that have a high risk for arc flash incidents, such as inserting draw-out circuit breakers on a live electrical bus. Remote racking systems are available which keep the operator outside the arc flash hazard zone.

Research

Both the Institute of Electrical and Electronics Engineers (IEEE) and the National Fire Protection Association (NFPA) have joined forces in an initiative to fund and support research and testing to increase the understanding of arc flash. The results of this collaborative project will provide information that will be used to improve electrical safety standards, predict the hazards associated with arcing faults and accompanying arc blasts, and provide practical safeguards for employees in the workplace.

Standards

- OSHA Standards 29-CFR, Part 1910. Occupational Safety and Health Standards. 1910 sub-part S (electrical) Standard number 1910.333 specifically addresses Standards for Work Practices and references NFPA 70E.
- The National Fire Protection Association (NFPA) Standard 70-2011 "The National Electrical Code" (NEC) contains requirements for warning labels.
- NFPA 70E 2009 provides guidance on implementing appropriate work practices that are required to safeguard workers from injury while working on or near exposed electrical conductors or circuit parts that could become energized.
- The Canadian Standards Association's CSA Z462 Arc Flash Standard is Canada's version of NFPA70E. Released in 2008.
- The Underwriters Laboratories of Canada's Standard on Electric Utility Workplace Electrical Safety for Generation, Transmission, and Distribution CAN/ULC_S801
- The Institute of Electronics and Electrical Engineers IEEE 1584 – 2002 Guide to Performing Arc-Flash Hazard Calculations.

Arc flash hazard software exists that allows businesses to comply with the myriad government regulations while providing their workforce with an optimally safe environment. Many software companies now offer arc flash hazard solutions. Few power services companies calculate safe flash boundaries.

AVALANCHE BREAKDOWN

Avalanche breakdown is a phenomenon that can occur in both insulating and semi-conducting materials. It is a form of electric current multiplication that can allow very large currents within materials which are otherwise good insulators. It is a type of electron avalanche. The avalanche process occurs when the carriers in the transition region are accelerated by the electric field to energies sufficient to free electron-hole pairs via collisions with bound electrons.

Explanation

Materials conduct electricity if they contain mobile charge carriers. There are two types of charge carrier in a semi-conductor : free electrons and electron holes. A fixed electron in a reverse-biased diode may break free due to its thermal energy, creating an electron-hole pair. If there is a voltage gradient in the semi-conductor, the electron will move towards the positive voltage while the hole will "move" towards the negative voltage. Most of the time, the electron and hole will just move to opposite ends of the crystal and stop. Under the right circumstances, however, (*i.e.* when the voltage is high enough) the free electron may move fast enough to knock other electrons free, creating more free-electron-hole pairs (*i.e.* more charge carriers), increasing the current. Fast-"moving" holes may also result in more electron-hole pairs being formed. In a fraction of a nanosecond, the whole crystal begins to conduct.

The large voltage drop and possibly large current during breakdown necessarily leads to the generation of heat. Therefore, a diode placed into a reverse blocking power application will usually be destroyed by breakdown, as the external circuit will be able to sustain a large current and dump excessive amounts of heat. In principle, however, avalanche breakdown only involves the passage of electrons, and intrinsically need not cause damage to the crystal. Avalanche diodes (commonly encountered as high voltage Zener diodes) are constructed to have a uniform junction that breaks down at a uniform voltage, to avoid current crowding during breakdown. These diodes can indefinitely sustain a moderate level of current while on the edge of breakdown.

The voltage at which the breakdown occurs is called the **breakdown voltage**. There is a hysteresis effect; once avalanche breakdown has occurred, the material will continue to conduct even if the voltage across it drops below the breakdown voltage. This is different from a Zener diode, which will stop conducting once the reverse voltage drops below the breakdown voltage.

Breakdown Voltage

The **breakdown voltage** of an insulator is the minimum voltage that causes a portion of an insulator to become electrically conductive.

The **breakdown voltage** of a diode is the minimum *reverse* voltage to make the diode conduct in reverse. Some devices (such as TRIACs) also have a *forward breakdown voltage*.

Fig. : High voltage dielectric breakdown within a block of plexiglas.

Solids

Breakdown voltage is a characteristic of an insulator that defines the maximum voltage difference that can be applied across the material before the insulator collapses and conducts. In solid insulating materials, this usually creates a weakened path within the material by creating permanent molecular or physical changes by the sudden current. Within rarefied gases found in certain types of lamps, **breakdown voltage** is also sometimes called the "striking voltage".

The breakdown voltage of a material is not a definite value because it is a form of failure and there is a statistical probability whether the material will fail at a given voltage. When a value is given it is usually the mean breakdown voltage of a large sample. Another term is also withstand voltage where the probability of failure at a given voltage is so low it is considered, when designing insulation, that the material will not fail at this voltage.

Two different breakdown voltage measurements of a material are the AC and impulse breakdown voltages. The AC voltage is the line frequency of the mains. The impulse breakdown voltage is simulating lightning strikes, and usually uses a 1.2 micro-second rise for the wave to reach 90% amplitude then drops back down to 50% amplitude after 50 micro-seconds.

Two technical standards governing performing these tests are ASTM D1816 and ASTM D3300 published by ASTM.

Gases and Vacuum

In standard conditions at atmospheric pressure, gas serves as an excellent insulator, requiring the application of a significant voltage before breaking down (*e.g.* lightning). In partial vacuum, this breakdown potential may decrease to an extent that two uninsulated surfaces with different potentials might induce the electrical breakdown of the surrounding gas. This has some useful applications in industry (*e.g.* the production of micro-processors) but in other situations may damage an apparatus, as breakdown is analogous to a short circuit.

In a gas, the breakdown voltage can be determined by Paschen's Law.

The breakdown voltage in a partial vacuum is represented as

$$V_b = \frac{Bpd}{\text{In } Apd - \text{In } (\text{In}(1 + \frac{1}{\gamma se}))}$$

where V_b is the breakdown potential in volts DC, A and B are constants that depend on the surrounding gas, p represents the pressure of the surrounding gas, d represents the distance in centimetres between the electrodes, and γse represents the Secondary Electron Emission Coefficient.

Diodes and Other Semi-conductors

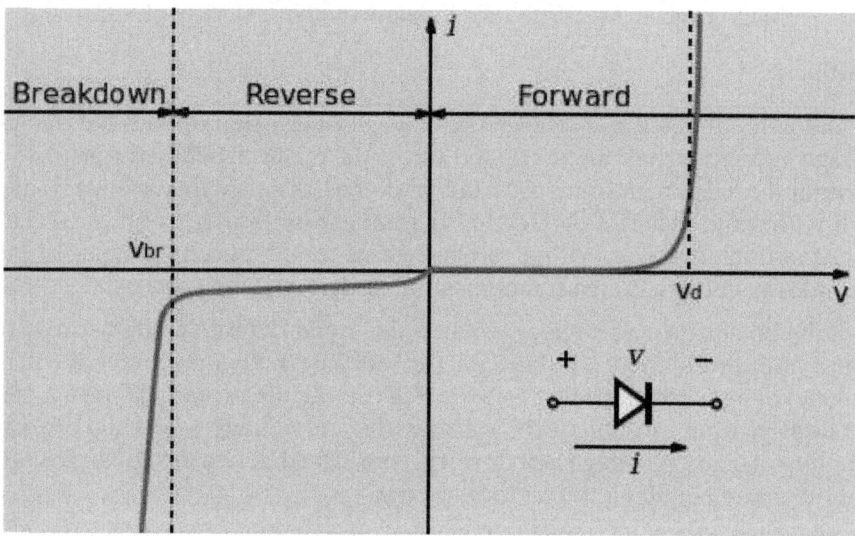

Fig. : Diode I-V diagram.

Breakdown voltage is a parameter of a diode that defines the largest reverse voltage that can be applied without causing an exponential increase in the current in the diode. As long as the current is limited, exceeding the breakdown voltage of a diode does no harm to the diode. In fact, Zener diodes are essentially just heavily doped normal diodes that exploit the breakdown voltage of a diode to provide regulation of voltage levels.

BRUSH DISCHARGE

A **brush discharge** is a type of corona discharge that takes place between two electrodes embedded in a non-conducting medium (*e.g.*, air at atmospheric pressure) and is characterized by non-sparking, faintly luciferous furcations composed of ionized particles.

Brush discharges can occur from charged insulating plastics (for example, polythene) to a conductor. The maximum energy associated with brush discharges is unlikely to exceed 4 mJ. Such discharges may be incendive but are less likely to cause ignition of a solvent - air mixture than an electrostatic discharge between two conductors.

COMPARATIVE TRACKING INDEX

The **Comparative Tracking Index** or **CTI** is used to measure the electrical breakdown (tracking) properties of an insulating material. To measure the tracking, 50 drops of 0.1% ammonium chloride solution are dropped on the material, and the voltage measured for a 3mm thickness is considered representative of the material performance. Also term PTI (Proof Tracking Index) is used : it means voltage at which during testing on five samples the samples pass the test with no failures.

Tracking is an electrical breakdown on the surface of an insulating material. A large voltage difference gradually creates a conductive leakage path across the surface of the material by forming a carbonized track. Testing method is specified in IEC standard 60112.

Performance Level Categories (PLC) were introduced to avoid excessive implied precision and bias.

The CTI value is used for electrical safety assessment of electrical apparatus, as for instance carried out by Underwriters Laboratory in USA and many other laboratories in the world. The minimum required creepage distances over an insulating material between electrically conducting parts in apparatus, especially between parts with a high voltage and parts that can be touched by human users, is dependent on the insulators CTI value. Also for internal distances in an apparatus by maintaining CTI based distances, the risk of fire is reduced.

Tracking Index (V)	PLC
600 and Greater	0
400 through 599	1
250 through 399	2
175 through 249	3
100 through 174	4
< 100	5

In design of medical products, the CTI is treated differently :

Comparative Tracking Index (CTI)	Material Group
600 <= CTI	I
400 <= CTI < 600	II
175 <= CTI < 400	IIIa
100 <= CTI < 175	IIIb

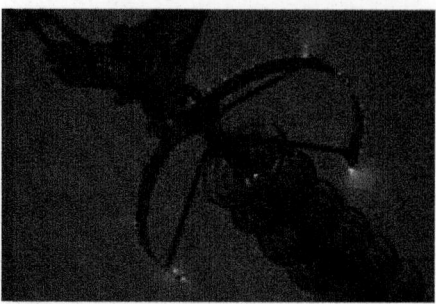

Fig. : Corona discharge on insulator string of a 500 kV overhead power line.
Corona discharges represent a significant power loss for electric utilities.

CORONA DISCHARGE

In electricity, a **corona discharge** is an electrical discharge brought on by the
ionization of a fluid surrounding a conductor that is electrically energized. The
discharge will occur when the strength (potential gradient) of the electric field
around the conductor is high enough to form a conductive region, but not high
enough to cause electrical breakdown or arcing to nearby objects. It is often seen
as a bluish (or other colour) glow in the air adjacent to pointed metal conductors
carrying high voltages. Spontaneous corona discharges are undesirable where
they waste power in high-voltage systems or where the high chemical activity
in a corona discharge creates objectionable or hazardous compounds, such as
ozone. Controlled corona discharges are used in a variety of filtration, printing
and other processes.

Introduction

Corona discharge is a process by which a current flows from an electrode with a
high potential into a neutral fluid, usually air, by ionizing that fluid so as to create
a region of plasma around the electrode. The ions generated eventually pass charge
to nearby areas of lower potential, or recombine to form neutral gas molecules.

When the potential gradient (electric field) is large enough at a point in the
fluid, the fluid at that point ionizes and it becomes conductive. If a charged object
has a sharp point, the electric field strength around that point will be much higher
than elsewhere. Air near the electrode can become ionized (partially conductive),
while regions more distant do not. When the air near the point becomes conductive, it has the effect of increasing the apparent size of the conductor. Since the
new conductive region is less sharp, the ionization may not extend past this local
region. Outside this region of ionization and conductivity, the charged particles
slowly find their way to an oppositely charged object and are neutralized.

If the geometry and gradient are such that the ionized region continues to
grow until it reaches another conductor at a lower potential, a low resistance
conductive path between the two will be formed, resulting in an electric arc.

Corona discharge usually forms at highly curved regions on electrodes, such as sharp corners, projecting points, edges of metal surfaces, or small diameter wires. The high curvature causes a high potential gradient at these locations, so that the air breaks down and forms plasma there first. In order to suppress corona formation, terminals on high voltage equipment are frequently designed with smooth large diameter rounded shapes like balls or toruses, and corona rings are often added to insulators of high voltage transmission lines.

Coronas may be *positive* or *negative*. This is determined by the polarity of the voltage on the highly-curved electrode. If the curved electrode is positive with respect to the flat electrode, it has a *positive corona*, if it is negative, it has a *negative corona*. The physics of positive and negative coronas are strikingly different. This asymmetry is a result of the great difference in mass between electrons and positively charged ions, with only the electron having the ability to undergo a significant degree of ionising inelastic collision at common temperatures and pressures.

An important reason for considering coronas is the production of ozone around conductors undergoing corona processes in air. A negative corona generates much more ozone than the corresponding positive corona.

Applications of Corona Discharge

Corona discharge has a number of commercial and industrial applications.

- Drag reduction over a flat surface
- Removal of unwanted electric charges from the surface of aircraft in flight and thus avoiding the detrimental effect of uncontrolled electrical discharge pulses on the performance of avionic systems
- Manufacture of ozone
- Sanitization of pool water
- Scrubbing particles from air in air-conditioning systems
- Removal of unwanted volatile organics, such as chemical pesticides, solvents, or chemical weapons agents, from the atmosphere
- Improvement of wetability or 'surface tension energy' of polymer films to improve compatibility with adhesives or printing inks
- Photocopying
- Air ionisers
- Production of photons for Kirlian photography to expose photographic film
- EHD thrusters, Lifters, and other ionic wind devices
- Nitrogen laser
- Surface treatment for tissue culture (polystyrene)
- Ionization of a gaseous sample for subsequent analysis in a mass spectrometer or an ion mobility spectrometer
- Solid-state cooling components for computer chips

Coronas can be used to generate charged surfaces, which is an effect used in electrostatic copying (photocopying). They can also be used to remove particulate matter from air streams by first charging the air, and then passing the charged stream through a comb of alternating polarity, to deposit the charged particles onto oppositely charged plates.

The free radicals and ions generated in corona reactions can be used to scrub the air of certain noxious products, through chemical reactions, and can be used to produce ozone.

Problems Caused by Corona Discharges

Coronas can generate audible and radio-frequency noise, particularly near electric power transmission lines. They also represent a power loss, and their action on atmospheric particulates, along with associated ozone and NOx production, can also be disadvantageous to human health where power lines run through built-up areas. Therefore, power transmission equipment is designed to minimise the formation of corona discharge.

Corona discharge is generally undesirable in :

- Electric power transmission, where it causes :
  - Power loss
  - Audible noise
  - Electromagnetic interference
  - Purple glow
  - Ozone production
  - Insulation damage.
- Electrical components such as transformers, capacitors, electric motors and generators.
  - Corona can progressively damage the insulation inside these devices, leading to equipment failure.
  - Elastomer items such as O-rings can suffer ozone cracking
  - Plastic film capacitors operating at mains voltage can suffer progressive loss of capacitance as corona discharges cause local vapourization of the metallization
- Situations where high voltages are in use, but ozone production is to be minimised
- Static electricity discharge.

In many cases coronas can be suppressed by corona rings, toroidal devices that serve to spread the electric field over larger area and decrease the field gradient below the corona threshold.

Mechanism of Corona Discharge

Corona discharge results when the electric field is strong enough to create a chain reaction : electrons in the air collide with atoms hard enough to ionize them, creating more electrons which ionize more atoms. The process is :

1. A neutral atom or molecule, in a region of strong electric field (such as the high potential gradient near the curved electrode) is ionized by a natural environmental event (for example, being struck by an ultra-violet photon or cosmic ray particle), to create a positive ion and a free electron.

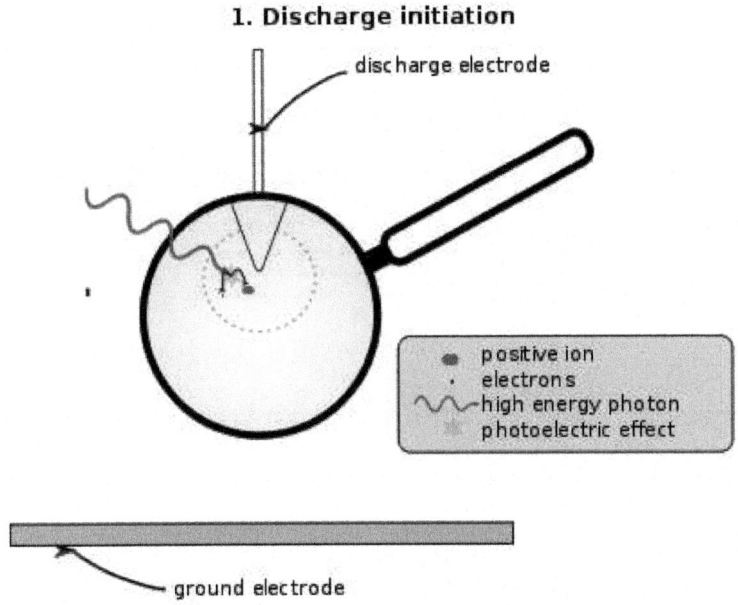

1. Discharge initiation

discharge electrode

positive ion
electrons
high energy photon
photoelectric effect

ground electrode

2. The electric field accelerates these oppositely charged particles in opposite directions, separating them, preventing their recombination, and imparting kinetic energy to each of them.

3. The electron has a much higher charge/mass ratio and so is accelerated to a higher velocity than the positive ion. It gains enough energy from the field that when it strikes another atom it ionizes it, knocking out another electron, and creating another positive ion. These electrons are accelerated and collide with other atoms, creating further electron/positive-ion pairs, and these electrons collide with more atoms, in a chain reaction process called an *electron avalanche*. Both positive and negative coronas rely on electron avalanches. In a positive corona all the electrons are attracted inward toward the nearby positive electrode and the ions are repelled outwards. In a negative corona the ions are attracted inward and the electrons are repelled outwards.

2. Electrical Breakdown

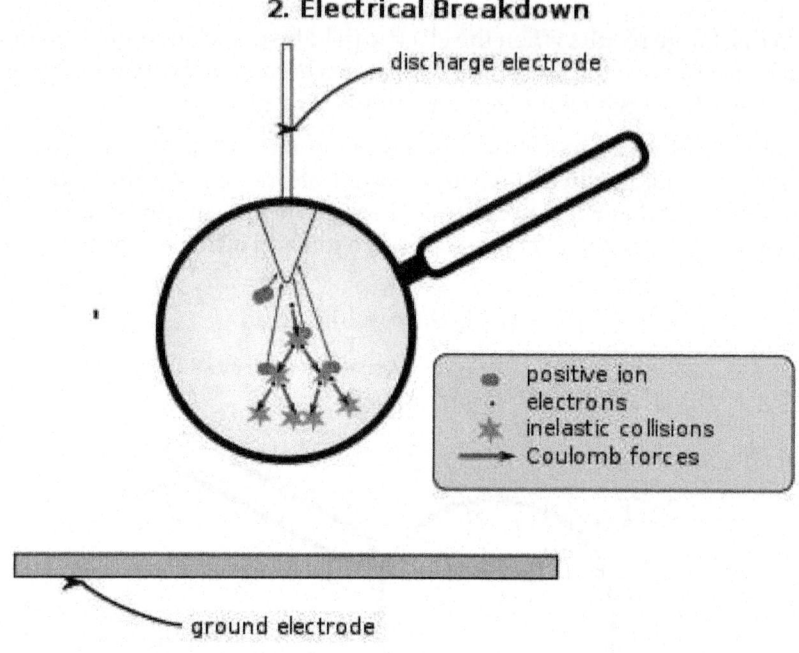

4. The glow of the corona is caused by electrons recombining with positive ions to form neutral atoms. When the electron falls back to its original energy level, it releases a photon of light. The photons serve to ionize other atoms, maintaining the creation of electron avalanches.

5. At a certain distance from the electrode, the electric field becomes low enough that it no longer imparts enough energy to the electrons to ionize atoms when they collide. This is the outer edge of the corona. Outside this the ions move through the air without creating new ions. The outward moving ions are attracted to the opposite electrode and eventually reach it and combine with electrons from the electrode to become neutral atoms again, completing the circuit.

Thermodynamically, a corona is a very *non-equilibrium* process, creating a non-thermal plasma. The avalanche mechanism does not release enough energy to heat the gas in the corona region generally and ionize it, as occurs in an electric arc or spark. Only a small number of gas molecules take part in the electron avalanches and are ionized, having energies close to the ionization energy of 1 - 3 ev, the rest of the surrounding gas is close to ambient temperature.

The onset voltage of corona or corona inception voltage (CIV) can be found with *Peek's law* (1929), formulated from empirical observations. Later papers derived more accurate formulas.

3. Recombination and upkeep of the discharge

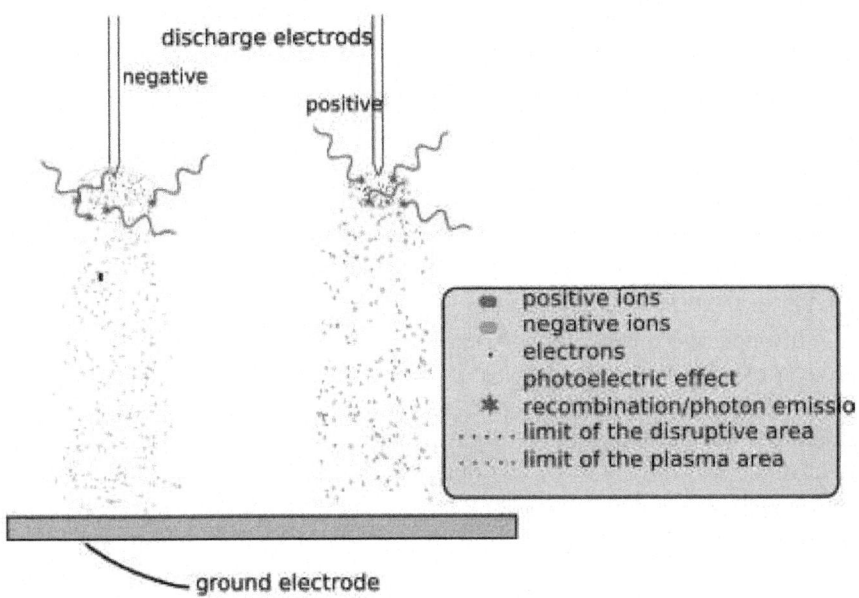

Positive Coronas

Properties

A positive corona is manifested as a uniform plasma across the length of a conductor. It can often be seen glowing blue/white, though many of the emissions are in the ultra-violet. With the same geometry and voltages, it appears a little smaller than the corresponding negative corona, owing to the lack of a non-ionising plasma region between the inner and outer regions.

A positive corona has much lower density of free electrons compared to a negative corona; perhaps a thousandth of the electron density, and a hundredth of the total number of electrons. However, the electrons in a positive corona are concentrated close to the surface of the curved conductor, in a region of high potential gradient (and therefore the electrons have a high energy), whereas in a negative corona many of the electrons are in the outer, lower-field areas. Therefore, if electrons are to be used in an application which requires a high activation energy, positive coronas may support a greater reaction constants than corresponding negative coronas; though the total number of electrons may be lower, the number of a very high energy electrons may be higher.

Coronas are efficient producers of ozone in air. A positive corona generates much less ozone than the corresponding negative corona, as the reactions which produce ozone are relatively low-energy. Therefore, the greater number of electrons of a negative corona leads to an increased production.

Beyond the plasma, in the *unipolar region*, the flow is of low-energy positive ions toward the flat electrode.

Mechanism

As with a negative corona, a positive corona is initiated by an exogenous ionisation event in a region of high potential gradient. The electrons resulting from the ionisation are attracted **toward** the curved electrode, and the positive ions repelled from it. By undergoing inelastic collisions closer and closer to the curved electrode, further molecules are ionized in an electron avalanche.

In a positive corona, secondary electrons, for further avalanches, are generated predominantly in the fluid itself, in the region outside the plasma or avalanche region. They are created by ionization caused by the photons emitted from that plasma in the various de-excitation processes occurring within the plasma after electron collisions, the thermal energy liberated in those collisions creating photons which are radiated into the gas. The electrons resulting from the ionisation of a neutral gas molecule are then electrically attracted back toward the curved electrode, attracted *into* the plasma, and so begins the process of creating further avalanches inside the plasma.

As can be seen, the positive corona is divided into two regions, concentric around the sharp electrode. The inner region contains ionising electrons, and positive ions, acting as a plasma, the electrons avalanche in this region, creating many further ion/electron pairs. The outer region consists almost entirely of the slowly migrating massive positive ions, moving toward the uncurved electrode along with, close to the interface of this region, secondary electrons, liberated by photons leaving the plasma, being re-accelerated into the plasma. The inner region is known as the **plasma** region, the outer as the **unipolar** region.

Negative Coronas

Properties

A negative corona is manifested in a non-uniform corona, varying according to the surface features and irregularities of the curved conductor. It often appears as tufts of corona at sharp edges, the number of tufts altering with the strength of the field. The form of negative coronas is a result of its source of secondary avalanche electrons. It appears a little larger than the corresponding positive corona, as electrons are allowed to drift out of the ionising region, and so the plasma continues some distance beyond it. The total number of electrons, and electron density is much greater than in the corresponding positive corona. However, they are of a predominantly lower energy, owing to being in a region of lower potential-gradient. Therefore, whilst for many reactions the increased electron density will increase the reaction rate, the lower energy of the electrons will mean that reactions which require a higher electron energy may take place at a lower rate.

Mechanism

Negative coronas are more complex than positive coronas in construction. As with positive coronas, the establishing of a corona begins with an exogenous ionization event generating a primary electron, followed by an electron avalanche.

Electrons ionized from the neutral gas are not useful in sustaining the negative corona process by generating secondary electrons for further avalanches, as the general movement of electrons in a negative corona is outward from the curved electrode. For negative corona, instead, the dominant process generating second-ary electrons is the photoelectric effect, from the surface of the electrode itself. The work function of the electrons (the energy required to liberate the electrons from the surface) is considerably lower than the ionization energy of air at standard temperatures and pressures, making it a more liberal source of secondary electrons under these conditions. Again, the source of energy for the electron-liberation is a high-energy photon from an atom within the plasma body relaxing after excitation from an earlier collision. The use of ionized neutral gas as a source of ionization is further diminished in a negative corona by the high-concentration of positive ions clustering around the curved electrode.

Under other conditions, the collision of the positive species with the curved electrode can also cause electron liberation.

The difference, then, between positive and negative coronas, in the matter of the generation of secondary electron avalanches, is that in a positive corona they are generated by the gas surrounding the plasma region, the new secondary electrons travelling inward, whereas in a negative corona they are generated by the curved electrode itself, the new secondary electrons travelling outward.

A further feature of the structure of negative coronas is that as the electrons drift outwards, they encounter neutral molecules and, with electronegative mol-ecules (such as oxygen and water vapour), combine to produce negative ions. These negative ions are then attracted to the positive uncurved electrode, completing the 'circuit'.

A negative corona can be divided into three radial areas, around the sharp electrode. In the inner area, high-energy electrons inelastically collide with neutral atoms and cause avalanches, whilst outer electrons (usually of a lower energy) combine with neutral atoms to produce negative ions. In the intermediate region, electrons combine to form negative ions, but typically have insufficient energy to cause avalanche ionization, but remain part of a plasma owing to the differ-ent polarities of the species present, and the ability to partake in characteristic plasma reactions. In the outer region, only a flow of negative ions and, to a lesser and radially-decreasing extent, free electrons toward the positive electrode takes place. The inner two regions are known as the corona *plasma*. The inner region is an *ionizing plasma*, the middle a *non-ionizing plasma*. The outer region is known as the *unipolar region*.

When an electrical transmission line is energized, the air surrounding the conductors is subjected to di-electric stress. At low voltage, nothing really occurs as the stress is too low to ionize the air outside. But when the voltage gradient around a conductor is higher than some threshold value, the air surrounding it experiences stress high enough to be dissociated into ions, making the atmosphere conducting. This results in electric discharge around the conductors due to the flow of these ions, giving rise to a faint luminescent glow, along with the hissing sound accompanied by the liberation of ozone, which is readily identified due to its characteristic odour. If the voltage across the lines is still increased the glow becomes more and more intense along with hissing noise, inducing very high power loss into the system.

Examples

Corona discharge may be seen around automotive spark plug wires that have become worn.

Corona discharge may arise from other conductors carrying alternating current at very high voltages, such as long-distance transmission lines and high-power short-wave transmitting antennae.

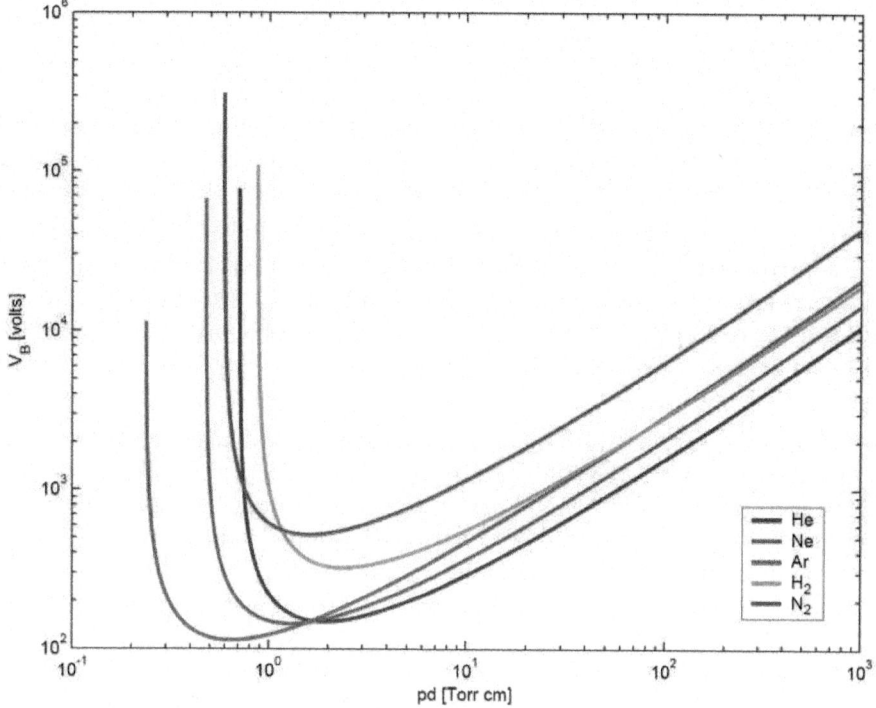

Fig. : Paschen curves obtained for helium, neon, argon, hydrogen and nitrogen, using the expression for the breakdown voltage as a function of the parameters A,B that interpolate the first Townsend coefficient.

PASCHEN'S LAW

Paschen's Law is an equation that gives the breakdown voltage, that is the voltage necessary to start a discharge or electric arc, between two electrodes in a gas as a function of pressure and gap length. It is named after Friedrich Paschen who discovered it empirically in 1889.

Paschen studied the breakdown voltage of various gases between parallel metal plates as the gas pressure and gap distance were varied. The voltage necessary to arc across the gap decreased as the pressure was reduced and then increased gradually, exceeding its original value. He also found that at normal pressure, the voltage needed to cause an arc reduced as the gap size was reduced but only to a point. As the gap was reduced further, the voltage required to cause an arc began to rise and again exceeded its original value. For a given gas, the voltage is a function only of the product of the pressure and gap length. The curve he found of voltage versus the pressure-gap length product *(right)* is called **Paschen's curve**. He found an equation that fit these curves, which is now called Paschen's law.

At higher pressures and gap lengths, the breakdown voltage is approximately *proportional* to the product of pressure and gap length, and the term Paschen's law is sometimes used to refer to this simpler relation. However this is only roughly true, over a limited range of the curve.

Paschen Curve

Early vacuum experimenters found a rather surprising behaviour. An arc would sometimes take place in a long irregular path rather than at the minimum distance between the electrodes. For example, in air, at a pressure of 10^{-3} atmospheres, the distance for minimum breakdown voltage is about 7.5 mm. The voltage required to arc this distance is 327 V which is insufficient to ignite the arcs for gaps that are either wider or narrower. For a 3.75 mm gap, the required voltage is 533 V, nearly twice as much. If 500 V were applied, it would not be sufficient to arc at the 2.85 mm distance, but would arc at a 7.5 mm distance.

It was found that breakdown voltage was described by the equation :

$$V = \frac{apd}{\ln{(pd)} + b}$$

Where V is the breakdown voltage in Volts, p is the pressure in Atmospheres or Bar, and d is the gap distance in meters. The constants a and b depend upon the composition of the gas. For air at standard atmospheric pressure of 101 kPa, $a = 4.36 \times 10^7$ V/(atm·m) and $b = 12.8$. The graph of this equation is the Paschen curve. By differentiating it with respect to pd and setting the derivative to zero, the minimum voltage can be found. This yields

$$pd = e^{1-b}$$

and predicts the occurrence of a minimum breakdown voltage for $pd = 7.5 \times 10^{-6}$ m·atm. This is 327 V in air at standard atmospheric pressure at a distance of 7.5 μm. The composition of the gas determines both the minimum arc voltage and the distance at which it occurs. For argon, the minimum arc voltage is 137 V at a larger 12 μm. For sulfur dioxide, the minimum arc voltage is 457 V at only 4.4 μm.

For air at STP, the voltage needed to arc a 1 meter gap is about 3.4 MV. The intensity of the electric field for this gap is therefore 3.4 MV/m. The electric field needed to arc across the minimum voltage gap is much greater than that necessary to arc a gap of one meter. For a 7.5 μm gap the arc voltage is 327 V which is 43 MV/m. This is about 13 times greater than the field strength for the 1 meter gap. The phenomenon is well verified experimentally and is referred to as the Paschen minimum. The equation loses accuracy for gaps under about 10 μm in air at one atmosphere and incorrectly predicts an infinite arc voltage at a gap of about 2.7 micrometers. Breakdown voltage can also differ from the Paschen curve prediction for very small electrode gaps when field emission from the cathode surface becomes important.

Physical Mechanism

The mean free path of a molecule in a gas is the average distance between its collision with other molecules. This is inversely proportional to the pressure of the gas. In air the mean free path of molecules is about 96 nm. Since electrons are much faster, their average distance between colliding with molecules is about 5.6 times longer or about 0.5 μm. This is a substantial fraction of the 7.5 μm spacing between the electrodes for minimum arc voltage. If the electron is in an electric field of 43 MV/m, it will be accelerated and acquire 21.5 electron volts of energy in 0.5 μm of travel in the direction of the field. The first ionization energy needed to dislodge an electron from nitrogen is about 15 eV. The accelerated electron will acquire more than enough energy to ionize a nitrogen atom. This liberated electron will in turn be accelerated which will lead to another collision. A chain reaction then leads to avalanche breakdown and an arc takes place from the cascade of released electrons.

More collisions will take place in the electron path between the electrodes in a higher pressure gas. When the pressure-gap product pd is high, an electron will collide with many different gas molecules as it travels from the cathode to the anode. Each of the collisions randomizes the electron direction, so the electron is not always being accelerated by the electric field — sometimes it travels back towards the cathode and is decelerated by the field.

Collisions reduce the electron's energy and make it more difficult for it to ionize a molecule. Energy losses from a greater number of collisions require larger voltages for the electrons to accumulate sufficient energy to ionize many gas molecules, which is required to produce an avalanche breakdown.

On the left side of the Paschen minimum, the pd product is small. The electron mean free path can become long compared to the gap between the electrodes. In this case, the electrons might gain lots of energy, but have fewer ionizing collisions. A greater voltage is therefore required to assure ionization of enough gas molecules to start an avalanche.

Derivation

Basics

To calculate the breakthrough voltage a homogeneous electrical field is assumed. This is the case in a parallel plate capacitor setup. The electrodes may have the distance d. The cathode is located at the point $x = 0$.

To get impact ionization the electron energy E_e must become greater than the ionization energy E_I of the gas atoms between the plates. Per length of path x a number of α ionizations will occur. α is known as the first Townsend coefficient as it was introduced by Townsend in. The increase of the electron current Γe can be described for the assumed setup as

$$\Gamma_e (x = d) = \Gamma_e (x = 0)e^{\alpha d} \tag{1}$$

(So the number of free electrons at the anode is equal to the number of free electrons at the cathode that were multiplied by impact ionization. The larger d and/or α the more free electrons are created.)

The number of created electrons is

$$\Gamma_e (d) - \Gamma_e (0) = \Gamma_e (0)\, (e^{\alpha d} - 1) \tag{2}$$

Neglecting possible multiple ionizations of the same atom, the number of created ions is the same as the number of created electrons :

$$\Gamma_i (0) - \Gamma_i (d) = \Gamma_e (0)\, (e^{\alpha d} - 1) \tag{3}$$

Γ_i is the ion current. To keep the discharge going on, free electrons must be created at the cathode surface. This is possible because the ions hitting the cathode release secondary electrons at the impact. (For very large applied voltages also field electron emission can occur.) Without field emission, we can write

$$\Gamma_e (0) = \gamma \Gamma_i (0) \tag{4}$$

where γ is the mean number of generated secondary electrons per ion. This is also known as the second Townsend coefficient. Assuming that $\Gamma_i (d) = 0$ one gets the relation between the Townsend coefficients by putting (4) into (3) and transforming :

$$\alpha d = \ln\left(1 + \frac{1}{\gamma}\right) \tag{5}$$

Impact Ionization

What is the amount of α? The number of ionization depends upon the probability that an electron hits an ion. This probability P is the relation of the cross-sectional area of a collision between electron and ion σ in relation to the overall area A that is available for the electron to fly through :

$$P = \frac{N\sigma}{A} = \frac{x}{\lambda} \tag{6}$$

As expressed by the second part of the equation, it is also possible to express the probability as relation of the path travelled by the electron x to the mean free path λ (distance at which another collision occurs).

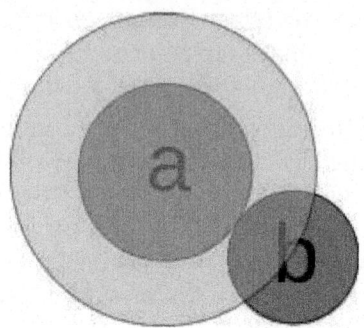

Fig.: Visualization of the cross-section σ : If the center of particle b penetrates the blue circle, a collision occurs with particle a. So the area of the circle is the cross-section and its radius r is the sum of the radii of the particles.

N is the number of electrons because every electron can hit. It can be calculated using the equation of state of the ideal gas

$$pV = Nk_BT \tag{7}$$

(p : pressure, V : volume, k_B : Boltzmann constant, T : temperature)

The adjoining sketch illustrates that $\sigma = \pi(r_a + r_b)^2$. As the radius of an electron can be neglected compared to the radius of an ion r_I it simplifies to $\sigma = \pi r^2 I$. Using this relation, putting (7) into (6) and transforming to λ one gets

$$\lambda = \frac{k_BT}{p\pi r_I^2} = \frac{1}{L \cdot p} \tag{8}$$

where the factor L was only introduced for a better overview.

The alteration of the current of not yet collided electrons at every point in the path x can be expressed as

$$d\Gamma_e(x) = -\Gamma_e(x)\frac{dx}{\lambda_e} \tag{9}$$

This differential equation can easily be solved :

$$\Gamma_e(x) = \Gamma_e(0)\exp\left(-\frac{x}{\lambda_e}\right) \tag{10}$$

The probability that $\lambda > x$ (that there was not yet a collision at the point x) is

$$P(\lambda > x) = \frac{\Gamma_e(x)}{\Gamma_e(0)} = \exp\left(-\frac{x}{\lambda_e}\right) \tag{11}$$

According to its definition α is the number of ionizations per length of path and thus the relation of the probability that there was no collision in the mean free path of the ions, and the mean free path of the electrons :

$$\alpha = \frac{P\,(\lambda > \lambda_I)}{\lambda_e} = \frac{1}{\lambda_e}\exp\left(-\frac{\lambda_I}{\lambda_e}\right) = \frac{1}{\lambda_e}\exp\left(-\frac{E_I}{E_e}\right) \tag{12}$$

It was hereby considered that the energy E that a charged particle can get between a collision depends on the electric field strength ε and the charge Q :

$$E = lQ\varepsilon \tag{13}$$

Breakdown Voltage

For the parallel-plate capacitor we have $\varepsilon = \dfrac{U}{d}$, where U is the applied voltage.

As a single ionization was assumed Q is the elementary charge e. We can now put (13) and (8) into (12) and get

$$\alpha = L \cdot p \exp\left(-\frac{L \cdot p \cdot d \cdot Er}{eU}\right) \tag{14}$$

Putting this into (5) and transforming to U we get the Paschen law for the breakdown voltage $U_{breakdown}$ that was first investigated by Paschen in and whose formula was first derived by Townsend in, :

$$U_{breakdown} = \frac{L \cdot p \cdot d \cdot E_I}{e(\ln(L \cdot p \cdot d) - \ln(\ln(1 + \gamma^{-1})))} \tag{15}$$

with $\qquad\qquad L = \dfrac{k_B T}{\pi r_I^2}$

Plasma Ignition

Plasma ignition in definition of Townsend (Townsend discharge) is a self-sustaining discharge, independent of an external source of free electrons. This means that electrons from the cathode can reach the anode in the distance d and ionize at least one atom on its way. So according to the definition of α this relation must be fulfilled :

$$\alpha d \geq 1 \tag{16}$$

If $\alpha d = 1$ is used instead of (5) one gets for the breakdown voltage

$$U_{brakdown\,Townsend} = \frac{L \cdot p \cdot d \cdot E_I}{\ln(L \cdot p \cdot d)} = \frac{d \cdot E_I}{\lambda_e \ln\left(\dfrac{d}{\lambda_e}\right)} \tag{17}$$

Conclusions / Validity

Paschen's law requires that

- There are already free electrons at the cathode ($\Gamma_e(x = 0) \neq 0$) which can be accelerated to trigger impact ionization. Such so-called *seed electrons* can be created by ionization by cosmic x-ray background.

- The creation of further free electrons is only achieved by impact ionization. Thus Paschen's law is not valid if there are external electron sources. This can for example be a light source creating secondary electrons via the photoelectric effect. This has to be considered in experiments.
- Each ionized atom leads to only one free electron. But multiple ionizations occur always in practice.
- Free electrons at the cathode surface are created by the impacting ions. The problem is that the number of thereby created electrons strongly depends on the material of the cathode, its surface (roughness, impurities) and the environmental conditions (temperature, humidity etc.). The experimental, reproducible determination of the factor γ is therefore nearly impossible.
- The electrical field is homogeneous.

Effects with Different Gases

Different gases will have different mean free paths for molecules and electrons. This is because different molecules have different diameters. Noble gases like helium and argon are monatomic and tend to have smaller diameters. This gives them a greater mean free path length.

Ionization potentials differ between molecules as well as the speed that they recapture electrons after they have been knocked out of orbit. All three effects change the number of collisions needed to cause an exponential growth in free electrons. These free electrons are necessary to cause an arc.

Chapter 4

MOTOR MAINTENANCE

Industrial and manufacturing facilities rely on motors to drive their processes. They are included in preventive maintenance programs because of investments, their critical importance to operations, and the cost of production downtime.

Motors are less visible in commercial facilities, where they are typically located behind closed doors in equipment rooms and closets. Nonetheless, they are very important because they drive fans, pumps, chillers, compressors, and other mechanical equipment that directly impacts facility operation and the well-being of its occupants.

INDUCTION MOTORS

Induction motors transform electrical energy into mechanical energy by induction, just as energy is transferred from the primary to the secondary of a transformer. The only difference between an induction motor and a transformer is that the "secondary" of an induction motor is the rotor, which rotates.

The "primary" is the stator winding, and it produces the rotating magnetic field that drives the rotor by inducing current. Most induction motors used in commercial applications are squirrel cage induction motors, which consist of bars embedded in a solid rotor and connected electrically through solid end rings. Only wound-rotor induction motors actually have windings on the rotor, and only some have connections to the rotor windings with brushes and slip rings for speed or torque control.

NEED FOR PREVENTIVE MAINTENANCE

Preventive maintenance involves identifying potential problems and correcting them before the equipment fails. Due to their construction, induction motors are very reliable, so many customers do not have effective preventive maintenance programs.

This is especially true for smaller commercial facilities with limited maintenance staff who lack the time and understanding to perform routine motor inspections and maintenance. These customers typically do not believe that any type of preventive maintenance needs to be performed and that motors just run until they fail and then are either rebuilt or replaced.

Customers should be informed that unnecessary expensive repairs, premature motor rebuilds and replacements, equipment and system downtime, and disruption to facility and occupant operations can be avoided by establishing a simple cost-effective preventive maintenance program.

WHEN DO MOTORS FAIL?

Motors tend to fail early in their service life due to manufacturing defects, damage before or during installation, improper installation, or misapplication. Similarly, motors have a high failure rate as they approach their rated life, which is typically determined by their insulation. In between these two endpoints, motors should experience low failure rates.

MOTOR PREVENTIVE MAINTENANCE

Motor preventive maintenance, especially for small induction motors, is more than just inspecting and maintaining the motor itself. If left running under rated conditions, an induction motor will provide many years of trouble-free service and may even outlast the building.

However, induction motors don't always operate in ideal conditions. Even in the best operating environments, they are occasionally subjected to adverse operating conditions due to spills, equipment failure, operator error, poor power quality, and other factors.

Motor failure is often symptomatic of external factors. A cost-effective preventive maintenance program for small induction motors should focus on external factors because service personnel can easily assess them and identify potential problems. A strategy for correcting the potential problem can be developed or additional tests can be performed on the motor to determine the gravity of the situation.

PREVENTIVE MAINTENANCE ACTIVITIES

Preventive maintenance activities should focus on inspection and limited testing. It should address the motor's physical environment, its condition, and the load that it serves. The physical environment is especially important because it reflects on the motor's insulation life, which in turn determines when the motor fails.

The ambient area temperature in which the motor is installed needs to be checked. Facilities change over time, and space may be reconfigured, ventilation may become restricted, and new motors and other heat-generating equipment may be installed near the motor.

Changes like these can result in an increase in ambient temperature, which can significantly reduce the motor's insulation life and must be corrected. A quick check of the motor's operating temperature can be made by using a contact thermometer or feeling the motor by hand.

Dirt and grease buildup on the motor or its windings can also result in insulation failure. Buildup on windings can reduce insulation levels and result in insulation failure that will require rewinding or replacement. Similarly, dirt and grease buildup affect the motor's ability to transfer excess heat to its surroundings, which will damage winding insulation.

Excessive dirt and grease need to be removed from the motor housing and ventilating openings. If found in the windings, they should be removed in accordance with manufacturer recommendations.

The electrical operation of the motor and its mechanical output can quickly be checked by measuring the current drawn while the motor is operating under load by using an ammeter, which measures true root mean squared (TRMS) current. Under normal operation, the phase currents should be balanced, and the current magnitude should be at or below the motor's nameplate rating.

Unbalanced phase currents may indicate a problem in the motor's windings, which can also be the electrical cause of excessive vibration, or slowness to reach starting rated speed. Similarly, excessively high currents can indicate winding problems in the motor, motor bearing problems, motor-load alignment, or a problem with the load itself. If any of these symptoms are encountered, additional investigation and testing needs to be performed to determine the root of the problem.

All rotating machinery vibrates when it operates, but excessive motor vibration can damage insulation and bearings. Electrical problems such as open circuits in the bars and end rings of a squirrel cage rotor can cause vibration.

More likely though, excessive mechanical vibration is caused by a mechanical problem in the motor such as a bearing failure, a failure to properly anchor the motor, misalignment between the motor and the driven load, or a mechanical problem in the load. The purpose of preventive maintenance in this case would be to note that there is excessive vibration by observing the operation of the motor, noting excessive noise, and/or simply placing a hand on the motor while in operation.

If excessive vibration is suspected, then additional testing is required..

ESTABLISHING AN EFFECTIVE MAINTENANCE PROGRAM

You must balance the costs of motor maintenance and failure. Maintenance takes time and is an expense for your customer. It would be impractical to perform the same tests, inspections, and maintenance on a 30 horsepower motor driving a fan as it would on a 300 horsepower motor driving a chiller. However, simple preventive maintenance activities are cost effective and can help reach their rated life by ensuring that these smaller motors are operating properly.

When establishing the preventive maintenance program, the motor's size and cost should not be the only factors considered. The motor's function and impact on the facility and its occupants should also be taken into account to determine the appropriate type of regular maintenance. In addition, any hazards that might result from a motor failure should also be factored into its maintenance program.

For example, a small motor driving an exhaust fan that removes fumes from a paint booth where artists work may require more attention than one that does not have the same safety implications for occupants.

If you already perform regular lighting maintenance for your customer, you could offer to perform routine motor maintenance. This should reduce the cost of performing the motor maintenance because service personnel are already on site. Similarly, other non-critical service work could be scheduled.

This type of preventive maintenance focuses on the customer's smaller motors and involves mainly inspection and simple tests to identify potential problems that need to be corrected in order to avoid premature failure. It does not focus on large motors that require specific maintenance, testing procedures, and scheduled downtime and personnel that specialize in motor maintenance and repair.

LARGE MOTOR MAINTENANCE : BASICS FOR MACHINE RELIABILITY

The true workhorses of industry, electric motors, provide the means to convert electrical energy into a meaningful and measurable output. Because they are so prevalent and critical to industry, the ability to accurately diagnose, predict and efficiently deal with motor problems is essential to maintenance, engineering and operations personnel.

One of the bigger challenges is being able to recognize, diagnose and remedy an evolving motor problem — *to the point that you can prevent an unexpected catastrophic event.* Understanding the basic visual, mechanical and electrical maintenance techniques will help you in this quest to keep large electric motors on line and producing.

The U.S. Department of Energy estimates that approximately 63% of all electricity used in industry is for process motor system energy.

(Refer to Fig. for a breakdown of the per cent averages by equipment type.) Industrial motor use accounts for approximately 25% of the total electricity usage in the United States. Thus, it makes sense that one would take all the necessary steps and precautions to assure long-term health and reliability of these vital machines, especially when they are of a larger horsepower.

Using guidance found within Inter National Electrical Testing Association (NETA) field-testing specifications, the ANSI/NETA MTS-2007 *Standard for Maintenance Testing Specifications for Electrical Power Distribution Equipment and Systems,* The focus will be primarily on medium-voltage AC induction (2.3 kV - 13.2 kV) machines.

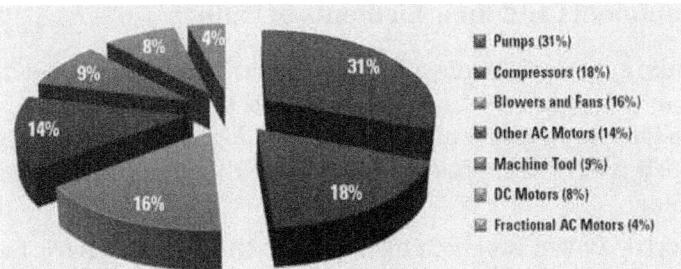

Fig. : Typical breakdown of energy usage, in percentage, by equipment type in an industrial facility (*Source* : U.S. DOE).

First and Foremost : Understand the Hazards!

Working on or near electrical equipment is by its very nature a hazardous task. Before any equipment is inspected or any maintenance is done, the person performing the tasks must be qualified and able to assess all hazards associated with the scope of work to be performed. If the person performing the work is not qualified, the end results could be significant equipment damage or possibly serious injury or death to personnel.

Some maintenance tasks require the work to be performed on energized equipment while it is in normal operation. Personnel should be familiar with – and comply with – the applicable OSHA, NFPA 70E, plant-specific and other electrical safety rules and regulations.

Whenever performing de-energized motor maintenance – *and before physically touching the motor* – one should be sure that the unit in question does not present a shock hazard to anyone that will be working on that particular motor. At a minimum, the affected parties should be notified, the machine should be locked out/tagged out, tried out, checked for the absence of voltage (live-dead-live), and the proper personnel protective grounds applied. Again, the key is to have a qualified worker perform these tasks to assure those in the area that an electrically safe condition has been obtained.

Electric Motor Basics

Motor Component Stresses				
Types of Stresses	Stator Winding	Rotor Assembly	Bearings	Shaft
Thermal	■	■	■	■
Electrical/Dielectric	■	■	■	
Mechanical	■	■	■	■
Dynamic		■	■	■
Shear				■
Vibration/Shock	■	■	■	■
Residual		■		■
Electromagnetic	■	■	■	■
Environmental	■	■	■	■

Fig. : The four basic components of electric motors can experience many different stresses that have the potential to cause failure. (*Source* : EASA).

Basic Components and their Elements of Failure

Understanding these elements of stress and the components they apply to is a key factor in being able to properly apply and prioritize the necessary maintenance decisions for large motors. The chart in Fig., on the next page, illustrates the relationship between types of stresses and the four basic components of an electric motor.

An Electric Power Research Institute (EPRI) study of electric motor failures indicated that 53% of electric motor failures are related to mechanical components and 47% to electrical faults (not including new and repaired equipment defects). Mechanical defects have traditionally been detected using vibration analysis and infrared thermography, while electrical defects have been detected with resistance tests, insulation tests, high-potential tests, surge comparison tests and partial discharge testing. The failure in Fig. was attributed to excessive vibration and heat from a defective fan shaft bearing.

Four Basic Electric Motor Components

While materials and insulation systems have changed, the basic principles and operation of an electric motor have not changed very much over the last 100+ years. An electric motor is made up of four basic components :

1. Stator winding
2. Rotor assembly
3. Bearings
4. Shaft.

Fig.

These components are exposed to stress conditions, failure of the motor can occur. The nine key elements of stress that can lead to motor failure are :

1. Thermal
2. Electrical/dielectric
3. Mechanical
4. Dynamic
5. Shear
6. Vibration/shock
7. Residual
8. Electromagnetic
9. Environmental.

To counteract these stress elements, one of the key parameters to an effective motor maintenance program is to establish test and inspection procedures that allow the owner to trend data over time. It is this trended data that helps in diagnosing the overall health of the machine. Let's look at the visual and mechanical tests/inspections and electrical tests that can be performed on large machines.

Visual and Mechanical Inspections

An important aspect of large machine maintenance is the visual and mechanical inspection.

1. Inspect the machine's physical and mechanical condition.
   o Check for signs of oil or water leakage.
   o Verify that air inlets are not plugged.
   o Check for abnormal sounds or smells.
   o Check the water and oil supply piping.
   o Check the drain piping.
   o Look at the condition of the foundation, grout, bed plates, anchor bolts, shaft extensions, couplings and guards.
   o Check the surroundings for any environmental issues that may affect performance or service life.
2. Inspect anchorage, alignment and grounding of the motor, driven equipment and base.
3. Inspect air baffles, filter media, cooling fans, slip rings, brushes and brush rigging.
4. Inspect bolted electrical connections for high resistance.
5. While the unit is under full load, perform a thermographic survey.
6. Perform special tests such as air-gap spacing and machine alignment, if applicable.

7. Verify the application of appropriate lubrication and lubrication systems.

    o Verify the bearing oil level.

    o Check for improper lubrication, oil of the wrong type, viscosity that is too heavy or too light.

    o Verify that there is sufficient oil in bearing bracket to cover bottom of rings.

    o Look for dirty oil or old oil (should be replaced and/or tested).

    o Verify that the oil rings are turning (especially at low temperatures).

    o Check for water or other contamination within the lubrication system.

    o Verify that the feed oil is connected to the correct ports when the bearing and seals are inspected the following should be considered :

    o Check for excessive bearing clearance.

    o Verify seal clearance and condition.

    o Make sure there is not improper seating of shaft journal in bearing or a bent shaft.

8. Verify the absence of unusual mechanical or electrical noise or signs of over-heating.

    o Check for pitting of bearing and journal surfaces due to bearing currents.

    o Verify integrity of bearing insulation.

    o Make sure there are no rough bearing surfaces due to corrosion or careless handling.

    o Verify that there is not excessive end thrust from the mechanical load.

    o Check for poor alignment.

    o Make sure that the bearing Babbitt has not been fractured or damaged due to impact or shock loading of the bearing journal.

9. Verify that resistance temperature detector (RTD) circuits conform to drawings and are functioning properly.

Electrical Tests for AC Induction Motors

As mentioned previously, the collection of valid test data and the trending of that data are vital if overall machinery health is to be determined. Electrical tests performed on large motors can yield significant information as to the overall health of the machine.

Some of the more common electrical tests and procedures include :

1. Resistance measurements taken through bolted connections with a low-resistance ohmmeter;

2. Insulation-resistance tests in accordance with ANSI/IEEE Standard 43;

3. DC overpotential tests on machines rated at 2300 volts and greater in accordance with ANSI/IEEE Standard 95;

4. Phase-to-phase stator resistance tests on machines 2300 volts and greater;

5. Insulation power-factor or dissipation-factor tests;

6. Power-factor tip-up tests;

7. Surge comparison tests;

8. Insulation-resistance tests on insulated bearings;

9. Testing and inspection of surge protection devices;

10. Testing and inspection of motor starters;

11. Resistance tests on resistance temperature detector (RTD) circuits;

12. Verification of machine space heater operation, if applicable;

13. Vibration testing of motor after it has started running.

THERMAL ELECTRIC MOTOR PROTECTION

The term thermally protected in motors, pertains to a thermal protector placed internallly within a motor or motor-compressor to protect the motor and motor windings components from dangerous overheating that can cause motor failure.

This overheating generally occurs when the motor is overloaded, a bearing seizes up, something locked the motor shaft from turning, or the motor simply fails to start properly. A failure to start may be caused by faulty start windings in a motor.

The thermal protector consists of one or more heat-sensing elements integral with the motror or motor-compressor and an external control device. The thermal protection is in place to turn the motor off when excessive heat is generated within the motor circuitry and keep it from burning up the motor.

These thermal protectors are reste when the motor cools to a safe operating temperature. There is usually a visible red button located on the wiring side of the motor, opposite of the motor shaft, but not always. In some cases, these buttons are reset manually, but not always.

Although having a motor shut down because it tripped a thermal limit is inconvenient, it certainly is better than having to replace a motor. Just think of it as a second chance to find out what the problem is with the motor or the connected devices or load attached to it.

You see, just because a motor fails to start or overheats during operation, doesn't mean that the motor is necessarily at fault. There could be an obstruction on the attached load to the motor, causing excessive load on the motor. This in turn causes excessive heat to build up in the motor and hopefully, the thermal protector trips the motor circuit to save the motor.

A good example of this is a sump pump, that probably everyone has used at one time or another. Imagine that the sump pump is pumping dirty water with sticks and who knows what all out of a sump pump hole. Suddenly, a piece of a stick gets caught in the impeller of the pump and locks the rotation of the pump motor. It keeps the motor from turning and causes the motor to heat very quickly. Suddenly the thermal protector trips and shuts the circuit down to the motor windings. This allows the motor to cool and saves the motor. Unfortunately, the water is no longer pumping, but it obviously wouldn't have anyway and you'd likely be changing the sump pump out, rather than just cleaning out the obstruction. That's not to say the pump motor hasn't gone bad, as they often fail from hours of use.

LUBRICATING THE FAN MOTOR ON CENTRAL AIR CONDITIONERS

Central air conditioners rely on fans to blow air through the condenser coils in order to dissipate heat. These fans are subject to the sun's heat, rain, snow, and other adverse conditions, including dirt and grime. To keep the fan in tiptop shape, these fan motors need maintenance in the form of regular oiling.

In order to oil the motor, you must first turn the power off to the system and remove the fan cage, which holds the fan motor in place. Turn the cage assembly upside down to access the oil ports on the motor. These ports will be located on the top of the motor, just below the fan blade of the motor.

Remove the oil plugs that protect the motor from debris. Each port should be oiled with three drops of all-purpose, three-in-one oil. Spin the fan blade slowly by hand to disperse the oil within the fan.

Now, replace the oil plugs and wipe any excess oil that might have spilled. This process should be repeated each season to ensure proper lubrication and long wear for your fan motor.

ELIMINATING CEILING FAN NOISE

Ceiling fans are not only a great addition to your home for circulating both heat and air conditioning around a room, they also add beauty to a room. Some ceiling fans are huggers that hug the ceiling. Others have long extension poles that allow them to hang down from angled roof, like "A"-frame houses have. More and more, people have made ceiling fans a popular choice when building or remodelling their home. Ceiling fans have been documented to reduce both heating and cooling costs when they are used in addition to the heating and cooling units. These fans sometimes get used day and night, but sometimes not much at all. They do require maintenance from time to time and these are some things that need to be addressed when ceiling fans become noisy.

One of first things to check on when you have a noisy ceiling fan is dirty blades. Dirty blades can cause a fan to wobble and shake. To clean, simply use a soft cloth with warm, soapy water or mild grease cleaner in extreme cases to clean them. Dry the blades with a towel and after allowing it to dry, test the fan to see if the problem disappeared.

Sometimes, however, the wobble may be something very simple. Check the bolts that hold the ceiling fan blades onto the fan motor. From time-to-time these blades can become loose and the blade will wobble, causing undue noise.

An intermittent noise may be the fan blade brackets dragging on the fan shroud. Often times, the shroud can be adjusted by simply loosening the four shroud screws and moving its position to clear the rubbing sound. Then, just tighten the screw back into place and the fan is as good as new.

And then we come to probably the biggest culprit of them all, the light fixtures and their lenses. Some light kits have four lights that each has three set screws that keep the lenses from falling off. These screws loosen as vibration increases and falling light fixture lenses can be dangerous.

To help cure rattling lenses, you can also place rubber lens guards that are placed between the glass lenses and the set screws. These are the equivalent of a rubber band, but they do work in reducing the noises with loose lenses.

Remember, shaking or noisy fans are a sign that something is terribly wrong with your ceiling fan. This should be addressed immediately! New ceiling fans come complete with clips to attach to your ceiling fan blades in an effort to balance them when things get a little out of whack. A smooth-running ceiling fan can reduce your heating and cooling so don't forget to take a few minutes to maintain the things that save you money throughout the year.

HOW TO CALCULATE SAFE ELECTRICAL LOAD CAPACITIES

We all have a mountain of electrical appliances around the house and may, if not all, of them have some sort of motor running them. These may include furnaces, dishwashers, sump pump, garbage disposal, and microwaves. These motorized gadgets need a dedicated circuit just for their own use. You see, they should not be on a shared circuit with anything else.

So how is one to know what sized circuit to put each of these items on? You see, the circuits are protected by either circuit breakers or fuses that limit the amount of amperage allowed to flow through that circuit. They watch over the circuit's power draw like a watchdog. But still, how do we determine the right size for these circuit breakers and fuses?

Motors have a nameplate rating that is listed on the side of the motor. It lists the type, serial number, voltage, whether it is AC or DC, the RPM's, and the amperage rating. If you know the voltage and amperage rating, you can determine the wattage or total capacity needed for the safe operation of the motor.

By using Ohm's Law, we can determine what the wattage of the motor is and determine what sized breaker or fuse is needed to protect it.

In order to do this calculation, you simply take the amperage (AMPS) times the voltage (VOLTS) to give you the power (WATTAGE). But we're not done yet.

A 15-amp circuit that is running on 120 volts has a total capacity of 1,800 watts. To determine the safe capacity, you need to multipy the 1,800 watts times 80% to give you 1,440 watts. The rating of your motor should not exceed this rating. so let's say your motor is 120 volts and 13 amps. 120V X 13A = 1,560 watts. Now consider a 20-amp circuit gives you 20A X 120V = 2,400 watts. 2,400 watts X 80% = 1,920 watts of safe capacity, more than enough for this installation. You can see that the 20-amp circuit will fit this installation well.

By reading the nameplate information, doing a little math, and sizing the ciruit protection properly, you can safely operate the motor.

DC MOTOR MAINTENANCE

A D.C. motor maintenance program is a preventative and corrective maintenance schedule that covers inspections, cleaning, testing, replacement and lubrication tasks that are necessary to ensure the proper operation of D.C. motors and associated equipment. A maintenance program is fairly easy to develop and implement. Once regular maintenance checks are incorporated into a shop's work schedule, they soon become transparent yet will reap cost-savings well beyond the investment in time and materials spent in maintaining them.

Overview

In brief, a D.C. motor maintenance program begins with reviewing a motor's service history. This review may reveal on-going problems that are both integral to the motor itself as well as external conditions (*e.g.*, overloads, unbalances, misapplications) that are adversely affecting the motor's normal operating condition. After a service history review, a visual inspection should be performed to identify any obvious wear, blockages to cooling fans, or environmental contamination (moisture or corrosion). This inspection should also be conducted after the motor has been disassembled to discover evidence of failed components, such as burnt windings, broken leads, etc. Next, the motor's windings should be tested. For most maintenance shops, this is a ground insulation test. Since the commutator and brush assembly are high-wear parts of a D.C. motor, extra time should be spent on inspecting, repairing or replacing these vital components. Finally, an inspection of the bearings should be performed; worn out or noisy bearings require replacement. In the event that sealed, non-lube bearings are not used, the motor should be lubricated and then reassembled. An operational test should be performed prior to the motor being shipped out of the repair shop.

Table 1 : D.C. Motor Maintenance Program

Wipe off dust, dirt, oil, etc.	Monthly
Clean vent screens and fans	Quarterly
Lubricate bearings (if applicable)	Semi-annual

Vacuum or blow out interior	Semi-annual
Check commutator, brushes leads	Semi-annual
Check brush spring tension	Semi-annual
Test field coils	Semi-annual
Test armature windings	Semi-annual
Check electrical connections	Semi-annual

Maintenance Guide Summary

This maintenance guide will discuss standard maintenance procedures for maintaining most D.C. motors. To determine the maintenance requirements of a specific motor, the maintenance technician should refer to the manufacturer's technical documentation prior to performing maintenance. This guide is divided into the following sub-sections :

- Reviewing the Service History
- Noise and Vibration Inspections
- Visual Inspection
- Windings Tests
- Brush and Commutator Maintenance
- Bearings and Lubrication.

Reviewing the Service History

D.C. motor maintenance, as with all types of industrial maintenance, requires pre-maintenance planning and scheduling. This starts with reviewing the motor's service history usually contained within an equipment maintenance log or, if the log is not available, interviewing the customer, operator or responsible party to determine what type of maintenance is required, preventative or corrective (failure repair).

The goal is to determine :

- What kind of maintenance is required.
- What maintenance personnel are needed to perform the maintenance (skill level).
- What parts are needed to complete the maintenance (*i.e.,* bearings, brushes, etc.)
- What kind of scheduling or co-ordination with other departments is required to perform the maintenance (downtime or off-hours scheduling)
- What kind of safety hazards exist that would interfere with the maintenance.
- If there are problems other than the motor itself that caused the motor breakdown.

Noise and Vibration Inspections

Prior to disconnecting the motor and sending it to the shop or a repair facility for maintenance, a noise and vibration inspection should be conducted. This requires the motor to be connected to its driven load, energized and operated normally (if possible). The existence of mechanical noises or vibrations can indicate a variety of problems, such as mechanical and/or electrical imbalances, misalignments, brush chattering, bad bearings, bent shafts, mechanically loose windings (shaken lose by excessive vibrations, for instance) or simply a loose cooling fan or something stuck inside the vents or shroud. If the windings are loose, after it is disassembled, check for insulation and lead damage. Vibrations can also be the root cause for excessive heat and brush sparking.

Noises and vibrations are not limited to mechanical problems or imbalances; electrical imbalances, such as open or shorted windings or uneven airgaps, can cause noises or vibrations. The easy way to troubleshoot an electrical from an mechanical imbalance is to first power up the motor, then disconnect power. If the noise/vibration exists while it is unpowered, the problem is mechanical; if the noise stops while power is disconnected, the problem is usually electrical.

Visual Inspections

Before disassembly, refer to the manufacturer's technical documentation on recommended inspection tests or procedures. This documentation will provide valuable information for conducting visual inspections.

A visual inspection is meant to observe and record anomalies about the physical condition of the motor in a de-energized state. A motor that appears dirty, corroded or has the "beat up" look indicates that it was operated in a rough environment and may have more problems than usual. This inspection should include the "smell" test. Is there a burned odour coming from the motor windings? The burnt smell is coming from the insulation varnish of the motor windings. If so, this suggests an overheating problem. Motor winding damage is possible under these conditions so winding tests should be conducted.

Overheating problems may not necessarily be internal to the motor; rather, they could be the result of mechanical overloads such as jams in the driven load or a cold oil that is being pumped via a motor drive, running the motor at low speeds such that there's inadequate cooling airflow, electrical noise from DC drives overheating the windings or it could simply be the result of a dirty environment. Dirt acts like a heat insulator and heat damage is the weakness of normal motor operation. Inspect the cooling fan and passages to ensure they are operational and free of blockage, respectively. Clean all surfaces with a rag and blow out or vacuum passages with a shop vac. Corrosion can damage motor windings as well as create high resistance wiring connections. If the corrosion is chronic, a motor rewind may be merited if the winding tests verify that winding damage has occurred. Re-lugging the motor connection box terminals also may be required.

Motor Winding Tests

Once the motor is disassembled, and a thorough inspection of internal components has been conducted, testing the motor windings is done. This is where a maintenance history can prove its value. What kind of service history is on record relative to the winding failures or abnormalities? Has the motor ever been rewound? If so, what was the cause of the failure? This information suggests what kind of motor winding tests are necessary. In some situations, motor winding tests beyond the ground insulation (megger) test may need to be conducted. Is there any evidence of overheating of the windings? This may appear as burn marks, cracks, or, if catastrophic, exposed wire. Severe damage would require rewinding the motor.

Once again, look at the physical condition of the windings. If they are dirty or corroded, clean the windings with a brush, hot water and detergents and a vacuum. Check the manufacturer's documentation before using any solvents or detergents to ensure they will not damage the insulation. Avoid using pressurized air because the force of the air may propel particles into the winding insulation and damage it. Is there moisture on the windings? If so, prior to conducting any winding tests, the windings must be thoroughly dried out. Moist or wet windings will generally give false readings when conducting insulation tests so the windings must be dried first. This is done by baking the motor windings in an oven until the insulation resistance is at least 10 megohms. Refer to the manufacturer's technical documentation on specific requirements. If this does not work, consider revarnishing the motor first. If the motor passes the insulation tests, this is an adequate solution. If not, a rewind will be necessary, a job that is beyond the capabilities of most general maintenance shops.

The standard way to test winding insulation is the megger test which applies a D.C. voltage, usually 500 or 1000 volts, to the motor and measures the resistance of the insulation. The minimum insulation resistance to ground is 1 megohm per kv of rating plus 1 megohm at 40 degrees Celsius ambient. Measurements of 50 megohms or more are common. Resistance readings depend on the motor size, type of wire, etc. Refer to the manufacturer's documentation for the specific values of ground insulation resistance. One caveat about testing ground insulation with a megger : the values can vary so conduct several tests over a period of time. Low readings indicate a problem that needs to be investigated. A ground insulation test is not a comprehensive motor insulation test; it does not, for instance, test the insulation's resistance between turns of the windings. To test coil-to-coil or turn-to-turn insulation failures, a high surge test, the Hipot test, would be required and requires special test equipment

Brush and Commutator Maintenance

The brushes and commutator are integral to the normal operation of a D.C. motor. The brushes ride or slide on the rotating commutator of the armature; there should be little brush noise, chatter or sparking when the motor is powered up. Excessive brush wear or chipping are signs that the motor is not commutating properly, which can be caused by a variety of factors. While de-energized, rotate the armature by hand to see if the brushes are free to ride on the commutator and there's adequate spring tension to keep them hugging the commutator. A good brush should have a polished surface which indicates that it has been seated properly. Check the brush connections to ensure they are tight and clean. Determine if the brushes are aligned properly. Misalignment from neutral can cause sparking (armature reaction). The brushes should have equidistant spacing around the commutator and parallel to the bars. Clean any debris around the brushes. Compare the brushes to a new set of brushes to gauge the amount of wear. If excessive or, if you don't think they will last until the next maintenance time, replace them.

The commutator should have a smooth, polished, brown appearance. There should be no grooves, scratches or scores. If there is any blackened, rough areas on the commutator, it's probably caused by brush sparking. If a commutator has a brassy appearance, there's excessive wear that could be caused by the wrong type of brush or the wrong spring tension. Check the manufacturer's technical documentation to verify the correct brushes are installed. Carbon dust and debris from the brushes can cause sparking and damage the commutator. If the commutator is rough and the bars are uneven, it will need to be turned on a lathe to restore its roundness. To clean the commutator, use a commutator cleaning brush (fiberglass) and some electric motor cleaner. Never use emery paper because it has metal particles in it that if rubbed off could cause electrical shorts. Remove the brush springs, slide the brush across the commutator hood and spray. When done, blow out the motor so it is dry and clean.

Bearings : Replacement and Lubrication

There are different types of bearings and the required maintenance on them will depend on the type of bearing, operating environment and the motor application. There are *lubed-for-life*, sealed bearings used in low horsepower motors that do not require lubrication.

Lubrication is only one of three maintenance tasks involved with motor bearings. Cleaning, removal and replacement are the other tasks. In the noise and vibration inspections, the bearings should have been inspected for abnormal noises, vibrations or hot bearings. The "feel" and "sound" tests are simple methods to gauge bearing condition. For the "feel" test, with the motor running, touch the bearing housing. If it is very hot to the touch, the bearing is probably malfunctioning. In the "sound" test, listen for thumping or grinding noises. If they exist, the bearings need a closer look and possible replacement. For most types, the sources of bearing failures are :

- Insufficient oil or grease.
- Too much grease causing churning and overheating.
- Worn bearings (*i.e.*, broken balls or rough races, etc.)
- Hot motor or external environment.

If the service history demonstrates repeated bearing failures, check the manufacturer's specifications to determine if the correct bearing has been installed. If that's not the case, then an external factor could be the cause. Prior to bearing removal or replacement, clean the housing with solvents or flushing oils. The bearings should be cleaned with a lint free rag. Take a lot of care to keep dirt out of the bearing. When bearings need to be replaced, remove them with the proper tool. Hammers should never be used since they can damage the bearing races. The bearing puller's claws should be attached to the sidewall of the inner ring or an adjacent part.

The lubrication schedule depends on the bearing and the motor application. Small-to-medium motors with ball bearings (except sealed) are greased every 3-6 years under normal conditions. A wet, corrosive or high temperature environment may require more frequent lubrication. The proper lubricant is critical to proper lubrication; check with the manufacturer on oil/grease recommendations. Prior to lubrication, remove the relief plug from bottom of the housing in order to prevent excessive pressure during lubrication. After completing greasing, run the motor 5-10 minutes until grease flowing out of the grease hole. This will expel the excess grease.

PREVENTIVE MAINTENANCE OF MOTORS AND CONTROLS

A well-planned preventive maintenance program is the key to dependable, long-life operation of motors and generators. In modern plant operations, unscheduled stoppage of production or long repair shutdowns are intolerable. The high cost of the resultant downtime eats deeply into profits. Although management probably realizes the value of a good preventive maintenance (PM) program, they sometimes resist.

A well-planned preventive maintenance program is the key to dependable, long-life operation of motors and generators.

In modern plant operations, unscheduled stoppage of production or long repair shutdowns are intolerable. The high cost of the resultant downtime eats deeply into profits. Although management probably realizes the value of a good preventive maintenance (PM) program, they sometimes resist the investment in proper tools, instruments, practices, or technical assistance. Therefore, it's very important that you show how a properly planned motor/generator PM program is justified.

The first step is to show that PM pays dividends. For example, illustrate the advantages gained by employing a motor maintenance program. You can do this by collecting case histories of motor breakdowns and the cost of resultant lost production. Show how budgeted PM costs are significantly less than the cost of lost production.

Second, select the best approach. Organizing and setting up the budget for a motor PM program is usually a difficult chore. The program must be effective and, at the same time, its cost must be kept to a practical minimum. Don't underestimate the importance of this initial planning. A PM program won't work if you don't have the proper test equipment and tools, along with trained men to properly apply them. Consider which equipment you'll need and the time required to perform inspections and keep accurate records. Determine which procedures are essential and whether they should be performed by facility electricians or a service organization geared to do the job.

Finally, select the best motor-maintenance techniques. For each type of motor, controller, or related equipment, a variety of maintenance methods may be selected. Choose the best methods and determine to what extent they should be applied. For example, should you check for possible bearing trouble on a motor simply by feeling components for over temperature and listening for unusual sounds, or should you install temperature monitoring devices and make inspections using a stethoscope or an infrared scanner?

Basic Guidelines to Motor Maintenance

Here are some valuable guidelines that you can use in your PM program.

Lubrication. Lubricate regularly according to manufacturer's instructions. On sleeve-bearing and other oil-lubricated machines, check oil reservoirs on a regular basis. In poor environments, change oil at least once a month. Never over-lubricate; excess grease or oil can get into windings and deteriorate insulation. Be sure to use only the lubricant specified for the machine in question. However, you should also check into the possibility of using modern lubricants that have excellent life and lubricating qualities.

Bearing inspection. Bearing failures are one of the most common causes of motor failures. Typical bearing problems include improper lubrication, misalignment of the motor with the load, replacement with the wrong type bearing, excessive loading, and harsh environments.

On essential motors or those that are heavily used or frequently duty cycled, you should check bearings daily using a stethoscope or infrared scanner (or camera, if appropriate). Check bearing surface temperature with a thermometer, electronic temperature sensing devices, or stick-on temperature indicating labels. Compare temperature of hot bearings with the temperatures of normally operating bearings. Check oil rings and watch for excessive end play.

Rotor/stator inspection. Check air gap between the rotor and stator with feeler gages at least annually. Measurements should be made at the top, bottom, and on both sides of the stator. Differences in readings obtained from year-to-year indicate bearing wear.

Belt inspection. Check belt tension; belts should have about 1 in. of play. Sheaves should be seated firmly with little or no play. Couplings should be tight, within tolerances, and should operate without excessive noise. An alignment check should be made on all motor-generator sets and on motor-load couplings when trouble is suspected.

Brush/commutator inspection. Inspect brushes and commutators of DC motors for excessive wear. Check brushes for proper type, hardness, conductivity, and fit in brush holders. Check holder spring pressure with a small scale. In most instances, pressure should be 2 to 2 1/2 lbs per sq in. of brush cross-sectional area. Call manufacturer or service company to solve recurring problems of brush chatter, excessive brush wear, and sparking, streaking, or threading of commutator.

Motor mount inspection. Check mounting bolts, steel base plates for possible warping, and concrete base for cracking or spalling.

Annually, perform vibration-analysis tests. Excessive vibration may be hard to detect by hand, but it could be enough to shorten motor life significantly. It can cause bearing failure, metal fatigue of parts, or failure of windings. The cause of vibration is usually mechanical in nature, such as excessive belt tension, defective sleeve or ball bearings, misalignment, or improper balance. The most common cause is the unbalance of a rotating member (the motor rotor, rotating load, or other drive train component). Simple testing of the motor is done by uncoupling the load or removing the belts and then running the motor. Electrical problems also can cause vibration.

Field vibration analysis can be accomplished by using a portable instrument that identifies vibrations and displays their amplitudes and frequencies.

Motor temperature control. Restricted ventilation will cause a motor to operate at a higher than desired temperature. Dirt, dust, chemicals, snow, oil, grass, weeds, etc., can clog ventilation passages of an open-frame motor. Keep motor clean and cool. In poor environments, blow out dirt with dry compressed air (no more than 50 lbs) as often as needed.

Open dripproof and totally enclosed motors are protected but must not be installed where air flow will be restricted or where excessive ambient temperatures might be encountered. In high-temperature locations, consider the use of energy-efficient motors that operate cooler than standard motors. Excessive ambient temperatures will shorten motor life.

Pull and disassemble important motors during summer shutdowns for thorough inspection, testing, cleaning, checking of bearings, couplings or accessories, or complete reconditioning.

Record keeping. Keep accurate records. Perform annual insulation-resistance (IR) and other appropriate tests. Important motors should also receive a thorough visual inspection, as well as voltage and current checks. All values should be recorded and compared each year. The trend of the readings will indicate the condition of the motor and offer a guide to its reliability.

Basic Guidelines to Control Maintenance

These guidelines will help you maintain motor controls.

Cleanliness. In poor environments, blow out dirt weekly; in normal environments, a quarterly or semi-annual cleaning should be adequate. Make sure that dust or contamination is kept off high-voltage equipment. This is important because dust may contain conducting materials that could form unwanted circuit paths, resulting in current leakage or possible grounds or short circuits.

Moving parts inspection. Moving parts should operate easily without excessive friction. Check operation of contactors and relays by hand, feeling for any binding or sticking. Look for loose pins, bolts, or bearings. If the control is dirty, it should be wiped or blown clean.

Contact inspection. Check contacts for pitting and signs of overheating, such as discolouration of metal, charred insulation, or odour. Be sure contact pressure is adequate and the same on all poles; verify with manufacturer's specification. Watch for frayed flexible leads.

Contact resistance testing. On essential controls, perform contact-resistance tests with a low-resistance ohmmeter on a regular basis. Proper contact resistance should be about 50 micro-ohms. Record readings for future comparison. This will indicate trends in the condition of contacts.

Overload relay inspection. Overload relays should receive a thorough inspection and cleaning. You also should check for proper setting. In general, maintenance requirements for these relays include checking that the rating or trip setting takes into account ambient temperature as well as the higher inrush currents of modern, energy-efficient motors. You also should verify that contacts are clean and free from oxidation and that the relay will operate dependably when needed. Relays should be tested and calibrated every one to three years. Special equipment such as an OL relay tester can be used.

Chapter 5

ELECTRIC GENERATOR

In electricity generation, an **electric generator** is a device that converts mechanical energy to electrical energy. A generator forces electric current to flow through an external circuit. The source of mechanical energy may be a reciprocating or turbine steam engine, water falling through a turbine or waterwheel, an internal combustion engine, a wind turbine, a hand crank, compressed air, or any other source of mechanical energy. Generators provide nearly all of the power for electric power grids.

The reverse conversion of electrical energy into mechanical energy is done by an electric motor, and motors and generators have many similarities. Many motors can be mechanically driven to generate electricity and frequently make acceptable generators.

HISTORY

Before the connection between magnetism and electricity was discovered, electrostatic generators were used. They operated on electrostatic principles. Such generators generated very high voltage and low current. They operated by using moving electrically charged belts, plates, and disks that carried charge to a high potential electrode. The charge was generated using either of two mechanisms :

- Electrostatic induction
- The triboelectric effect, where the contact between two insulators leaves them charged.

Because of their inefficiency and the difficulty of insulating machines that produced very high voltages, electrostatic generators had low power ratings, and were never used for generation of commercially significant quantities of electric power. The Wimshurst machine and Van de Graaff generator are examples of these machines that have survived.

In 1827, Hungarian Anyos Jedlik started experimenting with the electromagnetic rotating devices which he called electromagnetic self-rotors, now called the Jedlik's dynamo. In the prototype of the single-pole electric starter (finished between 1852 and 1854) both the stationary and the revolving parts were electromagnetic. He formulated the concept of the dynamo at least 6 years before Siemens and Wheatstone but didn't patent it as he thought he wasn't the first to realize this. In essence the concept is that instead of permanent magnets, two electromagnets opposite to each other induce the magnetic field around the rotor. It was also the discovery of the principle of self-excitation.

In the years of 1831–1832, Michael Faraday discovered the operating principle of electromagnetic generators. The principle, later called Faraday's law, is that an electromotive force is generated in an electrical conductor which encircles a varying magnetic flux. He also built the first electromagnetic generator, called the Faraday disk, a type of homopolar generator, using a copper disc rotating between the poles of a horseshoe magnet. It produced a small DC voltage.

This design was inefficient, due to self-cancelling counterflows of current in regions that were not under the influence of the magnetic field. While current was induced directly underneath the magnet, the current would circulate backwards in regions that were outside the influence of the magnetic field. This counterflow limited the power output to the pickup wires, and induced waste heating of the copper disc. Later homopolar generators would solve this problem by using an array of magnets arranged around the disc perimeter to maintain a steady field effect in one current-flow direction.

Another disadvantage was that the output voltage was very low, due to the single current path through the magnetic flux. Experimenters found that using multiple turns of wire in a coil could produce higher, more useful voltages. Since the output voltage is proportional to the number of turns, generators could be easily designed to produce any desired voltage by varying the number of turns. Wire windings became a basic feature of all subsequent generator designs.

The **dynamo** was the first electrical generator capable of delivering power for industry. The dynamo uses electromagnetic induction to convert mechanical rotation into direct current through the use of a commutator. The first dynamo was built by Hippolyte Pixii in 1832.

A dynamo machine consists of a stationary structure, which provides a constant magnetic field, and a set of rotating windings which turn within that field. On small machines the constant magnetic field may be provided by one or more permanent magnets; larger machines have the constant magnetic field provided by one or more electromagnets, which are usually called field coils.

Through a series of accidental discoveries, the dynamo became the source of many later inventions, including the DC electric motor, the AC alternator, the AC synchronous motor, and the rotary converter.

Alternating current generating systems were known in simple forms from the discovery of the magnetic induction of electric current. The early machines were developed by pioneers such as Michael Faraday and Hippolyte Pixii.

Faraday developed the "rotating rectangle", whose operation was *heteropolar* - each active conductor passed successively through regions where the magnetic field was in opposite directions. The first public demonstration of a more robust "alternator system" took place in 1886. Large two-phase alternating current generators were built by a British electrician, J.E.H. Gordon, in 1882. Lord Kelvin and Sebastian Ferranti also developed early alternators, producing frequencies between 100 and 300 Hz. In 1891, Nikola Tesla patented a practical "high-frequency" alternator (which operated around 15 kHz). After 1891, polyphase alternators were introduced to supply currents of multiple differing phases. Later alternators were designed for varying alternating-current frequencies between sixteen and about one hundred hertz, for use with arc lighting, incandescent lighting and electric motors.

Large power generation dynamos are now rarely seen due to the now nearly universal use of alternating current for power distribution. Before the adoption of AC, very large direct-current dynamos were the only means of power generation and distribution. AC has come to dominate due to the ability of AC to be easily transformed to and from very high voltages to permit low losses over large distances.

ELECTROMAGNETIC GENERATORS

Dynamo

A dynamo is an electrical generator that produces direct current with the use of a commutator. Dynamos were the first electrical generators capable of delivering power for industry, and the foundation upon which many other later electric-power conversion devices were based, including the electric motor, the alternating-current alternator, and the rotary converter. Today, the simpler alternator dominates large scale power generation, for efficiency, reliability and cost reasons. A dynamo has the disadvantages of a mechanical commutator. Also, converting alternating to direct current using power rectification devices (vacuum tube or more recently solid state) is effective and usually economic.

Alternator

Without a commutator, a dynamo becomes an alternator, which is a synchronous singly fed generator. Alternators produce alternating current with a frequency that is based on the rotational speed of the rotor and the number of magnetic poles.

Automotive alternators produce a varying frequency that changes with engine speed, which is then converted by a rectifier to DC. By comparison, alternators used to feed an electric power grid are generally operated at a speed very close to a specific frequency, for the benefit of AC devices that regulate their speed and performance based on grid frequency. Some devices such as incandescent lamps and ballast-operated fluorescent lamps do not require a constant frequency, but synchronous motors such as in electric wall clocks do require a constant grid frequency.

When attached to a larger electric grid with other alternators, an alternator will dynamically interact with the frequency already present on the grid, and operate at a speed that matches the grid frequency. If no driving power is applied, the alternator will continue to spin at a constant speed anyway, driven as a synchronous motor by the grid frequency. It is usually necessary for an alternator to be accelerated up to the correct speed and phase alignment before connecting to the grid, as any mismatch in frequency will cause the alternator to act as a synchronous motor, and suddenly leap to the correct phase alignment as it absorbs a large inrush current from the grid, which may damage the rotor and other equipment.

Typical alternators use a rotating field winding excited with direct current, and a stationary (stator) winding that produces alternating current. Since the rotor field only requires a tiny fraction of the power generated by the machine, the brushes for the field contact can be relatively small. In the case of a brushless exciter, no brushes are used at all and the rotor shaft carries rectifiers to excite the main field winding.

Induction Generator

An induction generator or asynchronous generator is a type of AC electrical generator that uses the principles of induction motors to produce power. Induction generators operate by mechanically turning their rotor faster than the synchronous speed, giving negative slip. A regular AC asynchronous motor usually can be used as a generator, without any internal modifications. Induction generators are useful in applications such as minihydro power plants, wind turbines, or in reducing high-pressure gas streams to lower pressure, because they can recover energy with relatively simple controls.

To operate an induction generator must be excited with a leading voltage; this is usually done by connection to an electrical grid, or sometimes they are self-excited by using phase correcting capacitors.

MHD Generator

A magnetohydrodynamic generator directly extracts electric power from moving hot gases through a magnetic field, without the use of rotating electromagnetic machinery. MHD generators were originally developed because the output of a plasma MHD generator is a flame, well able to heat the boilers of a steam power plant. The first practical design was the AVCO Mk. 25, developed in 1965. The U.S. government funded substantial development, culminating in a 25 MW demonstration plant in 1987. In the Soviet Union from 1972 until the late 1980s, the MHD plant U 25 was in regular commercial operation on the Moscow power system with a rating of 25 MW, the largest MHD plant rating in the world at that time. MHD generators operated as a topping cycle are currently (2007) less efficient than combined cycle gas turbines.

Other Rotating Electromagnetic Generators

Other types of generators, such as the asynchronous or induction singly fed generator, the doubly fed generator, or the brushless wound-rotor doubly fed generator, do not incorporate permanent magnets or field windings that establish a constant magnetic field, and as a result, are seeing success in variable speed constant frequency applications, such as wind turbines or other renewable energy technologies.

The full output performance of any generator can be optimized with electronic control but only the doubly fed generators or the brushless wound-rotor doubly fed generator incorporate electronic control with power ratings that are substantially less than the power output of the generator under control, a feature which, by itself, offers cost, reliability and efficiency benefits.

Homopolar Generator

A homopolar generator is a DC electrical generator comprising an electrically conductive disc or cylinder rotating in a plane perpendicular to a uniform static magnetic field. A potential difference is created between the center of the disc and the rim (or ends of the cylinder), the electrical polarity depending on the direction of rotation and the orientation of the field. It is also known as a **unipolar generator**, **acyclic generator**, **disk dynamo**, or **Faraday disc**. The voltage is typically low, on the order of a few volts in the case of small demonstration models, but large research generators can produce hundreds of volts, and some systems have multiple generators in series to produce an even larger voltage. They are unusual in that they can produce tremendous electric current, some more than a million amperes, because the homopolar generator can be made to have very low internal resistance.

Excitation

An electric generator or electric motor that uses field coils rather than permanent magnets requires a current to be present in the field coils for the device to be able to work. If the field coils are not powered, the rotor in a generator can spin without producing any usable electrical energy, while the rotor of a motor may not spin at all.

Smaller generators are sometimes *self-excited*, which means the field coils are powered by the current produced by the generator itself. The field coils are connected in series or parallel with the armature winding. When the generator first starts to turn, the small amount of remanent magnetism present in the iron core provides a magnetic field to get it started, generating a small current in the armature. This flows through the field coils, creating a larger magnetic field which generates a larger armature current. This "bootstrap" process continues until the magnetic field in the core levels off due to saturation and the generator reaches a steady state power output.

Very large power station generators often utilize a separate smaller generator to excite the field coils of the larger. In the event of a severe widespread power outage where islanding of power stations has occurred, the stations may need to perform a black start to excite the fields of their largest generators, in order to restore customer power service.

Fig. : Suppose that the conditions are as in the figure, with the segment A_1 positive and the segment B_1 negative. Now, as A_1 moves to the left and B_1 to the right, their potentials will rise on account of the work done in separating them against attraction. When A_1 and neighbouring sectors comes opposite the segment B_2 of the B plate, which is now in contact with the brush Y, they will cause a displacement of electricity along the conductor between Y and Y_1 bringing a negative charge, larger than the positive charge in A_1 alone, on Y and sending a positive charge to the segment touching Y_1. As A1 moves on, it passes near the brush Z and is partially discharged into the external circuit. It then passes on until, on touching the brush X, has a new charge, this time negative, driven into it by induction from B_2 and neighboring sectors. As the machine turns, the process causes exponential increases in the voltages on all positions, until sparking occurs limiting the increase.

ELECTROSTATIC GENERATOR

An **electrostatic generator**, or **electrostatic machine**, is a mechanical device that produces *static electricity*, or electricity at high voltage and low continuous current. The knowledge of static electricity dates back to the earliest civilizations, but for millennia it remained merely an interesting and mystifying phenomenon, without a theory to explain its behaviour and often confused with magnetism. By the end of the 17th Century, researchers had developed practical means of generating electricity by friction, but the development of electrostatic machines did not begin in earnest until the 18th century, when they became fundamental instruments in the studies about the new science of electricity. Electrostatic generators operate by using manual (or other) power to transform mechanical work into electric energy. Electrostatic generators develop electrostatic charges of opposite signs rendered to two conductors, using only electric forces, and work by using moving plates, drums, or belts to carry electric charge to a high potential electrode. The charge is generated by one of two methods : either the triboelectric effect (friction) or electrostatic induction.

Wimshurst Machine

The **Wimshurst influence machine** is an electrostatic generator, a machine for generating high voltages developed between 1880 and 1883 by British inventor James Wimshurst (1832–1903). It has a distinctive appearance with two large contra-rotating discs mounted in a vertical plane, two crossed bars with metallic brushes, and a spark gap formed by two metal spheres.

Van de Graaff Generator

A **Van de Graaff generator** is an electrostatic generator which uses a moving belt to accumulate very high voltages on a hollow metal globe on the top of the stand. It was invented by American physicist Robert J. Van de Graaff in 1929. The potential difference achieved in modern Van de Graaff generators can reach 5 megavolts. The Van de Graaff generator can be thought of as a constant-current source connected in parallel with a capacitor and a very large electrical resistance, so it can produce a visible electrical discharge to a nearby grounding surface which can potentially cause a "spark" depending on the voltage.

TERMINOLOGY

The two main parts of a generator or motor can be described in either mechanical or electrical terms.

Mechanical :

- Rotor : The rotating part of an electrical machine
- Stator : The stationary part of an electrical machine

Electrical :

- Armature : The power-producing component of an electrical machine. In a generator, alternator, or dynamo the armature windings generate the electric current. The armature can be on either the rotor or the stator.
- Field : The magnetic field component of an electrical machine. The magnetic field of the dynamo or alternator can be provided by either electromagnets or permanent magnets mounted on either the rotor or the stator.

Because power transferred into the field circuit is much less than in the armature circuit, AC generators nearly always have the field winding on the rotor and the stator as the armature winding. Only a small amount of field current must be transferred to the moving rotor, using slip rings. Direct current machines (dynamos) require a commutator on the rotating shaft to convert the alternating current produced by the armature to direct current, so the armature winding is on the rotor of the machine.

EQUIVALENT CIRCUIT

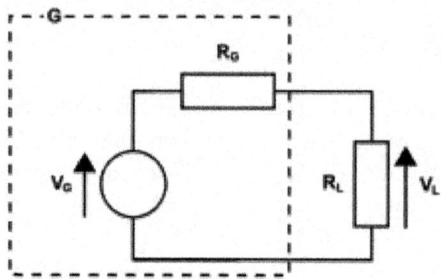

Equivalent circuit of generator and load.

G = generator

V_G = generator open-circuit voltage

R_G = generator internal resistance

V_L = generator on-load voltage

R_L = load resistance.

The equivalent circuit of a generator and load is shown in the diagram to the right. The generator's V_G and R_G parameters can be determined by measuring the winding resistance (corrected to operating temperature), and measuring the open-circuit and loaded voltage for a defined current load.

VEHICLE-MOUNTED GENERATORS

Early motor vehicles until about the 1960s tended to use DC generators with electromechanical regulators. These have now been replaced by alternators with built-in rectifier circuits, which are less costly and lighter for equivalent output. Moreover, the power output of a DC generator is proportional to rotational speed, whereas the power output of an alternator is independent of rotational speed. As a result, the charging output of an alternator at engine idle speed can be much greater than that of a DC generator. Automotive alternators power the electrical systems on the vehicle and recharge the battery after starting. Rated output will typically be in the range 50-100 A at 12 V, depending on the designed electrical load within the vehicle. Some cars now have electrically powered steering assistance and air conditioning, which places a high load on the electrical system. Large commercial vehicles are more likely to use 24 V to give sufficient power at the starter motor to turn over a large diesel engine. Vehicle alternators do not use permanent magnets and are typically only 50-60% efficient over a wide speed range. Motorcycle alternators often use permanent magnet stators made with rare earth magnets, since they can be made smaller and lighter than other types.

A magneto, like a dynamo, uses permanent magnets but generates alternating current like an alternator. Because of the limited field strength of permanent magnets, magneto generators are not used for high-power production applications, but have had specialist uses, particularly in light-houses as they are simple and reliable. This reliability is part of why they are still used as ignition magnetos in aviation piston engines.

Some of the smallest generators commonly found power bicycle lights. Called a bottle dynamo these tend to be 0.5 ampere, permanent-magnet alternators supplying 3-6 W at 6 V or 12 V. Being powered by the rider, efficiency is at a premium, so these may incorporate rare-earth magnets and are designed and manufactured with great precision. The maximum efficiency is around 80% for the best of these generators—60% is more typical—due in part to the rolling friction at the tire–generator interface, imperfect alignment, the small size of the generator, and bearing losses. Cheaper designs tend to be less efficient. Due to the use of permanent magnets, efficiency falls at high speeds because the magnetic field strength cannot be controlled in any way. Hub dynamos remedy many of these flaws since they are internal to the bicycle hub and do not require an interface between the generator and tire. The increasing use of LED lights, more efficient than incandescent bulbs, reduces the power needed for cycle lighting.

Sailing boats may use a water- or wind-powered generator to trickle-charge the batteries. A small propeller, wind turbine or impeller is connected to a low-power alternator and rectifier to supply currents of up to 12 A at typical cruising speeds.

Still smaller generators are used in micro-power applications.

ENGINE-GENERATOR

An *engine-generator* is the combination of an electrical generator and an engine (prime mover) mounted together to form a single piece of self-contained equipment. The engines used are usually piston engines, but gas turbines can also be used. And there are even hybrid diesel-gas units, called dual-fuel units. Many different versions of engine-generators are available-ranging from very small portable petrol powered sets to large turbine installations. The primary advantage of engine-generators is the ability to independently supply electricity, allowing the units to serve as backup power solutions.

HUMAN POWERED ELECTRICAL GENERATORS

A generator can also be driven by human muscle power (for instance, in field radio station equipment).

Human powered direct current generators are commercially available, and have been the project of some DIY enthusiasts. Typically operated by means of pedal power, a converted bicycle trainer, or a foot pump, such generators can be practically used to charge batteries, and in some cases are designed with an integral inverter. The average adult could generate about 125-200 watts on a pedal powered generator, but at a power of 200 W, a typical healthy human will reach complete exhaustion and fail to produce any more power after approximately 1.3 hours. Portable radio receivers with a crank are made to reduce battery purchase requirements. During the mid 20th century, pedal powered radios were used throughout the Australian outback, to provide schooling (School of the Air), medical and other needs in remote stations and towns.

LINEAR ELECTRIC GENERATOR

In the simplest form of linear electric generator, a sliding magnet moves back and forth through a solenoid - a spool of copper wire. An alternating current is induced in the loops of wire by Faraday's law of induction each time the magnet slides through. This type of generator is used in the Faraday flashlight. Larger linear electricity generators are used in wave power schemes Template : Linear electric generator.

TACHOGENERATOR

A tachogenerator is an electromechanical device which produce an output voltage proportional to its shaft speed. It may be used for a speed indicator or in a feedback speed control system. Two commonly used tachogenerators are DC and AC tachogenerators.

Tachogenerators are frequently used to power tachometers to measure the speeds of electric motors, engines, and the equipment they power. Generators generate voltage roughly proportional to shaft speed. With precise construction and design, generators can be built to produce very precise voltages for certain ranges of shaft speeds.

DIESEL GENERATOR

A **diesel generator** is the combination of a diesel engine with an electric generator (often an alternator) to generate electrical energy. This is a specific case of engine-generator.

Diesel generating sets are used in places without connection to the power grid, as emergency power-supply if the grid fails, as well as for more complex applications such as peak-lopping, grid support and export to the power grid. Sizing of diesel generators is critical to avoid low-load or a shortage of power and is complicated by modern electronics, specifically non-linear loads.

Diesel Generator Set

The packaged combination of a diesel engine, a generator and various ancillary devices (such as base, canopy, sound attenuation, control systems, circuit breakers, jacket water heaters and starting system) is referred to as a "generating set" or a "genset" for short.

Set sizes range from 8 to 30 kW (also 8 to 30 kVA single phase) for homes, small shops and offices with the larger industrial generators from 8 kW (11 kVA) up to 2,000 kW (2,500 kVA three phase) used for large office complexes, factories. A 2,000 kW set can be housed in a 40 ft (12 m) ISO container with fuel tank, controls, power distribution equipment and all other equipment needed to operate as a stand alone power station or as a standby backup to grid power. These units, referred to as power modules are gensets on large triple axle trailers weighing 85,000 pounds (38,555 kg) or more. A combination of these modules are used for small power stations and these may use from one to 20 units per power section and these sections can be combined to involve hundreds of power modules. In these larger sizes the power module (engine and generator) are brought to site on trailers separately and are connected together with large cables and a control cable to form a complete synchronized power plant. A number of options also exist to tailor specific needs, including control panels for autostart and mains paralleling, acoustic canopies for fixed or mobile applications, ventilation equipment, fuel supply systems, exhaust systems, etc. Diesel generators, sometimes as small as 200 kW (250 kVA) are widely used not only for emergency power, but also many have a secondary function of feeding power to utility grids either during peak periods, or periods when there is a shortage of large power generators.

Ships often also employ diesel generators, sometimes not only to provide auxiliary power for lights, fans, winches etc., but also indirectly for main propulsion. With electric propulsion the generators can be placed in a convenient position, to allow more cargo to be carried. Electric drives for ships were developed prior to World War I. Electric drives were specified in many warships built during World War II because manufacturing capacity for large reduction gears was in short supply, compared to capacity for manufacture of electrical equipment. Such a diesel-electric arrangement is also used in some very large land vehicles such as railroad locomotives.

Generator Size

Generating sets are selected based on the electrical load they are intended to supply, the electrical loads total characteristics kWe, kVA, var and harmonic content including starting currents (normally from motors) and non-linear loads. The expected duty, for example, emergency, prime or continuous power as well as environmental conditions such as altitude, temperature and emissions regulations must be taken into account as well.

Most of the larger generator set manufacturers offer software that will perform the complicated sizing calculations by simply inputting site conditions and connected electrical load characteristics.

Power Plants – Electrical "Island" Mode

One or more diesel generators operating without a connection to an electrical grid are referred to as operating in island mode, which features several parallel generators that provide the advantages of redundancy and better efficiency at partial loads. The plant brings generator sets online and takes them off line depending on the demands of the system at a given time. An islanded power plant intended for primary power source of an isolated community ("Prime Power") will often have at least three diesel generators, any two of which are rated to carry the required load. Groups of up to 20 are not uncommon.

Generators can be electrically connected together through the process of synchronization. Synchronization involves matching voltage, frequency and phase before connecting the generator to the system. Failure to synchronize before connection could cause a high short circuit current or wear and tear on the generator or its switchgear. The synchronization process can be done automatically by an auto-synchronizer module. The auto-synchronizer will read the voltage, frequency and phase parameters from the generator and busbar voltages, while regulating the speed through the engine governor or ECM (Engine Control Module).

Load can be shared among parallel running generators through load sharing. Load sharing can be achieved by using droop speed control controlled by the frequency at the generator, while it constantly adjusts the engine fuel control to shift load to and from the remaining power sources. A diesel generator will take more load when the fuel supply to its combustion system is increased, while load is released if fuel supply is decreased.

Supporting Main Utility Grids

In addition to their well known role as power supplies during power failures, diesel generator sets also routinely support main power grids worldwide in two distinct ways :

Grid Support

Emergency standby diesel generators, for example such as those used in hospitals, water plant, are, as a secondary function, widely used in the US and the UK (Short Term Operating Reserve) to support the respective national grids at times for a variety of reasons. In the UK for example, some 0.5 GWe of diesels are routinely used to support the National Grid, whose peak load is about 60 GW. These are sets in the size range 200 kW to 2 MW. This usually occurs during, for example, the sudden loss of a large conventional 660 MW plant, or a sudden unexpected rise in power demand eroding the normal spinning reserve available.

This is extremely beneficial for both parties - the diesels have already been purchased for other reasons; but to be reliable need to be fully load tested.

Grid paralleling is a convenient way of doing this. This method of operation is normally undertaken by a third party aggregator who manages the operation of the generators and the interaction with the system operator.

In this way the UK National Grid can call on about 2 GW of plant which is up and running in parallel as quickly as two minutes in some cases. This is far quicker than a base load power station which can take 12 hours from cold, and faster than a gas turbine, which can take several minutes. Whilst diesels are very expensive in fuel terms, they are only used a few hundred hours per year in this duty, and their availability can prevent the need for base load station running inefficiently at part load continuously. The diesel fuel used is fuel that would have been used in testing anyway.

A similar system operates in France known as EJP, where at times of grid extrema special tariffs can mobilize at least 5 GW of diesel generating sets to become available. In this case, the diesels prime function is to feed power into the grid.

During normal operation in synchronization with the electricity net, powerplants are governed with a five per cent droop speed control. This means the full load speed is 100% and the no load speed is 105%. This is required for the stable operation of the net without hunting and dropouts of power plants. Normally the changes in speed are minor. Adjustments in power output are made by slowly raising the droop curve by increasing the spring pressure on a centrifugal governor. Generally this is a basic system requirement for all powerplants because the older and newer plants have to be compatible in response to the instantaneous changes in frequency without depending on outside communication.

Cost of Generating Electricity

Typical Operating Costs

Fuel consumption is the major portion of diesel plant owning and operating cost for power applications, whereas capital cost is the primary concern for backup generators. Specific consumption varies, but a modern diesel plant will consume between 0.28 and 0.4 litres of fuel per kilowatt hour at the generator terminals.

However diesel engines can operate on a variety of different fuels, depending on configuration, though the eponymous diesel fuel derived from crude oil is most common. The engines can work with the full spectrum of crude oil distillates, from natural gas, alcohols, gasoline, wood gas to the *fuel oils* from diesel oil to residual fuels. This is implemented by introducing gas with the intake air and using a small amount of diesel fuel for ignition. Conversion to 100% diesel fuel operation can be achieved instantaneously.

Generator Sizing and Rating

Rating

Generators must provide the anticipated power required reliably and without damage and this is achieved by the manufacturer giving one or more ratings to a specific generator set model. A specific model of a generator operated as a standby generator may only need to operate for a few hours per year, but the same model operated as a prime power generator must operate continuously. When running, the standby generator may be operated with a specified - e.g. 10% overload that can be tolerated for the expected short running time. The same model generator will carry a higher rating for standby service than it will for continuous duty. Manufacturers give each set a *rating* based on internationally agreed definitions.

These standard rating definitions are designed to allow correct machine selection and valid comparisons between manufacturers to prevent them from misstating the performance of their machines, and to guide designers.

Generator Rating Definitions

Standby Rating based on Applicable for supplying emergency power for the duration of normal power interruption. No sustained overload capability is available for this rating. (Equivalent to Fuel Stop Power in accordance with ISO3046, AS2789, DIN6271 and BS5514). Nominally rated.

Typical application - emergency power plant in hospitals, offices, factories etc. Not connected to grid.

Prime (Unlimited Running Time) Rating : Should not be used for Construction Power applications. Output available with varying load for an unlimited time. Typical peak demand 100% of prime-rated ekW with 10% of overload capability for emergency use for a maximum of 1 hour in 12. A 10% overload capability is available for limited time. (Equivalent to Prime Power in accordance with ISO8528 and Overload Power in accordance with ISO3046, AS2789, DIN6271, and BS5514). This rating is not applicable to all generator set models.

Typical application - where the generator is the sole source of power for say a remote mining or construction site, fairground, festival etc.

Base Load (Continuous) Rating based on : Applicable for supplying power continuously to a constant load up to the full output rating for unlimited hours. No sustained overload capability is available for this rating. Consult authorized distributor for rating. (Equivalent to Continuous Power in accordance with ISO8528, ISO3046, AS2789, DIN6271, and BS5514). This rating is not applicable to all generator set models

Typical application - a generator running a continuous unvarying load, or paralleled with the mains and continuously feeding power at the maximum permissible level 8,760 hours per year. This also applies to sets used for peak shaving/ grid support even though this may only occur for say 200 hours per year.

As an example if in a particular set the Standby Rating were 1000 kW, then a Prime Power rating might be 850 kW, and the Continuous Rating 800 kW. However these ratings vary according to manufacturer and should be taken from the manufacturer's data sheet.

Often a set might be given all three ratings stamped on the data plate, but sometimes it may have only a standby rating, or only a prime rating.

Sizing

Typically however it is the size of the maximum load that has to be connected and the acceptable maximum voltage drop which determines the set size, not the ratings themselves. If the set is required to start motors, then the set will have to be at least three times the largest motor, which is normally started first. This means it will be unlikely to operate at anywhere near the ratings of the chosen set.

Many gen-set manufacturers have software programs that enable the correct choice of set for any given load combination. Sizing is based on site conditions and the type of appliances, equipment, and devices that will be powered by the generator set.

Installation

Manufacturers provide detailed installation guidelines to ensure correct functioning, reliability and low maintenance costs. Guidelines cover such things as :

- Sizing and selection
- Electrical factors
- Cooling
- Ventilation
- Fuel storage
- Noise
- Exhaust
- Starting systems.

Engine Damage

Diesel engines can suffer damage as a result of misapplication or misuse-namely internal glazing (occasionally referred to as bore glazing or piling) and carbon build-up. Ideally, diesel engines should be run at least 60% to 75% of their maximum rated load. Short periods of low load running are permissible providing the set is brought up to full load, or close to full load on a regular basis.

Internal glazing and carbon build-up is due to prolonged periods of running at low speeds or low loads. Such conditions may occur when an engine is left idling as a 'standby' generating unit, ready to run up when needed, (misuse); if the engine powering the set is over-powered (misapplication) for the load applied to it, causing the diesel unit to be under-loaded, or as is very often the case, when sets are started and run off load as a test (misuse).

Running an engine under low loads causes low cylinder pressures and consequent poor piston ring sealing since this relies on the gas pressure to force them against the oil film on the bores to form the seal. Low cylinder pressures causes poor combustion and resultant low combustion pressures and temperatures.

This poor combustion leads to soot formation and unburnt fuel residues which clogs and gums piston rings, causing a further drop in sealing efficiency and exacerbates the initial low pressure. Glazing occurs when hot combustion gases blow past the now poorly-sealing piston rings, causing the lubricating oil on the cylinder walls to 'flash burn', creating an enamel-like glaze which smooths the bore and removes the effect of the intricate pattern of honing marks machined into the bore surface which are there to hold oil and return it to the crankcase via the scraper ring.

Hard carbon also forms from poor combustion and this is highly abrasive and scrapes the honing marks on the bores leading to bore polishing, which then leads to increased oil consumption (blue smoking) and yet further loss of pressure, since the oil film trapped in the honing marks is intended to maintain the piston seal and pressures.

Unburnt fuel then leaks past the piston rings and contaminates the lubricating oil. Poor combustion causes the injectors to become clogged with soot, causing further deterioration in combustion and black smoking.

The problem is increased further with the formation of acids in the engine oil caused by condensed water and combustion by-products which would normally boil off at higher temperatures. This acidic build-up in the lubricating oil causes slow but ultimately damaging wear to bearing surfaces.

This cycle of degradation means that the engine soon becomes irreversibly damaged and may not start at all and will no longer be able to reach full power when required.

Under-loaded running inevitably causes not only white smoke from unburnt fuel but over time will be joined by blue smoke of burnt lubricating oil leaking past the damaged piston rings, and black smoke caused by damaged injectors. This pollution is unacceptable to the authorities and neighbours.

Once glazing or carbon build up has occurred, it can only be cured by stripping down the engine and re-boring the cylinder bores, machining new honing marks and stripping, cleaning and de-coking combustion chambers, fuel injector nozzles and valves. If detected in the early stages, running an engine at maximum load to raise the internal pressures and temperatures allows the piston rings to scrape glaze off the bores and allows carbon build-up to be burnt off. However, if glazing has progressed to the stage where the piston rings have seized into their grooves, this will not have any effect.

The situation can be prevented by carefully selecting the generator set in accordance with manufacturers printed guidelines. (the use off additional depth oil and fuel By Pass filtration, down to <3 micron level can prevent build up of the particulate or carbon build that contributes to the varnishing.)

For emergency only sets, it may be impractical to use the supported load for testing. A temporary or permanent load bank can be used testing. Sometimes the switchgear can be designed to allow the set to feed power into the grid for load testing.

Chapter 6

LOAD MANAGEMENT

Load management, also known as demand side management (DSM), is the process of balancing the supply of electricity on the network with the electrical load by adjusting or controlling the load rather than the power station output. This can be achieved by direct intervention of the utility in real time, by the use of frequency sensitive relays triggering circuit breakers (ripple control), by time clocks, or by using special tariffs to influence consumer behaviour. Load management allows utilities to reduce demand for electricity during peak usage times, which can, in turn, reduce costs by eliminating the need for peaking power plants. In addition, peaking power plants also often require hours to bring on-line, presenting challenges should a plant go off-line unexpectedly. Load management can also help reduce harmful emissions, since peaking plants or backup generators are often dirtier and less efficient than base load power plants. New load-management technologies are constantly under development — both by private industry and public entities.

BRIEF HISTORY

In 1972, Theodore George "Ted" Paraskevakos, while working for Boeing in Huntsville, Alabama, developed a sensor monitoring system which used digital transmission for security, fire, and medical alarm systems as well as meter-reading capabilities for all utilities. This technology was a spin-off of his patented automatic telephone line identification system, now known as caller ID. In, 1974, Mr. Paraskevakos was awarded a U.S. Patent for this technology.

At the request of the Alabama Power Company, Mr. Paraskevakos developed a load-management system along with automatic meter-reading technology. In doing so, he utilized the ability of the system to monitor the speed of the watt power meter disc and, consequently, power consumption. This information, along with the time of day, gave the power company the ability to instruct individual meters to manage water heater and air conditioning consumption in order to prevent peaks in usage during the high consumption portions of the day. For this approach, Mr. Paraskevakos was awarded multiple patents.

ADVANTAGES AND OPERATING PRINCIPLES

Since electrical energy is a form of energy that cannot be effectively stored in bulk, it must be generated, distributed, and consumed immediately. When the load on a system approaches the maximum generating capacity, network operators must either find additional supplies of energy or find ways to curtail the load, hence load management. If they are unsuccessful, the system will become unstable and blackouts can occur.

Long-term load management planning may begin by building sophisticated models to describe the physical properties of the distribution network (*i.e.* topology, capacity, and other characteristics of the lines), as well as the load behaviour. The analysis may include scenarios that account for weather forecasts, the predicted impact of proposed load-shed commands, estimated time-to-repair for off-line equipment, and other factors.

The utilization of load management can help a power plant achieve a higher **capacity factor**, a measure of average capacity utilization. Capacity factor is a measure of the output of a power plant compared to the maximum output it could produce. Capacity factor is often defined as *the ratio of average load to capacity* or *the ratio of average load to peak load in a period of time*. A higher load factor is advantageous because a power plant may be less efficient at low load factors, a high load factor means fixed costs are spread over more kWh of output (resulting in a lower price per unit of electricity), and a higher load factor means greater total output. If the power load factor is affected by non-availability of fuel, maintenance shut-down, unplanned breakdown, or reduced demand (as consumption pattern fluctuate throughout the day), the generation has to be adjusted, since grid energy storage is often prohibitively expensive.

Smaller utilities that buy power instead of generating their own find that they can also benefit by installing a load control system. The penalties they must pay to the energy provider for peak usage can be significantly reduced. Many report that a load control system can pay for itself in a single season.

COMPARISONS TO DEMAND RESPONSE

When the decision is made to curtail load, it is done so on the basis of system *reliability*. The utility in a sense "owns the switch" and sheds loads only when the stability or reliability of the electrical distribution system is threatened. The utility (being in the business of generating, transporting, and delivering electricity) will not disrupt their business process without due cause. Load management, when done properly, is non-invasive, and imposes no hardship on the consumer.

Demand response places the "on-off switch" in the hands of the consumer using devices such as a smart grid controlled load control switch. While many residential consumers pay a flat rate for electricity year-round, the utility's costs actually vary constantly, depending on demand, the distribution network, and composition of the company's electricity generation portfolio. In a free market, the wholesale price of energy varies widely throughout the day. Demand response programs such as those enabled by smart grids attempt to incentivize the consumer to limit usage based upon *cost* concerns. As costs rise during the day (as the system reaches peak capacity and more expensive peaking power plants are used), a free market economy should allow the price to rise. A corresponding drop in demand for the commodity should meet a fall in price. While this works for predictable shortages, many crises develop within seconds due to unforeseen equipment failures. They must be resolved in the same time-frame in order to avoid a power blackout. Many utilities who are interested in demand response have also expressed an interest in load control capability so that they might be able to operate the "on-off switch" before price updates could be published to the consumers.

The application of load control technology continues to grow today with the sale of both radio frequency and powerline communication based systems. Certain types of smart meter systems can also serve as load control systems.

The largest residential load control system in the world is found in Florida and is managed by Florida Power and Light. It utilizes 800,000 Load Control Transponders (LCTs) and controls 1,000 MW of electrical power (2,000 MW in an emergency). FPL has been able to avoid the construction of numerous new power plants due to their load management programs.

RIPPLE CONTROL

Ripple control is the most common form of load control, and is used in many countries around the world, including Australia, New Zealand, the United Kingdom, Germany, and South Africa. Ripple control involves superimposing a higher-frequency signal (usually between 100 and 1600 Hz) onto the standard 50–60 Hz of the main power signal. When receiver devices attached to non-essential residential or industrial loads receive this signal, they shut down the load until the signal is disabled or another frequency signal is received.

Early implementations of ripple control occurred during World War II in various parts of the world using a system that communicates over the electrical distribution system. Ripple control systems are generally paired with a two- (or more) tiered pricing system, whereby electricity is more expensive during peak times (evenings) and cheaper during low-usage times (early morning).

Affected residential devices will vary by region, but may include residential electric hot-water heaters, air conditioners, pool pumps, or crop-irrigation pumps. In a distribution network outfitted with load control, these devices are outfitted with communicating controllers that can run a program that limits the duty cycle of the equipment under control. Consumers are usually rewarded for participating in the load control program by paying a reduced rate for energy. Proper load management by the utility allows them to practice load shedding to avoid rolling blackouts and reduce costs.

EXAMPLES OF SCHEMES

In many countries, including United States, United Kingdom and France, the power grids routinely use privately held, emergency diesel generators in load management schemes

New Zealand

Since the 1950s, New Zealand has had a system of load management based on ripple control, allowing the electricity supply for domestic and commercial water storage heaters to be switched off and on, as well as allowing remote control of nightstore heaters and street lights. Ripple injection equipment located within each local distribution network signals to ripple control receivers at the customer's premises. Control may either done manually by the local distribution network company in response to local outages or requests to reduce demand from the transmission system operator (*i.e.* Transpower), or automatically when injection equipment detects mains frequency falling below 49.2 Hz. Ripple control receivers are assigned to one of several ripple channels to allow the network company to only turn off supply on part of the network, and to allow staged restoration of supply to reduce the impact of a surge in demand when power is restored to water heaters after a period of time off.

Depending on the area, the consumer may have two electricity meters, one for normal supply ("Anytime") and one for the load-managed supply ("Controlled"), with Controlled supply billed at a lower rate per kilowatt-hour than Anytime supply. For those with load-managed supply but only a single meter, electricity is billed at the "Composite" rate, priced between Anytime and Controlled.

France

France has an EJP tariff, which allows it to disconnect certain loads and to encourage consumers to disconnect certain loads. This tariff is no longer available for new clients (as of July 2009). The *Tempo* tariff also includes different types of days with different prices, but has been discontinued for new clients as well (as of July 2009). Reduced prices during nighttime are available for customers for a higher monthly fee.

United Kingdom

Rltec in the UK in 2009 reported that domestic refrigerators are being sold fitted with their dynamic load response systems. In 2011 it was announced that the Sainsbury supermarket chain will use dynamic demand technology on their heating and ventilation equipment.

In the UK, night storage heaters are used to increase the load by about 5 GW to accommodate the nuclear programme. There is also a programme that allows industrial loads to be disconnected using circuit breakers triggered automatically by frequency sensitive relays fitted on site. This operates in conjunction with Standing Reserve, a programme using diesel generators.

These can also be remotely switched using BBC Radio 4 Longwave Radio teleswitch.

Germany

Existing storage heaters in Germany are charged up with a lower tariff electricity, mainly during night and afternoon hours. The switch and the two-tariff electricity meter are remotely controlled, using either ripple control through the powerline or the European radio teleswitching system based on longwave radio signals.

Chapter 7

TRANSFORMER MAINTENANCE

TRANSFORMER OIL

Transformer oil or **insulating oil** is a highly refined mineral oil that is stable at high temperatures and has excellent electrical insulating properties. It is used in oil-filled transformers, some types of high voltage capacitors, fluorescent lamp ballasts, and some types of high voltage switches and circuit breakers. Its functions are to insulate, suppress corona and arcing, and to serve as a coolant.

Explanation

The oil helps cool the transformer. Because it also provides part of the electrical insulation between internal live parts, transformer oil must remain stable at high temperatures for an extended period. To improve cooling of large power transformers, the oil-filled tank may have external radiators through which the oil circulates by natural convection. Very large or high-power transformers (with capacities of thousands of kVA) may also have cooling fans, oil pumps and even oil-to-water heat exchangers.

Large, high voltage transformers undergo prolonged drying processes, using electrical self-heating, the application of a vacuum, or both to ensure that the transformer is completely free of water vapour before the cooling oil is introduced. This helps prevent corona formation and subsequent electrical breakdown under load.

Oil filled transformers with a conservator may have a gas detector relay (Buchholz relay). These safety devices detect the build up of gas inside the transformer due to corona discharge, overheating, or an internal electric arc. On a slow accumulation of gas, or rapid pressure rise, these devices can trip a protective circuit breaker to remove power from the transformer. Transformers without conservators are usually equipped with sudden pressure relays, which perform a similar function as the Buchholz relay.

Fig. : Oil transformer with air convection cooled heat exchangers in the front and at the side.

The flash point (min) and pour point (max) are 140 °C and −6 °C respectively. The dielectric strength of new untreated oil is 12 MV/m (RMS) and after treatment it should be >24 MV/m (RMS).

Large transformers for indoor use must either be of the dry type, that is, containing no liquid, or use a less-flammable liquid.

Recently, Research is underway in making transformer oil nano fluids by mixing insulating or semi-conducting particles with transformer oil to enhance its thermal conductivity and electrical breakdown strength.

Current Mineral Oil Alternatives

Today, most transformers use a fluid that achieves a much higher performance level than standard naphthenic mineral oil, with far less risk. Mineral oils invariably have an issue with corrosive sulphur that can render them problematic in service, and attempts to balance this out with copper passivators are insufficient compared to readily-available, safer alternatives.

Pentaerythritol tetra fatty acid natural and synthetic esters have emerged as an increasingly common mineral oil alternative. They offset all the main risks associated with mineral oil, such as high flammability, environmental impact and poor moisture tolerance. Esters are also non-toxic to aquatic life, readily biodegradable and provide a lower volatility and higher flash point.

Additionally, they have a high fire point of over 300°C and K-class fluids such as these are often used in high-risk transformer applications, such as indoors or offshore. They also have a lower pour point, greater moisture tolerance and improved function at high temperatures.

Silicone-based or fluorinated hydrocarbons, where the added expense of a fire-resistant offsets any additional costs of building a transformer vault, have also been presented as a viable mineral oil alternative. However, silicone has been proven to be much less biodegradable than esters in the event of a leak or spillage.

Vegetable-based oils have also been suggested, but these are unsuitable for use in cold climates or for voltages over 230kV. Some papers have also cited co-conut oil as a potential substitute for use in transformers.

Polychlorinated biphenyls (PCBs)

Well into the 1970s, polychlorinated biphenyls (PCBs) were often used as a dielec-tric fluid since they are not flammable. PCBs do not break down when released into the environment but accumulate in the tissues of plants and animals, where they can have hormone-like effects. When burned, PCBs can form highly toxic products, such as chlorinated dioxins and chlorinated dibenzofurans. Starting in the early 1970s, production and new uses of PCBs have been banned due to concerns about the accumulation of PCBs and toxicity of their byproducts. In many countries significant programs are in place to reclaim and safely destroy PCB contaminated equipment.

Polychlorinated biphenyls were banned in 1979 in the US. Since PCB and transformer oil are miscible in all proportions, and since sometimes the same equipment (drums, pumps, hoses, and so on) was used for either type of liquid, contamination of oil-filled transformers is possible. Under present regulations, concentrations of PCBs exceeding 5 parts per million can cause an oil to be clas-sified as hazardous waste in California.

Testing and Oil Quality

Transformer oils are subject to electrical and mechanical stresses while a trans-former is in operation. In addition there is contamination caused by chemical interactions with windings and other solid insulation, catalyzed by high operating temperature. The original chemical properties of transformer oil change gradu-ally, rendering it ineffective for its intended purpose after many years. Oil in large transformers and electrical apparatus is periodically tested for its electrical and chemical properties, to make sure it is suitable for further use. Sometimes oil condition can be improved by filtration and treatment. Tests can be divided into :

1. Dissolved gas analysis
2. Furan analysis
3. PCB analysis
4. General electrical & physical tests :
   o Colour & Appearance

- o Breakdown Voltage
- o Water Content
- o Acidity (Neutralization Value)
- o Dielectric Dissipation Factor
- o Resistivity
- o Sediments & Sludge
- o Flash Point
- o Pour Point
- o Density
- o Kinematic Viscosity.

The details of conducting these tests are available in standards released by IEC, ASTM, IS, BS, and testing can be done by any of the methods. The Furan and DGA tests are specifically not for determining the quality of transformer oil, but for determining any abnormalities in the internal windings of the transformer or the paper insulation of the transformer, which cannot be otherwise detected without a complete overhaul of the transformer. Suggested intervals for these test are :

- General and physical tests - bi-yearly
- Dissolved gas analysis - yearly
- Furan testing - once every 2 years, subject to the transformer being in operation for min 5 years.

TRANSFORMER OIL MAINTENANCE

Perhaps, the easiest way to explain the reasoning for proper transformer oil maintenance is to compare a transformer to a car. A car requires oil for lubrication while a transformer requires oil for insulation and cooling. Just as a car requires regular checks of both oil quality and oil level, a transformer requires the same proactive treatment. Oil in each instance degrades over time resulting in the potential for failure or costly repair bills. Regular maintenance and upgrading of the oil will add years to the life of both your car and your transformer.

Oil Quality : Transformer oil is a mineral based oil. It is commonly used in transformers because of its chemical properties and dielectric strength. Minimal breakdown of the oil occurs under normal operating conditions.

The insulation and cooling properties of the transformer are effected by the quality of the oil. The quality of the oil is reduced by oxidization and contamination. The results of each of these can be summarized briefly as follows :

a. Contamination commonly found in oil can be water and particulate. The presence of either contaminant will cause reduced insulation quality of the oil.

b. Oxidization is the acid that forms in the oil when it reacts to oxygen. This acid will form sludge which will settle on the windings of the transformer reducing the heat dissipation from the transformer. The heat transfer from the windings to the oil is limited thus causing the windings to run hotter. Sludge formation on the windings has a snowball effect on the transformer with more sludge creating more heat, creating more sludge etc. The high acid content together with the excessive temperatures will cause the deterioration of the transformer insulation to be accelerated and if left untreated the transformer will fail.

Testing

The first part of a preventative maintenance program requires establishing when remedial action is necessary. With transformer oil this is done through testing of the oil. Development of an ongoing maintenance program will help prevent costly failures. Initial testing will establish a base line and annual testing will plot any changes happening internally in the transformer. The following 5 part tests are a minimal requirement of a yearly maintenance program.

1. Dielectric breakdown : Dielectric strength is a measure of the voltage which the oil will conduct. Many contaminants conduct electricity better than oil therefore lowering the dielectric breakdown voltage.
2. Neutralization/Acid Number : Oxidation occurs in the oil causing the buildup of acid which will lead to the formation of sludge. The test indicates the level of acid present in the oil.
3. Interfacial tension : This test points to the presence of polar compounds, which indicate oxidation contaminates or deterioration from the transformer materials *i.e.* paint, varnish, paper.
4. Colour : Indicates the quality, aging, and the presence of contaminants.
5. Water : Is measured in parts per million. The presence of water will decrease the dielectric breakdown voltage.

Another useful tool in a maintenance program is a Dissolved Gas Analysis (DGA). By analyzing the gasses present in the oil it is possible to determine if there is a fault such as arcing, corona or overheated connections.

From the testing results, it can be determined whether remedial action needs to be taken. Predetermined limits must be set based on voltage class and KVA rating of the equipment. Any analysis showing test results lower than the parameters set indicate further investigation. A downward trend in any of the testing values warrants further testing and an evaluation of these results. Equipment should not be condemned by one test but by a series of tests.

If remedial action is required to the transformer oil, a recent PCB analysis is required in addition to the preceding testing. If the PCB level of the oil is less than 2 ppm, in most cases on-site reclamation is recommended based on the customer's preference. If the oil is over 2 ppm but less than 50 ppm, it can be shipped to a recycling facility and the transformer can be retrofilled with either new oil or recycled oil. In all cases, any results over the 50 ppm limit require special handling.

Remediation Treatment

If testing indicates a problem with the oil quality, the owner must determine if the cost of reclamation of the oil is warranted. Factors influencing this decision would include the replacement cost of the transformer or expenses associated with a failure of the equipment at a critical time.

After the decision has been made to upgrade the oil, the owner will need to evaluate if the oil is to be replaced or treated. Onsite reclamation includes fuller's earth processing and degasification. It is possible to restore oil to new oil specifications with a combination of these treatments. Any transformer oil can be brought back to new oil specifications, although some contamination levels may make it more economically viable to replace the oil. Antioxidant may be added at this time if an inhibited oil is required.

A regular maintenance program allows the transformer oil to be upgraded before sludge created by oxidization occurs. This results in a cost saving to the transformer owner. Once sludge occurs, the internal parts of the transformer require flushing with hot oil to remove the sediment. If not rinsed down, the core and coil assembly will hold the acids which will leach back into the oil over time. This will cause the re-deterioration of the oil to be accelerated. If reclamation is done in the early stages of the acid build up (approx. 0.1) before sludging occurs, the oil will retain its properties longer under normal operating conditions.

Reclamation of oil with a high acid content consists of fuller's earth treatment and degasification. Fuller's earth treatment will remove both acid and particulate in the oil. This process corrects the acid number as well as the colour. Degasification removes gasses and water. Transformer oil will hold water in suspension. The amount of water it will hold depends on the temperature of the oil. For example, if the oil temperature is 20C the maximum amount of water that can stay in suspension is approx. 53 ppm. It is likely that if oil is at its saturation point there will be free water at the bottom of the transformer. The ASTM limit for water is 30 ppm but the temperature must be considered. Due to the decreased dielectric strength with high water content it is important to complete a degasification on the oil. If the oil is especially high in water a hot oil dry out should be considered. Although it is more costly than just a degasification, it also removes any water that may be in the core and coil assembly.This is completed by circulating the oil through the degasification process and the transformer to heat the transformer. Then if the transformer is capable of vacuum the oil will be removed and will be held under vacuum for a determined number of hours. The oil is then degasified back in to the transformer.

If the decision has been made to replace the oil, either reclaimed or new oil can be used as retrofill material. If the tank is capable of a vacuum, the transformer should be filled under vacuum, based on manufacturer's recommendations. If the transformer cannot withstand a vacuum, the oil should be degasified into the transformer and circulated through the degasifier three times the volume of the transformer to help remove any moisture present in the insulation.

New oil often requires further degasification to remove air and moisture added during transportation and handling procedures. This further degasification will increase the life expectancy of the oil in the transformer.

Environmental Concerns

Mineral insulating oil is a valuable resource that can be recycled many times and returned to its original condition. The benefits of a regular maintenance program for your transformer include both economical and environmental factors. A failure of a transformer can result in significant environmental clean up costs.

In conclusion, a good preventative maintenance program for your transformers is money well spent. You will have peace of mind that comes from knowing that you have done due diligence in maintaining the continuing operation of your transformers.

RESONANT INDUCTIVE COUPLING

Resonant inductive coupling or **electrodynamic induction** is the near field wireless transmission of electrical energy between two coils that are tuned to resonate at the same frequency. The equipment to do this is sometimes called a **resonant** or **resonance transformer**. While many transformers employ resonance, this type has a high Q and is often air cored to avoid 'iron' losses. The two coils may exist as a single piece of equipment or comprise two separate pieces of equipment.

Resonant transfer works by making a coil *ring* with an oscillating current. This generates an oscillating magnetic field. Because the coil is highly resonant, any energy placed in the coil dies away relatively slowly over very many cycles; but if a second coil is brought near it, the coil can pick up most of the energy before it is lost, even if it is some distance away. The fields used are predominately non-radiative, near field (sometimes called evanescent waves), as all hardware is kept well within the ¼ wavelength distance they radiate little energy from the transmitter to infinity.

One of the applications of the resonant transformer is for the CCFL inverter. Another application of the resonant transformer is to couple between stages of a superheterodyne receiver, where the selectivity of the receiver is provided by tuned transformers in the intermediate-frequency amplifiers. Resonant transformers such as the Tesla coil can generate very high voltages with or without arcing, and are able to provide much higher current than electrostatic high-voltage generation machines such as the Van de Graaff generator. Resonant energy transfer is the operating principle behind proposed short range (up to 2 metre) wireless electricity systems such as WiTricity and systems that have already been deployed, such as Qi power transfer, passive RFID tags and contactless smart cards.

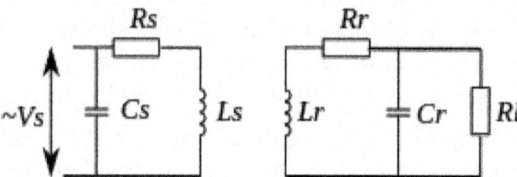

Fig. : Basic transmitter and receiver circuits, Rs and Rr are the resistances and losses in the associated capacitors and inductors. Ls and Lr are coupled by small coupling coefficient, k, usually below 0.2

Resonant Coupling

Non-resonant coupled inductors, such as typical transformers, work on the principle of a primary coil generating a magnetic field and a secondary coil subtending as much as possible of that field so that the power passing through the secondary is as close as possible to that of the primary. This requirement that the field be covered by the secondary results in very short range and usually requires a magnetic core. Over greater distances the non-resonant induction method is highly inefficient and wastes the vast majority of the energy in resistive losses of the primary coil.

Using resonance can help improve efficiency dramatically. If resonant coupling is used, each coil is capacitively loaded so as to form a tuned LC circuit. If the primary and secondary coils are resonant at a common frequency, it turns out that significant power may be transmitted between the coils over a range of a few times the coil diameters at reasonable efficiency.

Coupling Coefficient

The coupling coefficient is the fraction of the flux of the primary that cuts the secondary coil, and is a function of the geometry of the system. The coupling coefficient, k, is between 0 and 1.

Systems are said to be tightly coupled, loosely coupled, critically coupled or overcoupled. Tight coupling is when the coupling coefficient is around 1 as with conventional iron-core transformers. Overcoupling is when the secondary coil is so close that it tends to collapse the primary's field, and critical coupling is when the transfer in the passband is optimal. Loose coupling is when the coils are distant from each other, so that most of the flux misses the secondary, in Tesla coils around 0.2 is used, and at greater distances, for example for inductive wireless power transmission, it may be lower than 0.01.

Energy Transfer and Efficiency

The general principle is that if a given oscillating amount of energy (for example, a pulse or a series of pulses) is placed into a primary coil which is capacitively loaded, the coil will 'ring', and form an oscillating magnetic field. The energy will transfer back and forth between the magnetic field in the inductor and the electric field across the capacitor at the resonant frequency. This oscillation will die away at a rate determined by the gain-bandwidth (Q factor), mainly due to resistive and radiative losses. However, provided the secondary coil cuts enough of the field that it absorbs more energy than is lost in each cycle of the primary, then most of the energy can still be transferred.

The primary coil forms a series RLC circuit, and the Q factor for such a coil is :

$$Q = \frac{1}{R}\sqrt{\frac{L}{C}},$$

For R=10 ohm,C=1 micro farad and L=10 mH, Q is given as 10.

Because the Q factor can be very high, (experimentally around a thousand has been demonstrated with air cored coils) only a small percentage of the field has to be coupled from one coil to the other to achieve high efficiency, even though the field dies quickly with distance from a coil, the primary and secondary can be several diameters apart.

It can be shown that a figure of merit for the efficiency is :

$$U = k\sqrt{Q_1 Q_2}$$

Where Q_1 and Q_2 is the Q factor of the source and receiver coils.

And the maximum achievable efficiency is :

$$\eta_{opt} = \frac{U^2}{(1 + \sqrt{1 + U^2})^2}$$

Power Transfer

Because the Q can be very high, even when low power is fed into the transmitter coil, a relatively intense field builds up over multiple cycles, which increases the power that can be received — at resonance far more power is in the oscillating field than is being fed into the coil, and the receiver coil receives a percentage of that.

Voltage Gain

The voltage gain of resonantly coupled coils is directly proportional to the square root of the ratio of secondary and primary inductances.

Transmitter Coils and Circuitry

Unlike the multiple-layer secondary of a non-resonant transformer, coils for this purpose are often single layer solenoids (to minimise skin effect and give improved Q) in parallel with a suitable capacitor, or they may be other shapes such as wave-wound litz wire. Insulation is either absent, with spacers, or low permittivity, low loss materials such as silk to minimise dielectric losses.

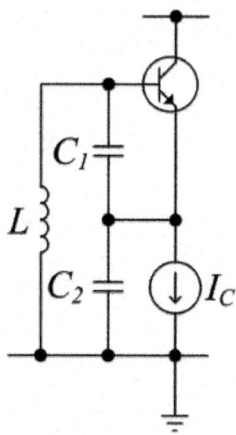

Fig. : Colpitts oscillator. In resonant energy transfer the inductor would be the transmitter coil and capacitors are used to tune the circuit to a suitable frequency.

To progressively feed energy/power into the primary coil with each cycle, different circuits can be used. One circuit employs a Colpitts oscillator.

In Tesla coils an intermittent switching system, a "circuit controller" or "break," is used to inject an impulsive signal into the primary coil; the secondary coil then rings and decays.

Receiver Coils and Circuitry

The secondary receiver coils are similar designs to the primary sending coils. Running the secondary at the same resonant frequency as the primary ensures that the secondary has a low impedance at the transmitter's frequency and that the energy is optimally absorbed.

Example receiver coil. The coil is loaded with a capacitor and two LEDs. The coil and the capacitor form a series LC circuit which is tuned to a resonant frequency that matches the transmission coil located inside of the brown matt. Power is transmitted over a distance of thirteen inches.

To remove energy from the secondary coil, different methods can be used, the AC can be used directly or rectified and a regulator circuit can be used to generate DC voltage.

History

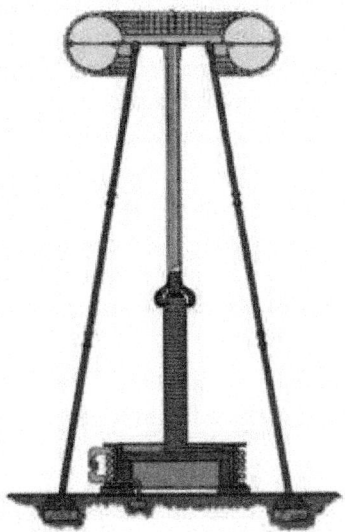

Fig. : This advanced Tesla coil was designed to implement wireless power by means of the *disturbed charge of ground and air method.*

In 1894 Nikola Tesla used resonant inductive coupling, also known as "electro-dynamic induction" to wirelessly light up phosphorescent and incandescent lamps at the 35 South Fifth Avenue laboratory, and later at the 46 E. Houston Street laboratory in New York City. In 1897 he patented a device called the high-voltage, resonance transformer or "Tesla coil." Transferring electrical energy from the primary coil to the secondary coil by resonant induction, a Tesla coil is capable of producing very high voltages at high frequency. The improved design allowed for the safe production and utilization of high-potential electrical currents, "without serious liability of the destruction of the apparatus itself and danger to persons approaching or handling it."

In the early 1960s resonant inductive wireless energy transfer was used successfully in implantable medical devices including such devices as pacemakers and artificial hearts. While the early systems used a resonant receiver coil, later systems implemented resonant transmitter coils as well. These medical devices are designed for high efficiency using low power electronics while efficiently accommodating some misalignment and dynamic twisting of the coils. The separation between the coils in implantable applications is commonly less than 20 cm. Today resonant inductive energy transfer is regularly used for providing electric power in many commercially available medical implantable devices.

Wireless electric energy transfer for experimentally powering electric automobiles and buses is a higher power application (>10 kW) of resonant inductive energy transfer. High power levels are required for rapid recharging and high energy transfer efficiency is required both for operational economy and to avoid negative environmental impact of the system. An experimental electrified roadway test track built circa 1990 achieved 80% energy efficiency while recharging the battery of a prototype bus at a specially equipped bus stop. The bus could be outfitted with a retractable receiving coil for greater coil clearance when moving. The gap between the transmit and receive coils was designed to be less than 10 cm when powered. In addition to buses the use of wireless transfer has been investigated for recharging electric automobiles in parking spots and garages as well.

Some of these wireless resonant inductive devices operate at low milliwatt power levels and are battery powered. Others operate at higher kilowatt power levels. Current implantable medical and road electrification device designs achieve more than 75% transfer efficiency at an operating distance between the transmit and receive coils of less than 10 cm.

In 1995, Professor John Boys and Prof. Grant Covic, of The University of Auckland in New Zealand, developed systems to transfer large amounts of energy across small air gaps.

In 1998, RFID tags were patented that were powered in this way.

In November 2006, Marin Soljačić and other researchers at the Massachusetts Institute of Technology applied this near field behaviour, well known in electromagnetic theory, the wireless power transmission concept based on strongly-coupled resonators. In a theoretical analysis, they demonstrate that, by designing electromagnetic resonators that suffer minimal loss due to radiation and absorption and have a near field with mid-range extent (namely a few times the resonator size), mid-range efficient wireless energy-transfer is possible. The reason is that, if two such resonant circuits tuned to the same frequency are within a fraction of a wavelength, their near fields (consisting of 'evanescent waves') couple by means of evanescent wave coupling. Oscillating waves develop between the inductors, which can allow the energy to transfer from one object to the other within times much shorter than all loss times, which were designed to be long, and thus with the maximum possible energy-transfer efficiency. Since the resonant wavelength is much larger than the resonators, the field can circumvent extraneous objects in the vicinity and thus this mid-range energy-transfer scheme does not require line-of-sight. By utilizing in particular the magnetic field to achieve the coupling, this method can be safe, since magnetic fields interact weakly with living organisms.

Apple Inc. applied for a patent on the technology in 2010, after WiPower did so in 2008.

Comparison with Other Technologies

Compared to inductive transfer in conventional transformers, except when the coils are well within a diameter of each other, the efficiency is somewhat lower (around 80% at short range) whereas tightly coupled conventional transformers may achieve greater efficiency (around 90-95%) and for this reason it cannot be used where high energy transfer is required at greater distances.

However, compared to the costs associated with batteries, particularly non-rechargeable batteries, the costs of the batteries are hundreds of times higher. In situations where a source of power is available nearby, it can be a cheaper solution. In addition, whereas batteries need periodic maintenance and replacement, resonant energy transfer can be used instead. Batteries additionally generate pollution during their construction and their disposal which is largely avoided.

Regulations and Safety

Unlike mains-wired equipment, no direct electrical connection is needed and hence equipment can be sealed to minimize the possibility of electric shock.

Because the coupling is achieved using predominantly magnetic fields; the technology may be relatively safe. Safety standards and guidelines do exist in most countries for electromagnetic field exposures (*e.g.*) Whether the system can meet the guidelines or the less stringent legal requirements depends on the delivered power and range from the transmitter.

Deployed systems already generate magnetic fields, for example induction cookers and contactless smart card readers.

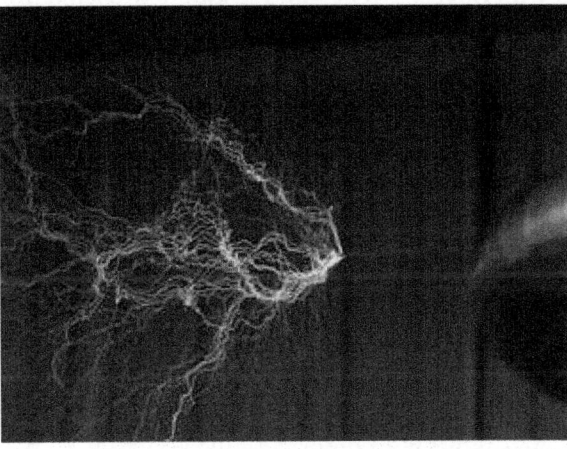

Fig. : High voltages may lead to electrical breakdown, resulting in an electrical discharge as illustrated by the plasma filaments streaming from a Tesla coil.

HIGH VOLTAGE

The term **high voltage** usually means electrical energy at voltages high enough to inflict harm or death upon living things. Equipment and conductors that carry high voltage warrant particular safety requirements and procedures. In certain industries, *high voltage* means voltage above a particular threshold. High voltage is used in electrical power distribution, in cathode ray tubes, to generate X-rays and particle beams, to demonstrate arcing, for ignition, in photomultiplier tubes, and in high power amplifier vacuum tubes and other industrial and scientific applications.

Definition

IEC voltage range	AC	DC	Defining risk
High voltage (supply system)	> 1000 V$_{rms}$	> 1500 V	electrical arcing
Low voltage (supply system)	50–1000 V$_{rms}$	120–1500 V	electrical shock
Extra-low voltage (supply system)	< 50 V$_{rms}$	< 120 V	low risk

The numerical definition of *high voltage* depends on context. Two factors considered in classifying a voltage as "high voltage" are the possibility of causing a spark in air, and the danger of electric shock by contact or proximity. The definitions may refer to the voltage between two conductors of a system, or between any conductor and ground.

In electric power transmission engineering, high voltage is usually considered any voltage over approximately 35,000 volts. This is a classification based on the design of apparatus and insulation.

The International Electrotechnical Commission and its national counterparts (IET, IEEE, VDE, etc.) define *high voltage* as above 1000 V for alternating current, and at least 1500 V for direct current — and distinguish it from low voltage (50–1000 V AC or 120–1500 V DC) and extra-low voltage (<50 V AC or <120 V DC) circuits. This is in the context of building wiring and the safety of electrical apparatus.

In the United States 2005 National Electrical Code (NEC), *high voltage* is any voltage over 600 V. British Standard BS 7671 : 2008 defines *high voltage* as any voltage difference between conductors that is higher than 1000 V AC or 1500 V ripple-free DC, or any voltage difference between a conductor and Earth that is higher than 600 V AC or 900 V ripple-free DC.

Electricians may only be licensed for particular voltage classes, in some jurisdictions. For example, an electrical license for a specialized sub-trade such as installation of HVAC systems, fire alarm systems, closed circuit television systems may be authorized to install systems energized up to only 30 volts between conductors, and may not be permitted to work on mains-voltage circuits. The general public may consider household mains circuits (100–250 V AC), which carry the highest voltages they normally encounter, to be *high voltage*.

Voltages over approximately 50 volts can usually cause dangerous amounts of current to flow through a human being who touches two points of a circuit — so safety standards, in general, are more restrictive around such circuits.. The definition of *extra high voltage* (EHV) again depends on context. In electric power transmission engineering, EHV refers to equipment that carries more than 345,000 volts between conductors. In electronics systems, a power supply that provides greater than 275,000 volts is called an *EHV Power Supply*, and is often used in experiments in physics.

The accelerating voltage for a television cathode ray tube may be described as *extra-high voltage* or *extra-high tension* (EHT), compared to other voltage supplies within the equipment. This type of supply ranges from >5 kV to about 50 kV.

In digital electronics, a logical *high* voltage is the one that represents a logic 1. It is typically represented by a voltage higher than the corresponding range for logic 0, but the difference may be less than a volt for some logic families. Older systems such as TTL used 5 volts, newer computers typically use 3.3 volts (LV-TTL) or even 1.8 volts.

Safety

Voltages greater than 50 V applied across dry unbroken human skin can cause heart fibrillation if they produce electric currents in body tissues that happen to pass through the chest area. The voltage at which there is the danger of electro-cution depends on the electrical conductivity of dry human skin. Living human tissue can be protected from damage by the insulating characteristics of dry skin up to around 50 volts. If the same skin becomes wet, if there are wounds, or if the voltage is applied to electrodes that penetrate the skin, then even voltage sources below 40 V can be lethal.

Accidental contact with high voltage supplying sufficient energy may result in severe injury or death. This can occur as a person's body provides a path for current flow, causing tissue damage and heart failure. Other injuries can include burns from the arc generated by the accidental contact. These burns can be espe-cially dangerous if the victim's airways are affected. Injuries may also be suffered as a result of the physical forces experienced by people who fall from a great height or are thrown a considerable distance.

Low-energy exposure to high voltage may be harmless, such as the spark produced in a dry climate when touching a doorknob after walking across a car-peted floor. The voltage can be in the thousand-volt range, but the amperage (the number of electrons involved) is low.

Safety equipment used by electrical workers includes insulated rubber gloves and mats. These protect the user from electric shock. Safety equipment is tested regularly to ensure it is still protecting the user. Test regulations vary according to country. Testing companies can test at up 300,000 volts and offer services from glove testing to Elevated Working Platform or EWP Truck testing.

Sparks in Air

The dielectric breakdown strength of dry air, at Standard Temperature and Pres-sure (STP), between spherical electrodes is approximately 33 kV/cm. This is only as a rough guide, since the actual breakdown voltage is highly dependent upon the electrode shape and size. Strong electric fields (from high voltages applied to small or pointed conductors) often produce violet-coloured corona discharges in air, as well as visible sparks. Voltages below about 500–700 volts cannot produce easily visible sparks or glows in air at atmospheric pressure, so by this rule these voltages are "low". However, under conditions of low atmospheric pressure (such as in high-altitude aircraft), or in an environment of noble gas such as argon or neon, sparks appear at much lower voltages. 500 to 700 volts is not a fixed minimum for producing spark breakdown, but it is a rule-of-thumb. For air at STP, the minimum sparkover voltage is around 327 volts, as noted by Friedrich Paschen.

While lower voltages don't, in general, jump a gap that is present before the voltage is applied, interrupting an existing current flow often produces a low-voltage spark or arc. As the contacts are separated, a few small points of contact become the last to separate. The current becomes constricted to these small *hot spots*, causing them to become incandescent, so that they emit electrons (through thermionic emission). Even a small 9 V battery can spark noticeably by this mechanism in a darkened room. The ionized air and metal vapour (from the contacts) form plasma, which temporarily bridges the widening gap. If the power supply and load allow sufficient current to flow, a self-sustaining arc may form. Once formed, an arc may be extended to a significant length before breaking the circuit. Attempting to open an inductive circuit often forms an arc, since the inductance provides a high-voltage pulse whenever the current is interrupted. AC systems make sustained arcing somewhat less likely, since the current returns to zero twice per cycle. The arc is extinguished every time the current goes through a zero crossing, and must reignite during the next half-cycle to maintain the arc.

Unlike an ohmic conductor, the resistance of an arc decreases as the current increases. This makes unintentional arcs in an electrical apparatus dangerous since even a small arc can grow large enough to damage equipment and start fires if sufficient current is available. Intentionally produced arcs, such as used in lighting or welding, require some element in the circuit to stabilize the arc's current/voltage characteristics.

Electrostatic Devices, Natural Static Electricity and Similar Phenomena

A high voltage is not necessarily dangerous if it cannot deliver substantial current. The common static electric sparks seen under low-humidity conditions always involve voltage well above 700 V. For example, sparks to car doors in winter can involve voltages as high as 20,000 V. Also, physics demonstration devices such as Van de Graaff generators and Wimshurst machines can produce voltages approaching one million volts, yet at worst they deliver a brief sting. That is because the number of electrons involved is not high. These devices have a limited amount of stored energy, so the current produced is low and usually for a short time. During the discharge, these machines apply high voltage to the body for only a millionth of a second or less. So a low-amperage current is applied for a very short time, and the number of electrons involved is very small.

The discharge may involve extremely high voltage over very short periods, but, to produce heart fibrillation, an electric power supply must produce a significant current (amperage) in the heart muscle continuing for many milli-seconds, and must deposit a total energy in the range of at least millijoules or higher. A current of relatively high amperage at anything more than about fifty volts can therefore be medically significant and potentially fatal.

Tesla coils are not electrostatic machines and can produce significant currents for a sustained interval. Although their appearance in operation is similar to high voltage static electricity devices, the current supplied to a human body will be relatively constant as long as contact is maintained, and the voltage will be much higher than the break-down voltage of human skin. Used correctly, the output of a Tesla coil of proper design can have useful therapeutic effects. Used incorrectly, the output can be dangerous or even fatal.

Power Lines

Electrical transmission and distribution lines for electric power always use voltages significantly higher than 50 volts, so contact with or close approach to the line conductors presents a danger of electrocution. Contact with overhead wires is a frequent cause of injury or death. Metal ladders, farm equipment, boat masts, construction machinery, aerial antennas, and similar objects are frequently involved in fatal contact with overhead wires. Digging into a buried cable can also be dangerous to workers at an excavation site. Digging equipment (either hand tools or machine driven) that contacts a buried cable may energize piping or the ground in the area, resulting in electrocution of nearby workers. A fault in a high-voltage transmission line or substation may result in high currents flowing along the surface of the earth, producing an earth potential rise that also presents a danger of electric shock.

Unauthorized persons climbing on power pylons or electrical apparatus are also frequently the victims of electrocution. At very high transmission voltages even a close approach can be hazardous, since the high voltage may spark across a significant air gap.

For high-voltage and extra-high-voltage transmission lines, specially trained personnel use "live line" techniques to allow hands-on contact with energized equipment. In this case the worker is electrically connected to the high-voltage line but thoroughly insulated from the earth so that he is at the same electrical potential as that of the line. Since training for such operations is lengthy, and still presents a danger to personnel, only very important transmission lines are subject to maintenance while live. Outside these properly engineered situations, insulation from earth does not guarantee that no current flows to earth — as grounding or arcing to ground can occur in unexpected ways, and high-frequency currents can burn even an ungrounded person. Touching a transmitting antenna is dangerous for this reason, and a high-frequency Tesla Coil can sustain a spark with only one endpoint.

Protective equipment on high-voltage transmission lines normally prevents formation of an unwanted arc, or ensures that it is quenched within tens of milliseconds. Electrical apparatus that interrupts high-voltage circuits is designed to safely direct the resulting arc so that it dissipates without damage. High voltage circuit breakers often use a blast of high pressure air, a special dielectric gas (such as SF_6 under pressure), or immersion in mineral oil to quench the arc when the high voltage circuit is broken.

Arc Flash Hazard

Depending on the prospective short circuit current available at a switchgear line-up, a hazard is presented to maintenance and operating personnel due to the possibility of a high-intensity electric arc. Maximum temperature of an arc can exceed 10,000 kelvin, and the radiant heat, expanding hot air, and explosive vapourization of metal and insulation material can cause severe injury to unprotected workers. Such switchgear line-ups and high-energy arc sources are commonly present in electric power utility substations and generating stations, industrial plants and large commercial buildings. In the United States, the National Fire Protection Association, has published a guideline standard NFPA 70E for evaluating and calculating *arc flash hazard*, and provides standards for the protective clothing required for electrical workers exposed to such hazards in the workplace.

Toxic Gases

Electrical discharges, including partial discharge and corona, can produce small quantities of toxic gases, which in a confined space can be a serious health hazard. These gases include ozone and various oxides of nitrogen.

Lightning

The largest-scale sparks are those produced naturally by lightning. An average bolt of negative lightning carries a current of 30 to 50 kiloamperes, transfers a charge of 5 coulombs, and dissipates 500 megajoules of energy (120 kg TNT equivalent, or enough to light a 100-watt light bulb for approximately 2 months). However, an average bolt of positive lightning (from the top of a thunderstorm) may carry a current of 300 to 500 kiloamperes, transfer a charge of up to 300 coulombs, have a potential difference up to 1 gigavolt (a billion volts), and may dissipate 300 GJ of energy (72 tons TNT, or enough energy to light a 100-watt light bulb for up to 95 years). A negative lightning stroke typically lasts for only tens of micro-seconds, but multiple strikes are common. A positive lightning stroke is typically a single event. However, the larger peak current may flow for hundreds of milliseconds, making it considerably hotter and more dangerous than negative lightning.

Hazards due to lightning obviously include a direct strike on persons or property. However, lightning can also create dangerous voltage gradients in the earth, as well as an electromagnetic pulse, and can charge extended metal objects such as telephone cables, fences, and pipelines to dangerous voltages that can be carried many miles from the site of the strike. Although many of these objects are not normally conductive, very high voltage can cause the electrical breakdown of such insulators, causing them to act as conductors. These transferred potentials are dangerous to people, livestock, and electronic apparatus. Lightning strikes also start fires and explosions, which result in fatalities, injuries, and property damage. For example, each year in North America, thousands of forest fires are started by lightning strikes.

Measures to control lightning can mitigate the hazard; these include lightning rods, shielding wires, and bonding of electrical and structural parts of buildings to form a continuous enclosure.

High-voltage lightning discharges in the atmosphere of Jupiter are thought to be the source of the planet's powerful radio frequency emissions.

VOLTAGE SPIKE

In electrical engineering, **spikes** are fast, short duration electrical transients in voltage (**voltage spikes**), current (**current spikes**), or transferred energy (**energy spikes**) in an electrical circuit.

Fast, short duration electrical transients (overvoltages) in the electric potential of a circuit are typically caused by :

- Lightning strikes
- Power outages
- Tripped circuit breakers
- Short circuits
- Power transitions in other large equipment on the same power line
- Malfunctions caused by the power company
- Electromagnetic pulses (EMP) with electromagnetic energy distributed typically up to the 100 kHz and 1 MHz frequency range.
- Inductive spikes

In the design of critical infrastructure and military hardware, one concern is of pulses produced by nuclear explosions, whose nuclear electromagnetic pulses distribute large energies in frequencies from 1 kHz into the gigahertz range through the atmosphere.

The effect of a voltage spike is to produce a corresponding increase in current (**current spike**). However some voltage spikes may be created by current sources. Voltage would increase as necessary so that a constant current will flow. Current from a discharging inductor is one example.

For sensitive electronics, excessive current can flow if this voltage spike exceeds a material's breakdown voltage, or if it causes avalanche breakdown. In semi-conductor junctions, excessive electric current may destroy or severely weaken that device. An avalanche diode, transient voltage suppression diode, transil, varistor, overvoltage crowbar, or a range of other overvoltage protective devices can divert (shunt) this transient current thereby minimizing voltage.

While generally referred to as a voltage spike, the phenomenon in question is actually an **energy spike**, in that it is measured not in volts but in joules; a transient response defined by a mathematical product of voltage, current, and time.

Voltage spikes may be created by a rapid buildup or decay of a magnetic field, which may induce energy into the associated circuit. However voltage spikes can also have more mundane causes such as a fault in a transformer or higher-voltage (primary circuit) power wires falling onto lower-voltage (secondary circuit) power wires as a result of accident or storm damage.

Voltage spikes may be longitudinal (common) mode or metallic (normal or differential) mode. Some equipment damage from surges and spikes can be prevented by use of surge protection equipment. Each type of spike requires selective use of protective equipment. For example, a common mode voltage spike may not even be detected by a protector installed for normal mode transients.

An uninterrupted voltage increase that lasts more than a few seconds is usually called a "voltage surge" rather than a spike. These are usually caused by malfunctions of the electric power distribution system.

POWER QUALITY

Power quality determines the fitness of electrical power to consumer devices. Synchronization of the voltage frequency and phase allows electrical systems to function in their intended manner without significant loss of performance or life. The term is used to describe electric power that drives an electrical load and the load's ability to function properly. Without the proper power, an electrical device (or load) may malfunction, fail prematurely or not operate at all. There are many ways in which electric power can be of poor quality and many more causes of such poor quality power.

The electric power industry comprises electricity generation (AC power), electric power transmission and ultimately electricity distribution to an electricity meter located at the premises of the end user of the electric power. The electricity then moves through the wiring system of the end user until it reaches the load. The complexity of the system to move electric energy from the point of production to the point of consumption combined with variations in weather, generation, demand and other factors provide many opportunities for the quality of supply to be compromised.

While "power quality" is a convenient term for many, it is the quality of the voltage — rather than power or electric current — that is actually described by the term. Power is simply the flow of

Introduction

The quality of electrical power may be described as a set of values of parameters, such as :

- Continuity of service
- Variation in voltage magnitude
- Transient voltages and currents
- Harmonic content in the waveforms for AC power.

It is often useful to think of power quality as a compatibility problem : is the equipment connected to the grid compatible with the events on the grid, and is the power delivered by the grid, including the events, compatible with the equipment that is connected? Compatibility problems always have at least two solutions : in this case, either clean up the power, or make the equipment tougher.

The tolerance of data-processing equipment to voltage variations is often characterized by the CBEMA curve, which give the duration and magnitude of voltage variations that can be tolerated.

Ideally, AC voltage is supplied by a utility as sinusoidal having an amplitude and frequency given by national standards (in the case of mains) or system specifications (in the case of a power feed not directly attached to the mains) with an impedance of zero ohms at all frequencies.

No real-life power source is ideal and generally can deviate in at least the following ways :

- Variations in the peak or RMS voltage are both important to different types of equipment.

- When the RMS voltage exceeds the nominal voltage by 10 to 80% for 0.5 cycle to 1 minute, the event is called a "swell".

- A "dip" (in British English) or a "sag" (in American English the two terms are equivalent) is the opposite situation : the RMS voltage is below the nominal voltage by 10 to 90% for 0.5 cycle to 1 minute.

- Random or repetitive variations in the RMS voltage between 90 and 110% of nominal can produce a phenomenon known as "flicker" in lighting equipment. Flicker is rapid visible changes of light level. Definition of the characteristics of voltage fluctuations that produce objectionable light flicker has been the subject of ongoing research.

- Abrupt, very brief increases in voltage, called "spikes", "impulses", or "surges", generally caused by large inductive loads being turned off, or more severely by lightning.

- "Under-voltage" occurs when the nominal voltage drops below 90% for more than 1 minute. The term "brownout" is an apt description for voltage drops somewhere between full power (bright lights) and a blackout (no power – no light). It comes from the noticeable to significant dimming of regular incandescent lights, during system faults or overloading etc., when insufficient power is available to achieve full brightness in (usually) domestic lighting. This term is in common usage has no formal definition but is commonly used to describe a reduction in system voltage by the utility or system operator to decrease demand or to increase system operating margins.

- "Over-voltage" occurs when the nominal voltage rises above 110% for more than 1 minute.

- Variations in the frequency.

- Variations in the wave shape – usually described as harmonics.

- Non-zero low-frequency impedance (when a load draws more power, the voltage drops).

- Non-zero high-frequency impedance (when a load demands a large amount of current, then stops demanding it suddenly, there will be a dip or spike in the voltage due to the inductances in the power supply line).

Each of these power quality problems has a different cause. Some problems are a result of the shared infrastructure. For example, a fault on the network may cause a dip that will affect some customers; the higher the level of the fault, the greater the number affected. A problem on one customer's site may cause a transient that affects all other customers on the same sub-system. Problems, such as harmonics, arise within the customer's own installation and may propagate onto the network and affect other customers. Harmonic problems can be dealt with by a combination of good design practice and well proven reduction equipment.

Power Conditioning

Power conditioning is modifying the power to improve its quality.

An uninterruptible power supply can be used to switch off of mains power if there is a transient (temporary) condition on the line. However, cheaper UPS units create poor-quality power themselves, akin to imposing a higher-frequency and lower-amplitude square wave atop the sine wave. High-quality UPS units utilize a double conversion topology which breaks down incoming AC power into DC, charges the batteries, then remanufactures an AC sine wave. This remanufactured sine wave is of higher quality than the original AC power feed.

A surge protector or simple capacitor or varistor can protect against most over-voltage conditions, while a lightning arrestor protects against severe spikes. Electronic filters can remove harmonics.

Smart Grids and Power Quality

Modern systems use sensors called phasor measurement units (PMU) distributed throughout their network to monitor power quality and in some cases respond automatically to them. Using such smart grids features of rapid sensing and automated self-healing of anomalies in the network promises to bring higher quality power and less downtime while simultaneously supporting power from intermittent power sources and distributed generation, which would if unchecked degrade power quality.

POWER QUALITY COMPRESSION ALGORITHM

A **power quality compression algorithm** is an algorithm used in power quality analysis. To provide high quality electric power service, it is essential to monitor the quality of the electric signals also termed as power quality (PQ) at different locations along an electrical power network. Electrical utilities carefully monitor waveforms and currents at various network locations constantly, to understand what lead up to any unforeseen events such as a power outage and blackouts. This is particularly critical at sites where the environment and public safety are at risk (institutions such as hospitals, sewage treatment plants, mines, etc.).

Power Quality Challenges

Engineers have at their disposal many meters, that are able to read and display electrical power waveforms and calculating parameters of the waveforms. These parameters may include, for example, current and voltage RMS, phase relationship between waveforms of a multi-phase signal, power factor, frequency, THD, active power (KW), reactive power (Kvar), apparent power (KVA) and active energy (KWh), reactive energy (Kvarh) and apparent energy (KVAh) and many more. In order to sufficiently monitor unforeseen events, Ribeiro *et. al.* explains that it is not enough to display these parameters, but to also capture voltage waveform data at all times. This is impracticable due to the large amount of data involved, causing what is known the "bottle effect". For instance, at a sampling rate of 32 samples per cycle, 1,920 samples are collected per second. For three-phase meters that measure both voltage and current waveforms, the data is 6-8 times as much. More practical solutions developed in recent years store data only when an event occurs (for example, when high levels of power system harmonics are detected) or alternatively to store the RMS value of the electrical signals. This data, however, is not always sufficient to determine the exact nature of problems.

Raw Data Compression

Nisenblat *et. al.* proposes the idea of power quality compression algorithm (similar to lossy compression methods) that enables meters to continuously store the waveform of one or more power signals, regardless whether or not an event of interest was identified. This algorithm referred to as PQZip empowers a processor with a memory that is sufficient to store the waveform, under normal power conditions, over a long period of time, of at least a month, two months or even a year. The compression is performed in real time, as the signals are acquired; it calculates a compression decision before all the compressed data is received. For instance should one parameter remain constant, and various others fluctuate, the compression decision retains only what is relevant from the constant data, and retains all the fluctuation data. It then decomposes the waveform of the power signal of numerous components, over various periods of the waveform. It concludes the process by compressing the values of at least some of these components over different periods, separately. This real time compression algorithm, performed independent of the sampling, prevents data gaps and has a typical 1000 : 1 compression ratio.

Aggregated Data Compression

A typical function of a common power quality analyzer is a generation of data archive aggregated over given interval. Most typically 10 minute or 1 minute interval is used as specified by the IEC/IEEE PQ standards. A significant archive sizes are created during an operation of such instrument. As Kraus *et. al.* have demonstrated the compression ratio on such archives using Lempel–Ziv–Markov chain algorithm, bzip or other similar lossless compression algorithms can be significant. By using prediction and modelling on the stored time series in the actual power quality archive the efficiency of post processing compression is usually further improved. This combination of simplistic techniques implies savings in both data storage and data acquisition processes.

SURGE PROTECTOR

A **surge protector** (or **surge suppressor**) is an appliance designed to protect electrical devices from voltage spikes. A surge protector attempts to limit the voltage supplied to an electric device by either blocking or by shorting to ground any unwanted voltages above a safe threshold.

The terms **surge protection device (SPD)**, or **transient voltage surge suppressor (TVSS)**, are used to describe electrical devices typically installed in power distribution panels, process control systems, communications systems, and other heavy-duty industrial systems, for the purpose of protecting against electrical surges and spikes, including those caused by lightning. Scaled-down versions of these devices are sometimes installed in residential service entrance electrical panels, to protect equipment in a household from similar hazards.

Many power strips have basic surge protection built in; these are typically clearly labelled as such. However, power strips that do *not* provide surge protection are sometimes erroneously referred to as "surge protectors".

Important Specifications

These are some of the most prominently featured specifications which define a surge protector for AC mains, as well as for some data communications protection applications.

Clamping Voltage

Also known as the **let-through voltage**. This specifies what spike voltage will cause the protective components inside a surge protector to divert unwanted energy from the protected line. A lower clamping voltage indicates better protection, but can sometimes result in a shorter life expectancy for the overall protective system. The lowest three levels of protection defined in the UL rating are 330 V, 400 V and 500 V. The standard let-through voltage for 120 V AC devices is 330 volts. The theoretical lowest possible let-through voltage for 120 V power lines was 180 V. New technology, high quality surge suppressors can now clamp voltage at 130 V.

Underwriters Laboratories (UL), a global independent safety science company, defines how a protector may be used safely. UL 1449, 3rd edition became compliance mandatory in September 2009 to increase safety compared to products conforming to 2nd edition. A Measured Limiting Voltage test, using six times higher current (and energy), defines a Voltage Protection Rating (VPR). For a specific protector, this voltage may be higher compared to a Suppressed Voltage Ratings (SVR) in previous editions that measured let-through voltage with less current. Due to non-linear characteristics of protectors, let-through voltages defined by 2nd edition and 3rd edition testing are not comparable.

A protector may be larger to obtain a same let-through voltage during 3rd edition testing. Therefore, a 3rd edition protector should provide superior safety with increased life expectancy.

Joules Rating

This number defines how much energy an MOV-based surge protector can theoretically absorb in a single event, without failure. Counter-intuitively, a lower number may indicate longer life expectancy if the device can divert more energy elsewhere and thus absorb less energy. In other words, a protective device offering a lower clamping voltage while diverting the same surge current will cause more of the surge energy to be dissipated elsewhere in that current's path. Better protectors exceed peak ratings of 1000 joules and 40,000 amperes.

It is often claimed that a lower joule rating is undersized protection, since the total energy in harmful spikes can be significantly larger than this. However, if properly installed, for every joule absorbed by a protector, another 4 to 30 joules may be dissipated harmlessly into ground. An MOV-based protector with a higher let-through voltage can receive a higher joule rating, even though it lets more surge energy through to the device to be protected.

The joule rating is a commonly quoted but very misleading parameter for comparing MOV-based surge protectors. A surge of any arbitrary ampere and voltage combination can occur in time, but surges commonly last only for nano-seconds to micro-seconds, and experimentally modelled surge energy has been far under 100 joules. Well-designed surge protectors should not rely on MOVs to *absorb* surge energy, but instead to survive the process of harmlessly *redirecting* it to ground.

Generally, more joules means an MOV absorbs less energy while diverting even more into ground.

Some manufacturers commonly design higher joule-rated surge protectors by connecting multiple MOVs in parallel. Since individual MOVs have slightly different non-linear responses when exposed to the same overvoltage, any given MOV might be more sensitive than others. This can cause one MOV in a group to conduct more (a phenomenon called **current-hogging**), leading to overuse and eventually premature failure of that component. If a single inline fuse is placed in series with the MOVs as a power-off safety feature, it will open and fail the surge protector even if remaining MOVs are intact. Thus, the *effective* surge energy absorption capacity of the entire system is dependent on the MOV with the lowest clamping voltage, and the additional MOVs do not provide any further benefit. This limitation can be surmounted by using carefully *matched sets* of MOVs, but this matching must be carefully coordinated with the original manufacturer of the MOV components.

According to industry standards, power line surges inside a building can be up to 6,000 volts, 3,000 amperes, and deliver up to 90 joules of energy, including surges from external sources.

Lightning and other high-energy transient voltage surges can be suppressed with a whole house surge protector. These products are more expensive than simple single-outlet surge protectors, and often need professional installation on the incoming electrical power feed; however, they promise whole house protection from surges via that path. Damage from direct lightning strikes via other paths must be controlled separately.

Response Time

Surge protectors don't operate instantaneously; a slight delay exists. The longer the response time, the longer the connected equipment will be exposed to the surge. However, surges don't happen instantly either. Surges usually take around a few micro-seconds to reach their peak voltage, and a surge protector with a nanosecond response time would kick in fast enough to suppress the most damaging portion of the spike.

Therefore, response time under standard testing is not a useful measure of a surge protector's ability when comparing MOV devices. All MOVs have response times measured in nanoseconds, while test waveforms usually used to design and calibrate surge protectors are all based on modelled waveforms of surges measured in micro-seconds. As a result, MOV-based protectors have no trouble producing impressive response-time specs.

Slower-responding technologies (notably, GDTs) may have difficulty protecting against fast spikes. Therefore, good designs incorporating slower but otherwise useful technologies usually combine them with faster-acting components, to provide more comprehensive protection.

Standards

Some frequently listed standards include :
- IEC 61643-1
- EN 61643-11 and 61643-21
- Telcordia Technologies Technical Reference TR-NWT-001011
- ANSI/IEEE C62.xx
- Underwriters Laboratories (UL) 1449.

Each standard defines different protector characteristics, test vectors, or operational purpose.

The UL1449 (3rd Edition) standard for SPDs is a major rewrite of previous editions, and has also been accepted as an ANSI standard for the first time.

EN 62305 and ANSI/IEEE C62.xx define what spikes a protector might be expected to divert. EN 61643-11 and 61643-21 specify both the product's performance and safety requirements. In contrast, the IEC only writes standards and does not certify any particular product as meeting those standards. IEC Standards are used by members of the CB Scheme of international agreements to test and certify products for safety compliance.

None of those standards guarantee that a protector will provide proper protection in a given application. Each standard defines what a protector should do or might accomplish, based on standardized tests that may or may not correlate to conditions present in a particular real-world situation. A specialized engineering analysis may be needed to provide sufficient protection, especially in situations of high lightning risk.

Primary Components

Systems used to reduce or limit high voltage surges can include one or more of the following types of electronic components. Some surge suppression systems use multiple technologies, since each method has its strong and weak points. The first six methods listed operate primarily by diverting unwanted surge energy away from the protected load, through a protective component connected in a *parallel* (or shunted) topology. The last two methods also block unwanted energy by using a protective component connected in *series* with the power feed to the protected load, and additionally may shunt the unwanted energy like the earlier systems.

Metal Oxide Varistor (MOV)

A metal oxide varistor consists of a bulk semi-conductor material (typically sintered granular zinc oxide) that can conduct large currents (effectively short-circuits) when presented with a voltage above its rated voltage. MOVs typically limit voltages to about 3 to 4 times the normal circuit voltage by diverting surge current elsewhere than the protected load. MOVs may be connected in parallel to increase current capability and life expectancy, providing they are *matched sets* (unmatched MOVs have a tolerance of approximately ±20% on voltage ratings, which is not sufficient).

MOVs have finite life expectancy and "degrade" when exposed to a few large transients, or many more smaller transients. As a MOV degrades, its triggering voltage falls lower and lower. If the MOV is being used to protect a low-power signal line, the ultimate failure mode typically is a partial or complete short circuit of the line, terminating normal circuit operation.

If used in a power filtering application, eventually the MOV behaves as a part-time effective short circuit on an AC (or DC) power line, which will cause it to heat up, starting a process called thermal runaway. As the MOV heats up, it may degrade further, causing a catastrophic failure that can result in a small explosion or fire, if the line current is not otherwise limited. An undersized MOV fails when "Absolute Maximum Ratings" in manufacturer's data-sheet are significantly exceeded.

MOVs are often connected in series with a thermal fuse, so that the fuse disconnects before catastrophic failure can happen. When this happens, only the MOV is disconnected. A failing MOV is a fire risk, which is an original reason for the National Fire Protection Association's (NFPA) UL1449 in 1986 and subsequent revisions in 1998 and 2009. NFPA's primary concern is protection from fire.

When used in power applications, MOVs usually are thermal fused or otherwise protected to avoid persistent short circuits and other fire hazards. In a typical power strip, the visible circuit breaker is distinct from the internal thermal fuse, which is not normally visible to the end user. The circuit breaker has no function related to disconnecting an MOV. A thermal fuse or some equivalent solution protects from MOV generated hazards.

If a surge current is so excessively large as to exceed the MOV parameters and blow the thermal fuse, then a light found on some protectors would indicate unacceptable failure. Even adequately sized MOV protectors will eventually degrade beyond acceptable limits, with or without a failure light indication. Therefore, all MOV-based protectors intended for long-term use should have an indicator that the protective components have failed, and this indication must be checked on a regular basis to insure that protection is still functioning.

Because of their good price/performance ratio, MOVs are the most common protector component in low-cost basic AC power protectors.

Transient Voltage Suppression (TVS) Diode

A TVS diode is a type of Zener diode, also called an avalanche diode or **silicon avalanche diode (SAD)**, which can limit voltage spikes. These components provide the fastest limiting action of protective components (theoretically in pico-seconds), but have a relatively low energy absorbing capability. Voltages can be clamped to less than twice the normal operation voltage. If current impulses remain within the device ratings, life expectancy is exceptionally long. If component ratings are exceeded, the diode may fail as a permanent short circuit; in such cases, protection may remain but normal circuit operation is terminated in the case of low-power signal lines. Due to their relatively limited current capacity, TVS diodes are often restricted to circuits with smaller current spikes. TVS diodes are also used where spikes occur significantly more often than once a year, since this component will not degrade when used within its ratings. A unique type of TVS diode (trade names Transzorb or Transil) contains reversed paired *series* avalanche diodes for bi-polar operation.

TVS diodes are often used in high-speed but low-power circuits, such as occur in data communications. These devices can be paired in *series* with another diode to provide low capacitance as required in communication circuits.

Thyristor Surge Protection Device (TSPD)

A Trisil is a type of **thyristor surge protection device (TSPD)**, a specialized solid-state electronic device used in crowbar circuits to protect against overvoltage conditions. A SIDACtor is another thyristor type device used for similar protective purposes.

These thyristor-family devices can be viewed as having characteristics much like a spark gap or a GDT, but can operate much faster. They are related to TVS diodes, but can "breakover" to a low clamping voltage analogous to an ionized and conducting spark gap. After triggering, the low clamping voltage allows large current surges to flow while limiting heat dissipation in the device.

Gas Discharge Tube (GDT)

A **gas discharge tube (GDT)** is a sealed glass-enclosed device containing a special gas mixture trapped between two electrodes, which conducts electric current after becoming ionized by a high voltage spike. GDTs can conduct more current for their size than other components. Like MOVs, GDTs have a finite life expectancy, and can handle a few very large transients or a greater number of smaller transients. The typical failure mode occurs when the triggering voltage rises so high that the device becomes ineffective, although lightning surges can occasionally cause a dead short.

GDTs take a relatively long time to trigger, permitting a higher voltage spike to pass through before the GDT conducts significant current. It is not uncommon for a GDT to let through pulses of 500 V or more of 100 ns in duration. In some cases, additional protective components are necessary to prevent damage to a protected load, caused by high-speed **let-through** voltage which occurs before the GDT begins to operate.

GDTs create an effective short circuit when triggered, so that if any electrical energy (spike, signal, or power) is present, the GDT will short this. Once triggered, a GDT will continue conducting (called **follow-on current**) until all electric current sufficiently diminishes, and the gas discharge quenches. Unlike other shunt protector devices, a GDT once triggered will continue to conduct at a voltage *less than* the high voltage that initially ionized the gas; this behaviour is called negative resistance. Additional auxiliary circuitry may be needed in DC (and some AC) applications to suppress follow-on current, to prevent it from destroying the GDT after the initiating spike has dissipated. Some GDTs are designed to deliberately short out to a grounded terminal when overheated, thereby triggering an external fuse or circuit breaker.

Many GDTs are light-sensitive, in that exposure to light lowers their triggering voltage. Therefore, GDTs should be shielded from light exposure, or opaque versions that are insensitive to light should be used. The CG2 SN series of surge arrestors formerly produced by C P Clare, are advertised as being non-radioactive, and the datasheet for that series states that some members of the CG/CG2 series (75-470V) are radioactive.

Due to their exceptionally low capacitance, GDTs are commonly used on high frequency lines, such as are used in telecommunications equipment. Because of their high current handling capability, GDTs can also be used to protect power lines, but the follow-on current problem must be controlled.

Selenium Voltage Suppressor

An "overvoltage clamping" bulk semi-conductor similar to an MOV, though it does not clamp as well. However, it usually has a longer life than an MOV. It is used mostly in high-energy DC circuits, like the exciter field of an alternator. It can dissipate power continuously, and it retains its clamping characteristics throughout the surge event, if properly sized.

Fig. : A telephone network connection point with spark-gap overvoltage suppressors. The two brass hex-head objects on the left cover the suppressors, which act to short overvoltage on the tip or ring lines to ground.

Carbon Block Spark Gap Overvoltage Suppressor

A spark gap is one of the oldest protective electrical technologies still found in telephone circuits, having been developed in the nineteenth century. A carbon rod electrode is held with an insulator a specific distance from a second electrode. The gap dimension determines the voltage at which a spark will jump between the two parts and short to ground. The typical spacing for telephone applications in North America is 0.076 mm (0.003"). Carbon block suppressors are similar to gas arrestors (GDTs) but with the two electrodes exposed to the air, so their behaviour is affected by the surrounding atmosphere, especially the humidity. Since their operation produces an open spark, these devices should *never* be installed where an explosive atmosphere may develop.

Quarter-wave Coaxial Surge Arrestor

Used in RF signal transmission paths, this technology features a tuned quarter-wavelength short-circuit stub that allows it to pass a bandwidth of frequencies, but presents a short to any other signals, especially down towards DC. The passbands can be narrowband (about ±5% to ±10% bandwidth) or wideband (above ±25% to ±50% bandwidth). Quarter-wave coax surge arrestors have coaxial terminals, compatible with common coax cable connectors (especially N or 7-16 types). They provide the most rugged available protection for RF signals above 400 MHz; at these frequencies they can perform much better than the gas discharge cells typically used in the universal/broadband coax surge arrestors. Quarter-wave arrestors are useful for telecommunications applications, such as Wi-Fi at 2.4 or

5 GHz but less useful for TV/CATV frequencies. Since a quarter-wave arrestor shorts out the line for low frequencies, it is not compatible with systems which send DC power for a LNB up the coaxial downlink.

Series Mode (SM) Surge Suppressors

These devices are not rated in joules because they operate differently from the earlier suppressors, and they do not depend on materials that inherently wear out during repeated surges. SM suppressors are primarily used to control transient voltage surges on electrical power feeds to protected devices. They are essentially heavy-duty low-pass filters connected so that they allow 50/60 Hz line voltages through to the load, while blocking and diverting higher frequencies. This type of suppressor differs from others by using banks of inductors, capacitors and resistors that suppress voltage surges and in rush current to the neutral wire, whereas other designs shunt to the ground wire. Surges are not diverted but actually suppressed. The inductors slow down the energy. Since the inductor in series with the circuit path slows the current spike, the peak surge energy is spread out in the time domain and harmlessly absorbed and slowly released from a capacitor bank.

Experimental results show that most surge energies occur at under 100 Joules, so exceeding the SM design parameters is unlikely. SM suppressors do not present a fire risk should the absorbed energy exceed design limits of the dielectric material of the components because the surge energy is also limited via arc-over to ground during lightning strikes, leaving a surge remnant that often does not exceed a theoretical maximum (such as 6000 V at 3000 A with a modelled shape of 8 x 20 micro-second waveform specified by IEEE/ANSI C62.41). Because SM work on both the current rise and the voltage rise, they can safely operate in the worst surge environments.

SM suppression focuses its protective philosophy on a *power supply input*, but offers nothing to protect against surges appearing between the input of an SM device and *data lines*, such as antennae, telephone or LAN connections, or multiple such devices cascaded and linked to the primary devices. This is because they do not divert surge energy to the ground line. Data transmission requires the ground line to be clean in order to be used as a reference point. In this design philosophy, such events are already protected against by the SM device before the power supply. NIST reports that "Sending them [surges] down the drain of a grounding conductor only makes them reappear within a micro-second about 200 meters away on some other conductor." So having protection on a data transmission line is only required if surges are diverted to the ground line.

In comparison to devices relying on 10 cent components that operate only briefly (such as MOVs or GDTs), SM devices tend to be bulkier and heavier than those simpler spike shunting components. The initial costs of SM filters are higher, typically 130 USD and up, but a long service life can be expected if they are used properly. In-field installation costs can be higher, since SM devices are installed in *series* with the power feed, requiring the feed to be cut and reconnected. But since the SM devices do not wear out and are not required to be replaced every few years, the overall cost of ownership is much lower.

SURGE ARRESTER

A **surge arrester** is a product installed near the end of any conductor which is long enough before the conductor lands on its intended electrical component. The purpose is to divert damaging lightning-induced transients safely to ground through property changes to its varistor in parallel arrangement to the conductor inside the unit. Also called a surge protection device (SPD) or transient voltage surge suppressor (TVSS), they are only designed to protect against electrical transients resulting from the lightning flash, not a direct lightning termination to the conductors.

Lightning termination to earth results in ground currents which pass over buried conductors and induce a transient that propagates outward towards the ends of the conductor. The same induction happens in overhead and above ground conductors which experience the passing energy of an atmospheric EMP caused by the flash. These devices only protect against induced transients characteristic of a lightning discharge's rapid rise-time and will not protect against electrification caused by a direct termination to the conductor. Transients similar to lightning-induced, such as from a high voltage system's switch faulting, may be safely diverted to ground, however, continuous overcurrents are not protected by these devices. The energy in the transient is infinitesimally small in comparison to that of a lightning discharge; however it is still of sufficient quantity to cause arcing between different circuit pathways within today's microprocessors.

Without very thick insulation, which is generally cost prohibitive, most conductors running for any length whatsoever, say greater than about 50 feet, will experience lightning-induced transients some time. Because the transient is usually initiated at some point between the two ends of the conductor, most applications install a surge arrestor just before the conductor lands in each device to be protected. Each conductor must be protected, as each will have its own transient induced, and each SPD must provide a pathway to earth to safely divert the transient away from the protected component, be it instrument or computer, etc. The one notable exception where they are not installed at both ends is in high voltage distribution systems. In general, the induced voltage is not sufficient to do damage at the electric generation end of the lines; however, installation at the service entrance to a building is key to protecting downstream products that are not as robust.

Types

- **Low-voltage surge arrester** : Apply in Low-voltage distribution system, exchange of electrical appliances protector, low-voltage distribution transformer windings

- **Distribution arrester** : Apply in 3KV, 6KV, 10KV AC power distribution system to protect distribution transformers, cables and power station equipment.

- **The station type of common valve arrester** : Used to protect the 3 ~ 220KV transformer station equipment and communication system.

- **Magnetic blow valve station arrester** : Use to 35 ~ 500KV protect communication systems, transformers and other equipment.

- **Protection of rotating machine using magnetic blow valve arrester** : Used to protect the AC generator and motor insulation.

- **Line Magnetic blow valve arrester** : Used to protect 330KV and above communication system circuit equipment insulation.

- **DC or blowing valve-type arrester** : Use to protect the DC system's insulation of electrical equipment.

- **Neutral protection arrester** : Apply in motor or the transformer's neutral protection.

- **Fiber-tube arrester** : Apply in the power station's wires and the weaknesses protection in the insulated.

- **Plug-in Signal Arrester** : Used to twisted-pair transmission line in order to protect communications and computer systems.

- **High-frequency feeder arrester** : Used to protect the microwave, mobile base stations satellite receiver, etc.

- **Receptacle-type surge arrester** : Use to Protect the terminal Electronic equipment.

- **Signal Arrester** : Apply in MODEM, DDN line, fax, phone, process control signal circuit etc.

- **Network arrester** : Apply in servers, workstations, interfaces etc.

- **Coaxial cable lightning arrester** : Used on the coaxial cable to protect the wireless transmission and receiving system.

ELECTRICAL EQUIPMENT IN HAZARDOUS AREAS

In electrical engineering, a **hazardous location** is defined as a place where concentrations of flammable gases, vapours, or dusts occur. Electrical equipment that must be installed in such locations is especially designed and tested to ensure it does not initiate an explosion, due to arcing contacts or high surface temperature of equipment.

For example, a household light switch may emit a small, harmless visible spark when switching; in an ordinary atmosphere this arc is of no concern, but if a flammable vapour is present, the arc might start an explosion. Electrical equipment intended for use in a chemical factory or refinery is designed either to contain any explosion within the device, or is designed not to produce sparks with sufficient energy to trigger an explosion.

Many strategies exist for safety in electrical installations. The simplest strategy is to minimize the amount of electrical equipment installed in a hazardous area, either by keeping the equipment out of the area altogether or by making the area less hazardous by process improvements or ventilation with clean air. Intrinsic safety, or non-incendive equipment and wiring methods, is a set of practices for apparatus designed with low power levels and low stored energy. Insufficient energy is available to produce an arc that can ignite the surrounding explosive mixture. Equipment enclosures can be pressurized with clean air or inert gas and designed with various controls to remove power or provide notification in case of supply or pressure loss of such gases. Arc-producing elements of the equipment can also be isolated from the surrounding atmosphere by encapsulation, immersion in oil, sand, etc. Heat producing elements such as motor winding, electrical heaters, including heat tracing and lighting fixtures are often designed to limit their maximum temperature below the autoignition temperature of the material involved. Both external and internal temperatures are taken into consideration.

As in most fields of electrical installation, different countries have approached the standardization and testing of equipment for hazardous areas in different ways. As world trade becomes more important in distribution of electrical products, international standards are slowly converging so that a wider range of acceptable techniques can be approved by national regulatory agencies.

Area classification is required by governmental bodies, for example by the U.S. Occupational Safety and Health Administration and compliance is enforced.

Documentation requirements are varied. Often an area classification plan-view is provided to identify equipment ratings and installation techniques to be used for each classified plant area. The plan may contain the list of chemicals with their group and temperature rating, and elevation details shaded to indicate Class, Division (Zone) and group combination. The area classification process would require the participation of operations, maintenance, safety, electrical and instrumentation professionals, the use of process diagrams and material flows, MSDS and any pertinent documents, information and knowledge to determine the hazards and their extent and the countermeasures to be employed. Area classification documentations are reviewed and updated to reflect process changes.

History

Soon after the introduction of electric power into coal mines, it was discovered that lethal explosions could be initiated by electrical equipment such as lighting, signals, or motors. The hazard of fire damp or methane accumulation in mines was well known by the time electricity was introduced, along with the danger of suspended coal dust. At least two British mine explosions were attributed to an electric bell signal system. In this system, two bare wires were run along the length of a drift, and any miner desiring to signal the surface would momentarily touch the wires to each other or bridge the wires with a metal tool. The inductance of the signal bell coils, combined with breaking of contacts by exposed metal surfaces, resulted in sparks which could ignite methane, causing an explosion.

Gas Divisions or Zones

In an industrial plant such as a refinery or chemical process plant, handling of large quantities of flammable liquids and gases creates a risk of leaks. In some cases the gas, ignitable vapour or dust is present all the time or for long periods. Other areas would have a dangerous concentration of flammable substances only during process upsets, equipment deterioration between maintenance periods, or during an incident. Refineries and chemical plants are then divided into areas of risk of release of gas, vapour or dust known as divisions or zones. The process of determining the type and size of these hazardous areas is called area classification. Guidance on assessing the extent of the hazard is given in the NFPA 497 Standard, or API 500 and according to their adaptation by other areas gas zones is given in the current edition of IEC 60079-10-1. For hazardous dusts, the guiding standard is IEC 60079-10-2.

Typical gas hazards are from hydrocarbon compounds, but hydrogen and ammonia are common industrial gases that are flammable.

Non-Hazardous Area :

An area such as a residence or office would be classed as Non-Hazardous (safe area), where the only risk of a release of explosive or flammable gas would be such things as the propellant in an aerosol spray. The only explosive or flammable liquid would be paint and brush cleaner. These are classed as very low risk of causing an explosion and are more of a fire risk (although gas explosions in residential buildings do occur). Non hazardous areas on chemical and other plant are present where it is absolutely certain that the hazardous gas is diluted to a concentration below 25% of its lower flammability limit (or lower explosive limit (LEL).

Division 2 or Zone 2 area :

This is a step up from the safe area. In this zone the gas, vapour or mist would only be present under abnormal conditions (most often leaks under abnormal conditions). As a general guide for Zone 2, unwanted substances should only be present under 10 hours/year or 0–0.1% of the time.

Division 1 or Zone 1 area :

Gas, vapour or mist will be present or expected to be present for long periods of time under normal operating conditions. As a guide for Zone 1, this can be defined as 10–1000 hours/year or 0.1–10% of the time.

Zone 0 area :

Gas or vapour is present all of the time. An example of this would be the vapour space above the liquid in the top of a tank or drum. The ANSI/NEC classification method consider this environment a Division 1 area. As a guide for Zone 0, this can be defined as over 1000 hours/year or >10% of the time.

Dust Zones

Flammable dusts when suspended in air can explode. An old system of area classification to a British standard used a system of letters to designate the zones. This has been replaced by a European numerical system, as set out in directive 1999/92/EU implemented in the UK as the Dangerous Substances and Explosives Atmospheres Regulations 2002.

The boundaries and extent of these three-dimensional zones should be decided by a competent person. There must be a site plan drawn up of the factory with the zones marked on.

The zone definitions are :

Zone 22

A place in which an explosive atmosphere in the form of a cloud of combustible dust in air is not likely to occur in normal operation but, if it does occur, will persist for a short period only

Zone 21

A place in which an explosive atmosphere in the form of a cloud of combustible dust in air is likely to occur, occasionally, in normal operation.

Zone 20

A place in which an explosive atmosphere in the form of a cloud of combustible dust in air is present continuously, or for long periods or frequently.

Gas Groups

Explosive gases, vapours and dusts have different chemical properties that affect the likelihood and severity of an explosion. Such properties include flame temperature, minimum ignition energy, upper and lower explosive limits, and molecular weight. Empirical testing is done to determine parameters such as the maximum experimental safe gap, minimum ignition current, explosion pressure and time to peak pressure, spontaneous ignition temperature, and maximum rate of pressure rise. Every substance has a differing combination of properties but it is found that they can be ranked into similar ranges, simplifying the selection of equipment for hazardous areas.

Flammability of combustible liquids are defined by their flash-point. The flash-point is the temperature at which the material will generate sufficient quantity of vapour to form an ignitable mixture. The flash point determines if an area needs to be classified. A material may have a relatively low autoignition temperature yet if its flash-point is above the ambient temperature, then the area may not need to be classified. Conversely if the same material is heated and handled above its flash-point, the area must be classified.

Each chemical gas or vapour used in industry is classified into a gas group.

Group	Representative Gases
I	All Underground Coal Mining. Firedamp (methane)
IIA	Industrial methane, propane, petrol and the majority of industrial
IIB	Ethylene, coke oven gas and other industrial gases
IIC	Hydrogen, acetylene, carbon disulphide

Apparatus marked IIB can also be used for IIA gases. IIC marked equipment can be used for both IIA and IIB. If a piece of equipment has just II and no A, B, or C after then it is suitable for any gas group.

A list must be drawn up of every chemical gas or vapour that is on the refinery/chemical complex and included in the site plan of the classified areas. The above groups are formed in order of how volatile the gas or vapour would be if it was ignited, IIC being the most volatile and IIA being the least. The groups also indicate how much energy is required to ignite the gas by spark ignition, Group IIA requiring the most energy and IIC the least.

Equipment Protection Level

In recent years also the Equipment Protection Level (EPL) is specified for several kinds of protection. The required Protection level is linked to the intended use in the zones described below :

Group	Ex risk	Zone	EPL	Minimum type of protection
I (mines)	energized		Ma	
I (mines)	de-energized in presence of Ex atmosphere		Mb	
II (gas)	explosive atmosphere > 1000 hrs/yr	0	Ga	ia, ma
II (gas)	explosive atmosphere between 10 and 1000 hrs/yr	1	Gb	ib, mb, px, py, e, o, q, s
II (gas)	explosive atmosphere between 1 and 10 hrs/yr	2	Gc	n, ic, pz
III (dust)	explosive surface > 1000 hrs/yr	20	Da	ia
III (dust)	explosive surface between 10 and 1000 hrs/yr	21	Db	ib
III (dust)	explosive surface between 1 and 10 hrs/yr	22	Dc	ic

Temperature Classification

Another important consideration is the temperature classification of the electrical equipment. The surface temperature or any parts of the electrical equipment that may be exposed to the hazardous atmosphere should be tested that it does not exceed 80% of the auto-ignition temperature of the specific gas or vapour in the area where the equipment is intended to be used.

The temperature classification on the electrical equipment label will be one of the following (in degree Celsius) :

USA °C		UK °C	Germany °C Continuous - Short Time
T1 - 450	T3A - 180	T1 - 450	G1 : 360 – 400
T2 - 300	T3B - 165	T2 - 300	G2 : 240 - 270
T2A - 280	T3C - 160	T3 - 200	G3 : 160 - 180
T2B - 260	T4 - 135	T4 - 135	G4 : 110 - 125
T2C - 230	T4A - 120	T5 - 100	G5 : 80 - 90
T2D - 215	T5 - 100	T6 - 85	
T3 - 200	T6 - 85		

The above table tells us that the surface temperature of a piece of electrical equipment with a temperature classification of T3 will not rise above 200 °C.

Auto-ignition Temperatures

The auto-ignition temperature of a liquid, gas or vapour is the temperature at which the substance will ignite without any external heat source. The exact temperature value determined depends on the laboratory test conditions and apparatus. Such temperatures for common substances are :

Gas	Temperature
Methane	580 °C
Hydrogen	560 °C
Propane	493 °C
Ethylene	425 °C
Acetylene	305 °C
Naphtha	290 °C
Carbon disulfide	102 °C

The surface of a high pressure steam pipe may be above the autoignition temperature of some fuel/air mixtures.

Auto-ignition Temperatures (Dust)

The auto-ignition temperature of a dust is usually higher than that of vapours & gases. Examples for common materials are :

Substance	Temperature
Sugar	460 °C
Wood	340 °C

Substance	Temperature
Flour	340 °C
Grain dust	300 °C
Tea	300 °C

Type of Protection

To ensure safety in a given situation, equipment is placed into protection level categories according to manufacture method and suitability for different situations. Category 1 is the highest safety level and Category 3 the lowest. Although there are many types of protection, a few are detailed

	Ex Code	Description	Standard	Location	Use
Flameproof	d	Equipment construction is such that it can withstand an internal explosion and provide relief of the external pressure via flamegap(s) such as the labyrinth created by threaded fittings or machined flanges. The escaping (hot) gases must sufficiently cool down along the escape path that by the time they reach the outside of the enclosure not to be a source of ignition of the outside, potentially ignitable surroundings. Equipment has flameproof gaps (max 0.006" (150 um) propane/ethylene, 0.004" (100 um) acetylene/hydrogen)	IEC/EN 60079-1	Zone 1 if gas group & temp. class correct	Motors, lighting, junction boxes, electronics
Increased Safety	e	Equipment is very robust and components are made to a high quality	IEC/EN 60079-7	Zone 2 or Zone 1	Motors, lighting, junction boxes
Oil Filled	o	Equipment components are completely submerged in oil	IEC/EN 60079-6	Zone 2 or Zone 1	switchgear
Sand/Powder/Quartz Filled	q	Equipment components are completely covered with a layer of Sand, powder or quartz	IEC/EN 60079-5	Zone 2 or Zone 1	Electronics, telephones, chokes
Encapsulated	m	Equipment components of the equipment are usually encased in a resin type material	IEC/EN 60079-18	Zone 1 (Ex mb) or Zone 0 (Ex ma)	Electronics (no heat)

Pres-surised/purged	p	Equipment is pressurised to a positive pressure relative to the surrounding atmosphere with air or an inert gas, thus the surrounding ignitable atmosphere can not come in contact with energized parts of the apparatus. The overpressure is monitored, maintained and controlled.	IEC/EN 60079-2	Zone 1 (px or py), or zone 2 (pz)	Ana-lysers, motors, control boxes, comput-ers
Intrinsi-cally safe	i	Any arcs or sparks in this equipment has insufficient energy (heat) to ignite a vapour Equipment can be installed in ANY housing provided to IP54. A 'Zener Barrier' or 'opto isol' or 'galvanic' unit may be used to assist with certification. A special standard for instrumentation is IEC/EN 60079-27, describing requirements for Fieldbus Intrinsically Safe Concept (FISCO) (zone 0, 1 or 2)	IEC/EN 60079-25 IEC/EN 60079-11 IEC/EN60079-27	'ia' : Zone 0 & 'ib' : Zone 1 'ic : zone 2	Instrumenta-tion, meas-urement, control
Non-Incen-dive	n	Equipment is non-incendive or non-sparking. A special standard for instrumentation is IEC/EN 60079-27, describing requirements for Fieldbus Non-Incendive Concept (FNICO) (zone 2)	IEC/EN 60079-15 IEC/EN 60079-27	Zone 2	Motors, lighting, junction boxes, electronic equipment
Special Protection	s	This method, being by definition special, has no specific rules. In effect it is any method which can be shown to have the required degree of safety in use. Much early equipment having Ex s protection was designed with encapsulation and this has now been incorporated into IEC 60079-18 [Ex m]. Ex s is a coding referenced in IEC 60079-0. The use of EPL and ATEX Category directly is an alternative for "s" marking. The IEC standard EN 60079-33 is made public and is expected to become effective soon, so that the normal Ex certification will also be possible for Ex-s	IEC/EN 60079-33	Zone depending upon Manufacturers Certification.	As its certification states

The types of protection are sub-divided into several sub-classes, linked to EPL : ma and mb, px, py and pz, ia, ib and ic. The a sub-divisions have the most stringent safety requirements, taking into account more the one independent component faults simultaneously.

Multiple Protection

Many items of EEx rated equipment will employ more than one method of protection in different components of the apparatus. These would be then labelled with each of the individual methods. For example, a socket outlet labeled EEx'de' might have a case made to EEx 'e' and switches that are made to EEx 'd'.

ANSI/NFPA Areas Description

Class I, Div. 1 - Where ignitable concentrations of flammable gases, vapours or liquids are present continuously or frequently within the atmosphere under normal operation conditions.

Class I, Div. 2 : Where ignitable concentrations of flammable gases, vapours, or liquids are present within the atmosphere under abnormal operating conditions.

Class II, Div. 1 : Where ignitable concentrations of combustible dusts are present within the atmosphere under normal operation conditions.

Class II, Div. 2 : Where ignitable concentrations of combustible dust are present within the atmosphere under abnormal operating conditions.

Class III, Div. 1 : Where easily ignitable fibers or materials producing combustible flyings are present within the atmosphere under normal operation conditions.

Class III, Div. 2 : Where easily ignitable fibers or materials producing combustible flyings are present within the atmosphere under abnormal operating conditions.

Common Materials within Associated Class & Group Ratings, such as "Class I, Division 1, Group A" :

Class I Areas : Group A : Acetylene / Group B : Hydrogen / Group C : Propane and Ethylene / Group D : Benzene, Butane, Methane & Propane

Class II Areas : Group E : Metal Dust / Group F : Carbon & Charcoal / Group G : Flour, Starch, Wood & Plastic

Class III Areas : NO GROUP : Cotton & Sawdust.

Equipment Category

The equipment category indicates the level of protection offered by the equipment.

- Category 1 equipment may be used in zone 0, zone 1 or zone 2 areas.
- Category 2 equipment may be used in zone 1 or zone 2 areas.

• Category 3 equipment may only be used in zone 2 areas.

Labelling

All equipment certified for use in hazardous areas must be labelled to show the type and level of protection applied.

Europe

Fig. : Mark for ATEX certified electrical equipment for explosive atmospheres.

In Europe the label must show the CE mark and the code number of the certifying body. The CE marking is complemented with the Ex mark, followed by the indication of the Group, Category and, if group II equipment, the indication relating to gases (G) or dust (D). For example : Ex II 1 G (Explosion protected, Group 2, Category 1, Gas) Specific type or types of protection being used will be marked.

• EEx ia IIC T4. (Type ia, Group 2C gases, Temperature category 4).
• EEx nA II T3 X (Type n, non-sparking, Group 2 gases, Temperature category 3, special conditions apply).

Industrial electrical equipment for hazardous area has to conform to appropriate parts of standard EN 60079 and in some cases, certified as meeting that standard. Independent test houses (known as Notified Bodies) are established in most European countries, and a certificate from any of these will be accepted across the EU. In the United Kingdom, the DTI appoint and maintain a list of Notified Bodies within the UK, of which Sira and Baseefa are the most well known.

North America

In North America the suitability of equipment for the specific hazardous area must be tested by a Nationally Recognized Testing Laboratory. Such institutes are UL, MET, FM, CSA or Intertek (ETL), for example.

The label will always list the Class(es), Division(s) and may list the Group(s) and temperature Code. Directly adjacent on the label one will find the mark of the listing agency.

Some manufacturers claim "suitability" or "built-to" hazardous areas in their technical literature, but in effect lack the testing agency's certification and thus unacceptable for the AHJ (Authority Having Jurisdiction) to permit operation of the electrical installation/system.

All equipment in Division 1 areas must have an approval label, but certain materials, such as rigid metallic conduit, does not have a specific label indicating the Cl./Div.1 suitability and their listing as approved method of installation in the NEC serves as the permission. Some equipment in Division 2 areas do not require a specific label, such as standard 3 phase induction motors that do not contain normally arcing components.

Also included in the marking are the manufacturers name or trademark and address, the apparatus type, name and serial number, year of manufacture and any special conditions of use. The NEMA enclosure rating or IP code may also be indicated, but it is usually independent of the Classified Area suitability.

INTRINSIC SAFETY

Intrinsic safety (IS) is a protection technique for safe operation of electrical equipment in hazardous areas by limiting the energy available for ignition. In signal and control circuits that can operate with low currents and voltages, the intrinsic safety approach simplifies circuits and reduces installation cost over other protection methods. Areas with dangerous concentrations of flammable gases or dust are found in applications such as petrochemical refineries and mines. As a discipline, it is an application of inherent safety in instrumentation. High-power circuits such as electric motors or lighting cannot use intrinsic safety methods for protection.

Fig. : An intrinsically-safe walkie talkie radio, which has been designed and tested to not become an ignition source in a flammable atmosphere. Test standards may specify combinations of internal failures which may be present while still passing the ignition test.

Operating and Design Principles

In normal use, electrical equipment often creates internal tiny sparks in switches, motor brushes, connectors, and in other places. Compact electrical equipment generates heat as well, which under some circumstances can become an ignition source. Arcing is also a consideration.

There are multiple ways to make equipment explosion-proof, or safe for use in ex-hazardous areas. Intrinsic Safety is one of a few methods available for ex-hazardous areas. Others include Explosion Proof Enclosures, Venting, Oil Immersion, Powder and Sand Filling, and Hermetic Sealing. For handheld electronics, intrinsic safety is the only realistic method that allows a functional device to be explosion-proof. A device termed intrinsically safe is designed to be incapable of producing heat or spark sufficient to ignite an explosive atmosphere.

There are several considerations in designing intrinsically safe electronics devices : reducing or eliminating internal sparking, controlling component temperatures, and eliminating component spacing that would allow dust to short a circuit. Elimination of spark potential within components is accomplished by limiting the stored energy in any given circuit and the system as a whole. Temperature, under certain fault conditions such as an internal short in a semi-conductor device, becomes an issue as the temperature of a component can rise to a level that can ignite some explosive gasses, even in normal use. Safeguards, such as current limiting by resistors and fuses, must be employed to ensure that in no circumstance can a component reach a temperature that could cause autoignition of a combustible atmosphere. In the highly compact electronic devices used today PCB's often have component spacing that create the possibility of an arc between components if dust or other particulate matter works into the circuitry, thus component spacing, siting and isolation become important to the design.

The primary concept behind intrinsic safety is the restriction of available electrical and thermal energy in the system so that ignition of a hazardous atmosphere (explosive gas or dust) cannot occur. This is achieved by ensuring that only low voltages and currents enter the hazardous area, and that no significant energy storage is possible.

One of the most common methods for protection is to limit electrical current by using multiple series resistors (assuming that resistors always fail open); and limit the voltage with multiple zener devices to ground (assuming diode always fail shorted). Sometimes an alternative type of barrier known as a galvanic isolation barrier may be used. Certification standards for intrinsic safety designs, which vary by device type, generally require that the barrier not exceed approved levels of voltage and current with specified damage to limiting components.

Equipment or instrumentation for use in a hazardous area will be designed to operate with low voltage and current, and will be designed without any large capacitors or inductors that could discharge in a spark. The instrument will be connected, using approved wiring methods, back to a control panel in a non-hazardous area that contains safety barriers. The safety barriers ensure that, no matter what accidental contact occurs between the instrument circuit and outside power sources, no more than the approved voltage or current enters the hazardous area.

For example, during marine transfer operations when flammable products are transferred between the marine terminal and tanker ships or barges, two-way radio communication needs to be constantly maintained in case the transfer needs to stop for unforeseen reasons such as a spill. The United States Coast Guard requires that the two way radio must be certified as intrinsically safe.

Another example is intrinsically safe or explosion-proof mobile phones used in explosive atmospheres, such as refineries. Intrinsically safe mobile phones must meet special battery design criteria in order to achieve UL, ATEX directive, or IECEx certification for use in explosive atmospheres.

No single field device or wiring is intrinsically safe by itself (except for properly designed battery-operated, self-contained devices), but is intrinsically safe only when employed in a properly designed IS system. Such systems are usually provided with detailed instructions to ensure safe use and maintenance.

Certifying Agencies

Several different agencies develop standards for intrinsic safety, and evaluate products for compliance with standards. Agencies may be run by governments or may be composed of members from insurance companies, manufacturers, and industries with an interest in safety standards. Certifying agencies allow manufacturers to affix a label or mark to identify that the equipment has been designed to the relevant product safety standards. Examples of such agencies in North America are the Factory Mutual Research Corporation, which certifies radios, Underwriters Laboratories (UL) that certifies mobile phones, and in Canada the Canadian Standards Association. In the EU the standard for intrinsic safety certification is the ATEX Directive, while in other countries around the world the IECEx standards are followed. To facilitate world trade, standards agencies around the world engage in harmonization activity so that intrinsically safe equipment manufactured in one country eventually might be approved for use in another without redundant, expensive, testing and documentation.

ARMATURE

In electrical engineering, an **armature** generally refers to one of the two principle electrical components of an electromechanical machine — generally in a motor or generator — but it may also mean the pole piece of a permanent magnet or electromagnet, or the moving iron part of a solenoid or relay.

The other component is the field winding or field magnet. The role of the "field" component is simply to create a magnetic field (magnetic flux) for the armature to interact with, thus the field component can comprise either permanent magnets, or electromagnets formed by a conducting coil.

The armature, in contrast, must carry current so it is always a conductor or a conductive coil, oriented normal to both the field and to the direction of motion, torque (rotating machine), or force (linear machine). The armature's role is twofold. The first is to carry current crossing the field, thus creating shaft torque in a rotating machine or force in a linear machine. The second role is to generate an electromotive force (EMF).

In the armature, an electromotive force is created by the relative motion of the armature and the field. When the machine acts in the motor mode, this EMF opposes the armature current, and the armature converts electrical power to mechanical torque, and power, unless the machine is stalled, and transfers it to the load via the shaft. When the machine acts in the generator mode, the armature EMF drives the armature current, and shaft mechanical power is converted to electrical power and transferred to the load. In an induction generator, these distinctions are blurred, since the generated power is drawn from the stator, which would normally be considered the field.

A growler is used to check the armature for shorts, opens and grounds.

Terminology

The parts of an alternator or related equipment can be expressed in either mechanical terms or electrical terms. Although distinctly separate these two sets of terminology are frequently used interchangeably or in combinations that include one mechanical term and one electrical term. This may cause confusion when working with compound machines like brushless alternators, or in conversation among people who are accustomed to work with differently configured machinery.

In alternating current machines, the armature is usually stationary, and is named the stator winding <Need Source>. In DC rotating machines other than brushless DC machines, it is usually rotating, and is known as the rotor <Need Source>. The pole piece of a permanent magnet or electromagnet and the moving, iron part of a solenoid, especially if the latter acts as a switch or relay, may also be referred to as armatures.

Mechanical

Rotor : The rotating part of an alternator, generator, dynamo or motor.
Stator : The stationary part of an alternator, generator, dynamo or motor

Electrical

Armature : The power-producing component of an alternator, generator, dynamo or motor. The armature can be on either the rotor or the stator.
Field : The magnetic field component of an alternator, generator, dynamo or motor. The field can be on either the rotor or the stator and can be either an electromagnet or a permanent magnet.

Armature Reaction in a DC Machine

In a DC machine, the main field is produced by field coils. In both the generating and motoring modes, the armature carries current and a magnetic field is established, which is called the armature flux. The effect of armature flux on the main field is called the armature reaction.

The armature reaction :

1. demagnetizes the main field, and
2. cross magnetizes the main field.

The demagnetizing effect can be overcome by adding extra ampere-turns on the main field. The cross-magnetizing effect can be reduced by having common poles.

Armature reaction is essential in Amplidyne rotating amplifiers.

Armature reaction drop is the effect of a magnetic field on the distribution of the flux under main poles of a generator.

Since an armature is wound with coils of wire, a magnetic field is set up in the armature whenever a current flows in the coils. This field is at right angles to the generator field, and is called cross-magnetization of the armature. The effect of the armature field is to distort the generator field and shift the neutral plane. The neutral plane is the position where the armature windings are moving parallel to the magnetic flux lines. This effect is known as armature reaction and is proportional to the current flowing in the armature coils.

The brushes of a generator must be set in the neutral plane; that is, they must contact segments of the commutator that are connected to armature coils having no induced emf. If the brushes were contacting commutator segments outside the neutral plane, they would short-circuit "live" coils and cause arcing and loss of power.

Armature reaction causes the neutral plane to shift in the direction of rotation, and if the brushes are in the neutral plane at no load, that is, when no armature current is flowing, they will not be in the neutral plane when armature current is flowing. For this reason it is desirable to incorporate a corrective system into the generator design.

These are two principal methods by which the effect of armature reaction is overcome. The first method is to shift the position of the brushes so that they are in the neutral plane when the generator is producing its normal load current. in the other method, special field poles, called interpoles, are installed in the generator to counteract the effect of armature reaction.

The brush-setting method is satisfactory in installations in which the generator operates under a fairly constant load. If the load varies to a marked degree, the neutral plane will shift proportionately, and the brushes will not be the correct position at all times. The brush-setting method is the most common means of correcting for armature reaction in small generators (those producing approximately 1000 W or less). Larger generators require the use of interpoles.

Winding Materials

The electrical energy efficiency of a motor can be improved by reducing the electrical losses in the coil (*e.g.*, by using materials with higher electrical conductivities). Armature wiring is made from copper or aluminum. Copper armature wiring enhances electrical efficiencies due to its higher electrical conductivity. Aluminum armature wiring is lighter and less expensive than copper.

BALANCING MACHINE

A **balancing machine** is a measuring tool used for balancing rotating machine parts such as rotors for electric motors, fans, turbines, disc brakes, disc drives, propellers and pumps. The machine usually consists of two rigid pedestals, with suspension and bearings on top supporting a mounting platform. The unit under test is bolted to the platform and is rotated either with a belt-, air-, or end-drive. As the part is rotated, the vibration in the suspension is detected with sensors and that information is used to determine the amount of unbalance in the part. Along with phase information, the machine can determine how much and where to add weights to balance the part.

Hard-Bearing vs. Soft-Bearing

There are two main types of balancing machines, **hard-bearing** and **soft-bearing**. The difference between them, however, is in the suspension and not the bearings.

In a hard-bearing machine, balancing is done at a frequency lower than the resonance frequency of the suspension. In a soft-bearing machine, balancing is done at a frequency higher than the resonance frequency of the suspension. Both types of machines have various advantages and disadvantages. A hard-bearing machine is generally more versatile and can handle pieces with greatly varying weights, because hard-bearing machines are measuring centrifugal effects and require only a one-time calibration. Only five geometric dimensions need to be fed into the measuring unit and the machine is ready for use. Therefore, it works very well for low- and middle-size volume production and in repair workshops.

A soft-bearing machine is not so versatile with respect to amount of rotor weight to be balanced. The preparation of a soft-bearing machine for individual rotor types is more time consuming, because it needs to be calibrated for different part types. It is very suitable for high-production volume and high-precision balancing tasks.

Hard- and soft-bearing machines can be automated to remove weight automatically, such as by drilling or milling, but hard-bearing machines are more robust and reliable. Both machine principles can be integrated into a production line and loaded by a robot arm or gantry, requiring very little human control.

How It Works

With the rotating part resting on the bearings, a vibration sensor is attached to the suspension. In most soft-bearing machines, a velocity sensor is used. This sensor works by moving a magnet in relation to a fixed coil that generates voltage proportional to the velocity of the vibration. Accelerometers, which measure acceleration of the vibration, can also be used.

A photocell (sometimes called a phaser), proximity sensor, or encoder is used to determine the rotational speed, as well as the relative phase of the rotating part. This phase information is then used to filter the vibration information to determine the amount of movement, or force, in one rotation of the part. Also, the time difference between the phase and the vibration peak gives the angle at which the unbalance exists. Amount of unbalance and angle of unbalance give an unbalance vector.

Calibration is performed by adding a known weight at a known angle. In a soft-bearing machine, trial weights must be added in correction planes for each part. This is because the location of the correction planes along the rotational axis is unknown, and therefore it is unknown how much a given amount of weight will affect the balance. By using trial weights, you are adding a known weight at a known angle and getting the unbalance vector caused by it.

Other Balancing Machine Types

Historically any balancing necessary for slower operating machinery would be done simply by rotating the component between centres and slackening the centres slightly until vibrations became apparent as the heavy side of the component flew out in a greater circle where its obvious eccentricity could be seen and marked with chalk. A small piece of plasticine or putty would be applied to the side opposite the chalk and the test repeated until the component semed to run with acceptable smoothness.

Static balancing machines differ from hard- and soft-bearing machines in that the part is not rotated to take a measurement. Rather than resting on its bearings, the part rests vertically on its geometric center. Once at rest, any movement by the part away from its geometric center is detected by two perpendicular sensors. Static balancers are often used to balance parts with a diameter much larger than their length, such as fans. The advantages of using a static balancer are speed and price. However a static balancer can only correct in one plane, so its accuracy is limited.

A **blade balancing machine** attempts to balance a part in assembly, so minimal correction is required later on. Blade balancers are used on parts such as fans, propellers, and turbines. On a blade balancer, each blade to be assembled is weighed and its weight entered into a balancing software package. The software then sorts the blades and attempts to find the blade arrangement with the least amount of unbalance.

Portable balancing machines are used to balance parts that cannot be taken apart and put on a balancing machine, usually parts that are currently in operation such as turbines, pumps, and motors. Portable balancers come with displacement sensors, such as accelerometers, and a photocell, which are then mounted to the pedestals or enclosure of the running part. Based on the vibrations detected, they calculate the part's unbalance. Many times these devices contain a spectrum analyzer so the part condition can be monitored without the use of a photocell and non-rotational vibration can be analyzed.

COMMUTATOR

A **commutator** is the moving part of a rotary electrical switch in certain types of electric motors or electrical generators that periodically reverses the current direction between the rotor and the external circuit. Commutators have two or more softer metallic brushes in contact with them to complete the other half of the switch. In a motor, it applies power to the best location on the rotor, and in a generator, picks off power similarly. As a switch, it has exceptionally long life, considering the number of circuit makes and breaks that occur in normal operation.

A commutator is a common feature of direct current rotating machines. By reversing the current direction in the moving coil of a motor's armature, a steady rotating force (torque) is produced. Similarly, in a generator, reversing of the coil's connection to the external circuit provides unidirectional (*i.e.* direct) current to the external circuit. The first commutator-type direct current machine was built by Hippolyte Pixii in 1832, based on a suggestion by André-Marie Ampère.

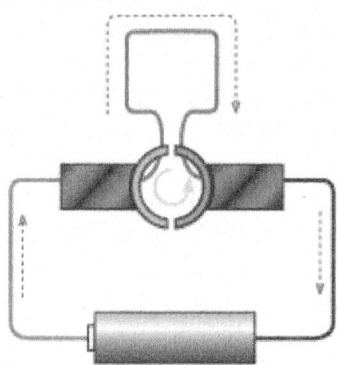

Principle of Operation

A commutator consists of a set of contact bars fixed to the rotating shaft of a machine, and connected to the armature windings. As the shaft rotates, the commutator reverses the flow of current in a winding. For a single armature winding, when the shaft has made one-half complete turn, the winding is now connected so that current flows through it in the opposite of the initial direction. In a motor, the armature current causes the fixed magnetic field to exert a rotational force, or a torque, on the winding to make it turn. In a generator, the mechanical torque applied to the shaft maintains the motion of the armature winding through the stationary magnetic field, inducing a current in the winding. In both the motor and generator case, the commutator periodically reverses the direction of current flow through the winding so that current flow in the circuit external to the machine continues in only one direction.

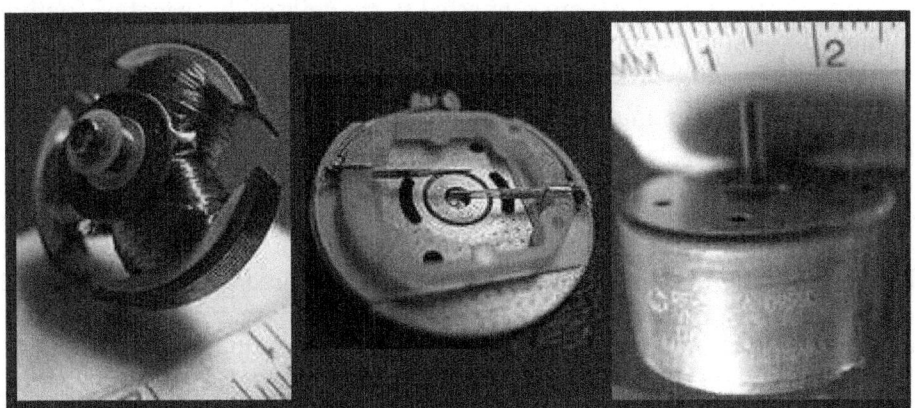

Simplest Practical Commutator

Practical commutators have at least three contact segments, to prevent a "dead" spot where two brushes simultaneously bridge only two commutator segments. Brushes are made wider than the insulated gap, to ensure that brushes are always in contact with an armature coil. For commutators with at least three segments, although the rotor can potentially stop in a position where two commutator segments touch one brush, this only de-energizes one of the rotor arms while the others will still function correctly. With the remaining rotor arms, a motor can produce sufficient torque to begin spinning the rotor, and a generator can provide useful power to an external circuit.

Ring/Segment Construction

A commutator consists of a set of copper segments, fixed around the part of the circumference of the rotating machine, or the rotor, and a set of spring loaded brushes fixed to the stationary frame of the machine. Two (or more) fixed brushes connect to the external circuit, either a source of current for a motor or a load for a generator.

Each conducting segment on the armature of the commutator is insulated from adjacent segments. Mica was used on early machines and is still used on large machines. Many other insulating materials are used to insulate smaller machines; plastics allow quick manufacture of an insulator, for example. The segments are held onto the shaft using a dovetail shape on the edges or underside of each segment, using insulating wedges around the perimeter of each commutation segment. For small appliance and tool motors the segments are typically crimped permanently in place and cannot be removed; when the motor fails it is discarded and replaced. On very large industrial machines (say, from several kilowatts to thousands of kilowatts in rating) it is economical to replace individual damaged segments, and so the end-wedge can be unscrewed and individual segments removed and replaced.

Commutator segments are connected to the coils of the armature, with the number of coils (and commutator segments) depending on the speed and voltage of the machine. Large motors may have hundreds of segments.

Friction between the segments and the brushes eventually causes wear to both surfaces. Carbon brushes, being made of a softer material, wear faster and may be designed to be replaced easily without dismantling the machine. Older copper brushes caused more wear to the commutator, causing deep grooving and notching of the surface over time. The commutator on small motors (say, less than a kilowatt rating) is not designed to be repaired through the life of the device. On large industrial equipment, the commutator may be re-surfaced with abrasives, or the rotor may be removed from the frame, mounted in a large metal lathe, and the commutator resurfaced by cutting it down to a smaller diameter. The largest of equipment can include a lathe turning attachment directly over the commutator.

Brush Construction

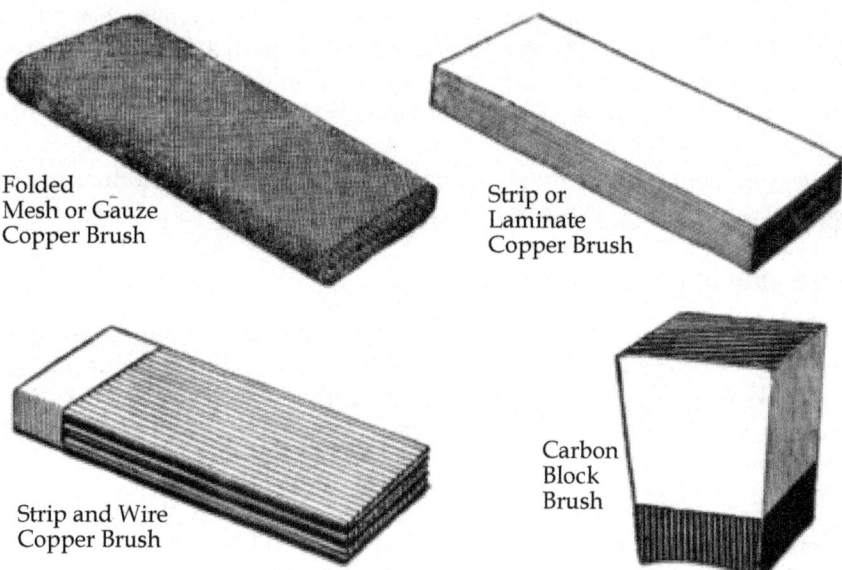

Folded Mesh or Gauze Copper Brush

Strip or Laminate Copper Brush

Strip and Wire Copper Brush

Carbon Block Brush

Fig. : Various types of copper and carbon brushes.

Early in the development of dynamos and motors, brushes made from strands of copper wire were used to contact the surface of the commutator. However, these hard metal brushes tended to scratch and groove the smooth commutator segments, eventually requiring resurfacing of the commutator. As the copper brushes wear away, the dust and pieces of the brush could wedge between commutator segments, shorting them and reducing the efficiency of the device. Fine copper wire mesh or gauze provided better surface contact with less segment wear, but gauze brushes were more expensive than strip or wire copper brushes.

Motors and generators suffer from a phenomenon known as 'armature reaction', one of the effects of which is to change the position at which the current reversal through the windings should ideally take place as the loading varies. Early machines had the brushes mounted on a ring that was provided with a handle. During operation, it was necessary to adjust the position of the brush ring to adjust the commutation to minimise the sparking at the brushes. This process was known as 'rocking the brushes'.

Various developments took place to automate the process of adjusting the commutation and minimizing the sparking at the brushes. One of these was the development of 'high resistance brushes', or brushes made from a mixture of copper powder and carbon. Although described as high resistance brushes, the resistance of such a brush was of the order of milliohms, the exact value dependent on the size and function of the machine. Also, the high resistance brush was not constructed like a brush but in the form of a carbon block with a curved face to match the shape of the commutator.

The high resistance or carbon brush is made large enough that it is significantly wider than the insulating segment that it spans (and on large machines may often span two insulating segments). The result of this is that as the commutator segment passes from under the brush, the current passing to it ramps down more smoothly than had been the case with pure copper brushes where the contact broke suddenly. Similarly the segment coming into contact with the brush has a similar ramping up of the current. Thus, although the current passing through the brush was more or less constant, the instantaneous current passing to the two commutator segments was proportional to the relative area in contact with the brush.

The introduction of the carbon brush had convenient side effects. Carbon brushes tend to wear more evenly than copper brushes, and the soft carbon causes far less damage to the commutator segments. As already noted : there is less sparking with carbon as compared to copper, and as the carbon wears away, the higher resistance of carbon results in fewer problems from the dust collecting on the commutator segments.

The ratio of copper to carbon are each better suited for a particular purpose. Brushes with higher copper content perform better with very low voltages and high current, while brushes with a higher carbon content are better for high voltage and low current. High copper content brushes typically carry 150 to 200 amperes per square inch of contact surface, while higher carbon content only carries 40 to 70 amperes per square inch. The higher resistance of carbon also results in a greater voltage drop of 0.8 to 1.0 volts per contact, or 1.6 to 2.0 volts across the commutator.

Modern rotating machines with commutators almost exclusively use carbon brushes, which may have copper powder mixed in to improve conductivity. Metallic copper brushes can be found in toy or very small motors, such as the one illustrated above, and some motors which only operate very intermittently, such as automotive starter motors.

Brush Holders

A spring is typically used with the brush, to maintain constant contact with the commutator. As the brush and commutator wear down, the spring steadily pushes the brush downwards towards the commutator. Eventually the brush wears small and thin enough that steady contact is no longer possible or it is no longer securely held in the brush holder, and so the brush must be replaced.

It is common for a flexible power cable to be directly attached to the brush, because current flowing through the support spring would cause heating, which may lead to a loss of metal temper and a loss of the spring tension.

When a commutated motor or generator uses more power than a single brush is capable of conducting, an assembly of several brush holders is mounted in parallel across the surface of the very large commutator.

This parallel holder distributes current evenly across all the brushes, and permits a careful operator to remove a bad brush and replace it with a new one, even as the machine continues to spin fully powered and under load.

High power, high current commutated equipment is now uncommon, due to the less complex design of alternating current generators that permits a low current, high voltage spinning field coil to energize high current fixed-position stator coils. This permits the use of very small singular brushes in the alternator design. In this instance, the rotating contacts are continuous rings, called slip rings, and no switching happens.

Modern devices using carbon brushes usually have a maintenance-free design that requires no adjustment throughout the life of the device, using a fixed-position brush holder slot and a combined brush-spring-cable assembly that fits into the slot. The worn brush is pulled out and a new brush inserted.

Brush Contact Angle

The different brush types make contact with the commutator in different ways. Because copper brushes have the same hardness as the commutator segments, the rotor cannot be spun backwards against the ends of copper brushes without the copper digging into the segments and causing severe damage. Consequently strip/laminate copper brushes only make tangential contact with the commutator, while copper mesh and wire brushes use an inclined contact angle touching their edge across the segments of a commutator that can spin in only one direction.

The softness of carbon brushes permits direct radial end-contact with the commutator without damage to the segments, permitting easy reversal of rotor direction, without the need to reorient the brush holders for operation in the opposite direction. Although never reversed, common appliance motors that use wound rotors, commutators and brushes have radial-contact brushes. In the case of a reaction-type carbon brush holder, carbon brushes may be reversely inclined with the commutator so that the commutator tends to push against the carbon for firm contact.

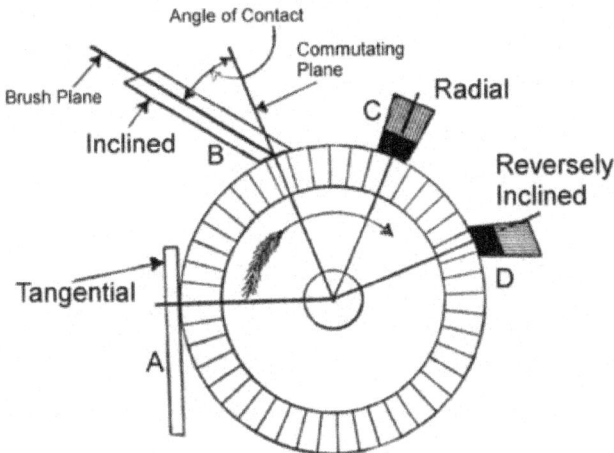

Fig. : Brush angle definitions.

The Commutating Plane

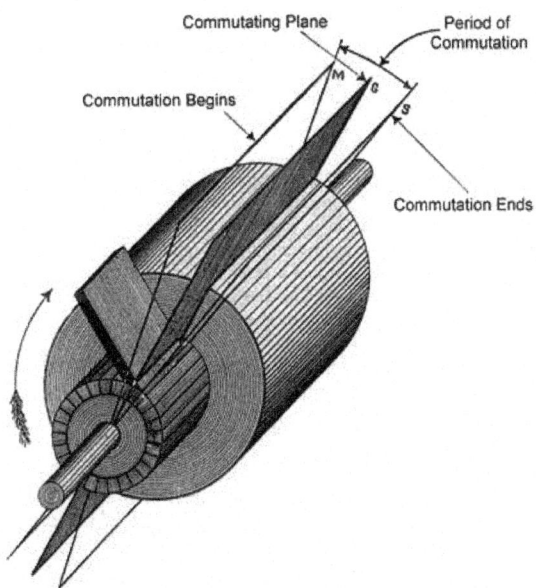

Fig. : Commutating plane definitions.

The contact point where a brush touches the commutator is referred to as the *commutating plane*. To conduct sufficient current to or from the commutator, the brush contact area is not a thin line but instead a rectangular patch across the segments. Typically the brush is wide enough to span 2.5 commutator segments. This means that two adjacent segments are electrically connected by the brush when it contacts both.

Compensation for Stator Field Distortion

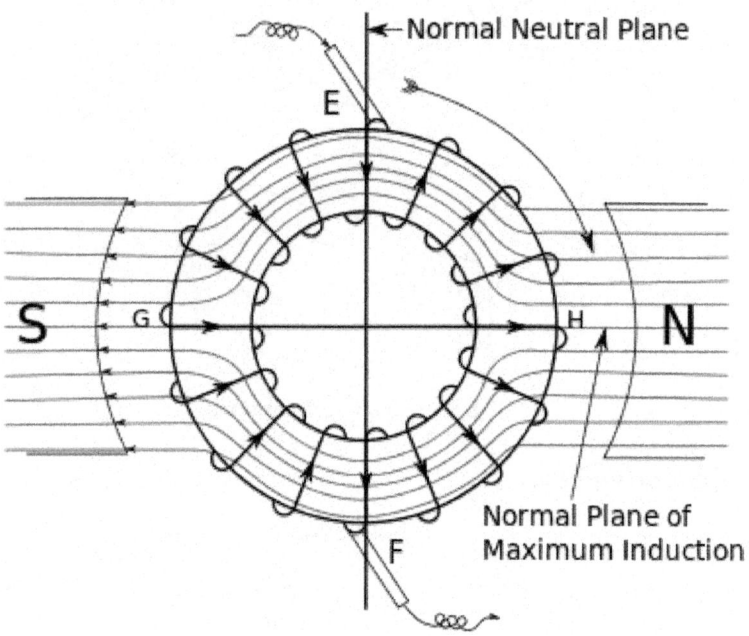

Fig. : Centered position of the commutating plane if there were no field distortion effects.

Most introductions to motor and generator design start with a simple two-pole device with the brushes arranged at a perfect 90-degree angle from the field. This ideal is useful as a starting point for understanding how the fields interact but it is not how a motor or generator functions in actual practice.

In a real motor or generator, the field around the rotor is never perfectly uniform. Instead, the rotation of the rotor induces field effects which drag and distort the magnetic lines of the outer non-rotating stator.

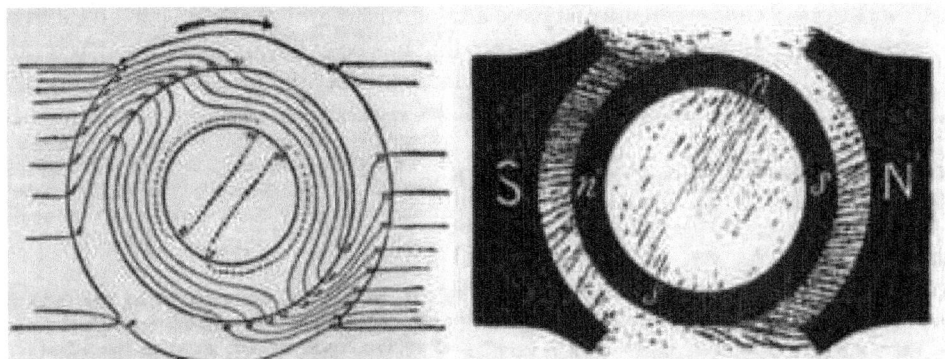

Fig. : On the left is an exaggerated example of how the field is distorted by the rotor. On the right, iron filings show the distorted field across the rotor.

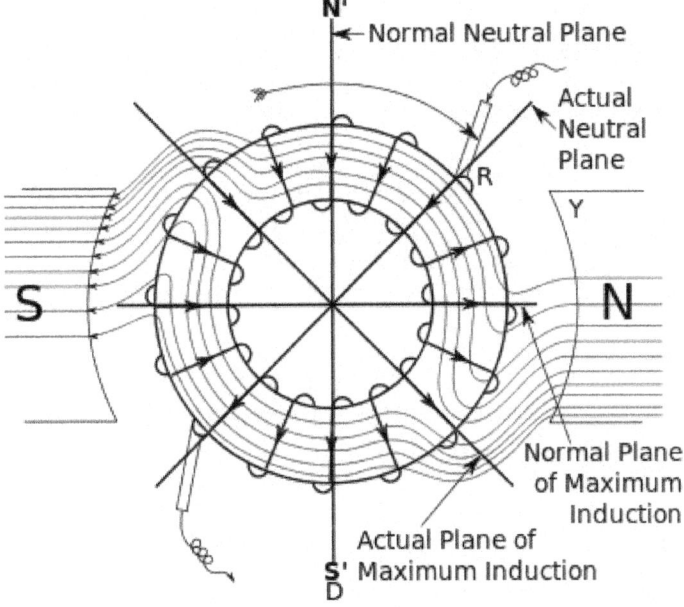

Fig. : Actual position of the commutating plane to compensate for field distortion.

The faster the rotor spins, the further this degree of field distortion. Because a motor or generator operates most efficiently with the rotor field at right angles to the stator field, it is necessary to either retard or advance the brush position to put the rotor's field into the correct position to be at a right angle to the distorted field.

These field effects are reversed when the direction of spin is reversed. It is therefore difficult to build an efficient reversible commutated dynamo, since for highest field strength it is necessary to move the brushes to the opposite side of the normal neutral plane.

The effect can be considered to be analogous to timing advance in an internal combustion engine. Generally a dynamo that has been designed to run at a certain fixed speed will have its brushes permanently fixed to align the field for highest efficiency at that speed.

Further Compensation for Self-induction

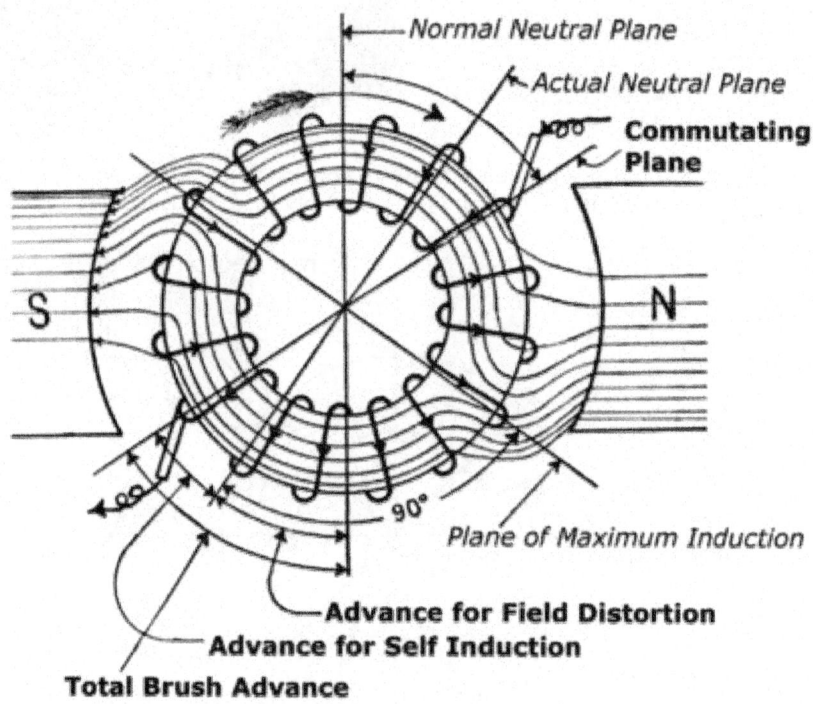

Fig. : Brush advance for Self-induction.

Self-induction – The magnetic fields in each coil of wire join and compound together to create a magnetic field that resists changes in the current, which can be likened to the current having inertia.

In the coils of the rotor, even after the brush has been reached, currents tend to continue to flow for a brief moment, resulting in a wasted energy as heat due to the brush spanning across several commutator segments and the current short-circuiting across the segments.

Spurious resistance is an apparent increase in the resistance in the armature winding, which is proportional to the speed of the armature, and is due to the lagging of the current.

To minimize sparking at the brushes due to this short-circuiting, the brushes are advanced a few degrees further yet, beyond the advance for field distortions. This moves the rotor winding undergoing commutation slightly forward into the stator field which has magnetic lines in the opposite direction and which oppose the field in the stator. This opposing field helps to reverse the lagging self-inducting current in the stator.

So even for a rotor which is at rest and initially requires no compensation for spinning field distortions, the brushes should still be advanced beyond the perfect 90-degree angle as taught in so many beginners textbooks, to compensate for self-induction.

Limitations and Alternatives

While commutators are widely applied in direct current machines, up to several thousand kilowatts in rating, they have limitations.

Brushes and copper segments wear. On small machines the brushes may last as long as the product (small power tools, appliances, etc.) but larger machines will require regular replacement of brushes and occasional resurfacing of the commutator. Brush-type motors may not be suitable for long service on aerospace equipment where maintenance is not possible.

The efficiency of direct current machines is limited by the "brush drop" due to the resistance of the sliding contact. This may be several volts, making low-voltage direct-current machines very inefficient. The friction of the brush on the commutator also absorbs some of the energy of the machine. By contrast to direct current motors, induction motors which do not use commutators or brushes are much more energy efficient.

Lastly, the current density in the brush is limited and the maximum voltage on each segment of the commutator is also limited. Very large direct current machines, say, more than several megawatts rating, cannot be built with commutators. The largest motors and generators, of hundreds of megawatt ratings, are all alternating-current machines.

With the widespread availability of power semi-conductors, it is now economical to provide electronic switching of the current in the motor windings. These "brushless direct current" motors eliminate the commutator; these can be likened to AC machines with a built-in DC to AC inverter. In these motors, rotor position determines when the stator windings switch polarity. Operating life is limited only by bearing wear, if other factors are not adverse.

Repulsion Induction Motors

These are single-phase AC-only motors with higher starting torque than could be obtained with split-phase starting windings, before high-capacitance (non-polar, relatively high-current electrolytic) starting capacitors became practical. They have a conventional wound stator as with any induction motor, but the wire-wound rotor is much like that with a conventional commutator. Brushes opposite each other are connected to each other (not to an external circuit), and transformer action induces currents into the rotor that develop torque by repulsion.

One variety, notable for having an adjustable speed, runs continuously with brushes in contact, while another uses repulsion only for high starting torque and in some cases lifts the brushes once the motor is running fast enough. In the latter case, all commutator segments are connected together as well, before the motor attains running speed.

Once at speed, the rotor windings become functionally equivalent to the squirrel-cage structure of a conventional induction motor, and the motor runs as such.

Laboratory Commutators

Commutators were used as simple forward-off-reverse switches for electrical experiments in physics laboratories. There are two well-known historical types :

Ruhmkorff Commutator

This is similar in design to the commutators used in motors and dynamos. It was usually constructed of brass and ivory (later ebonite).

Pohl Commutator

This consisted of a block of wood or ebonite with four wells, containing mercury, which were cross-connected by copper wires. The output was taken from a pair of curved copper wires which were moved to dip into one or other pair of mercury wells. Instead of mercury, ionic liquids or other liquid metals could be used.

Chapter 8

ELECTRIC POWER DISTRIBUTION

Electricity distribution is the final stage in the delivery of electricity to end users. A distribution system's network carries electricity from the transmission system and delivers it to consumers. Typically, the network would include medium-voltage (2kV to 34.5kV) power lines, sub-stations and pole-mounted transformers, low-voltage (less than 1 kV) distribution writing and sometimes meters.

HISTORY

First Commercial Distrib of Electric Power

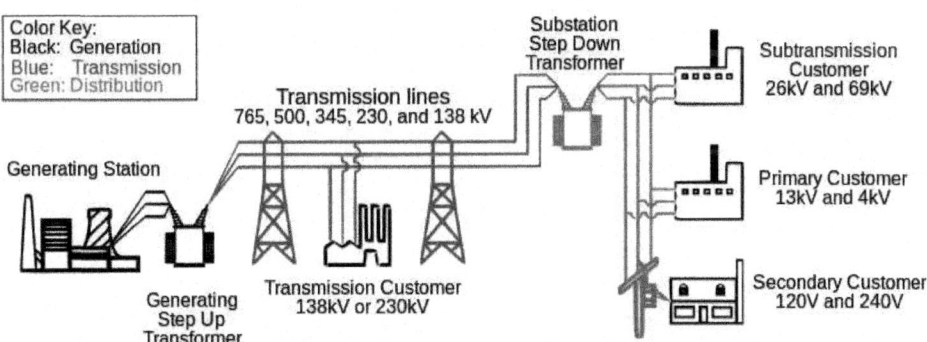

Fig.: Simplified diagram of AC electricity distribution from generation stations to consumers. Transmission system elements are shown in blue, distribution system elements are in green.

In the very early days of electricity distribution (for example, Thomas Edison's Pearl Street Station), direct current (DC) generators were connected to loads at the same voltage. The generation, transmission and loads had to be of the same voltage because there was no way of changing DC voltage levels, other than inefficient motor-generator sets. Low DC voltages (around 100 volts) were used since that was a practical voltage for incandescent lamps, which were the primary electrical load. Low voltage also required less insulation for safe distribution within buildings. The loss in a cable is proportional to the square of the current, and the resistance of the cable. A higher transmission voltage would reduce the copper size to transmit a given quantity of power, but no efficient method existed to change the voltage of DC power circuits. To keep losses to an economically practical level the Edison DC system needed thick cables and local generators. Early DC generating plants needed to be within about 1.5 miles (2.4 km) of the farthest customer to avoid excessively large and expensive conductors.

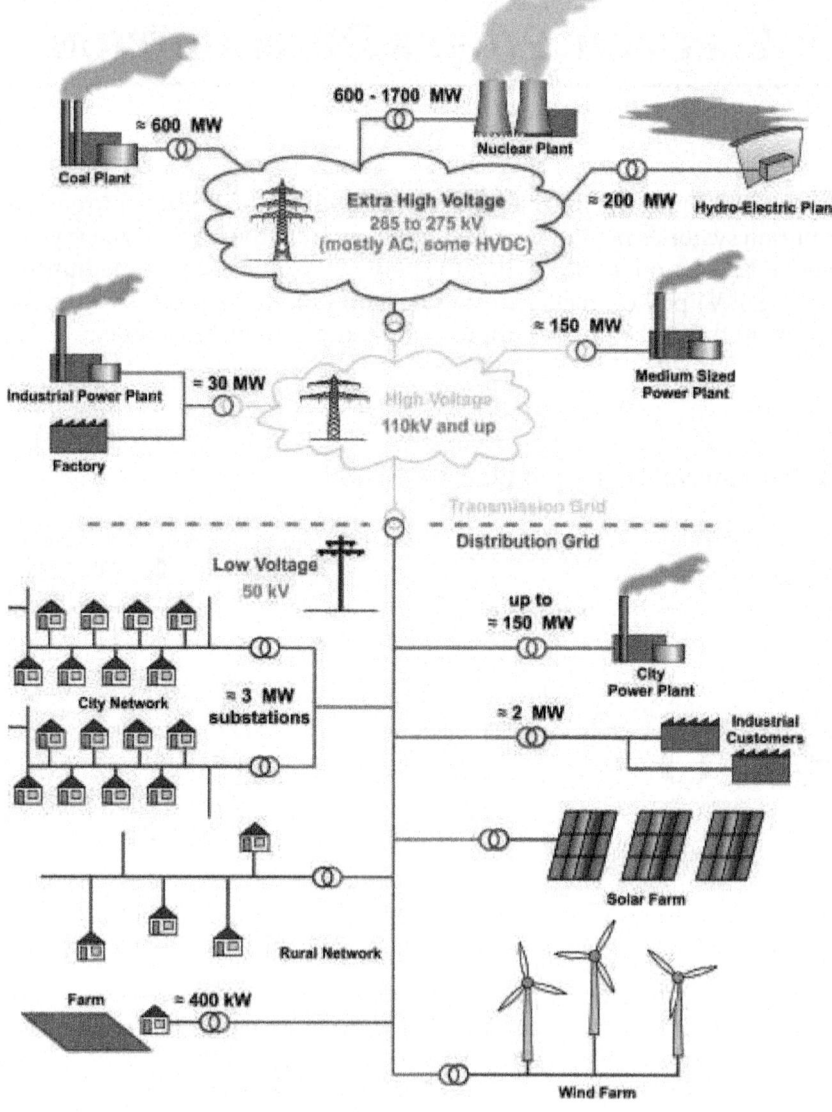

Fig. : General layout of electricity networks.

Introduction of Alternating Current

The competition between the direct current (DC) and alternating current (AC) (in the U.S. backed by Thomas Edison and George Westing-house respectively) was known as the War of Currents. At the conclusion of their campaigning, AC became the dominant form of transmission of power. Power transformers, installed at power stations, could be used to raise the voltage from the generators, and transformers at local sub-stations could reduce voltage to supply loads. Increasing the voltage reduced the current in the transmission and distribution lines and hence the size of conductors and distribution losses. This made it more economical to distribute power over long distances. Generators (such as hydroelectric sites) could be located far from the loads.

Variations

North American and European power distribution systems differ in that North American systems tend to have a greater number of low-voltage step-down transformers located close to customers' premises. For example, in the US a pole-mounted transformer in a sub-urban setting may supply 7-11 houses, whereas in the UK a typical urban or sub-urban low-voltage sub-station would normally be rated between 315 kVA and 1 MVA and supply a whole neighbourhood. This is because the higher domestic voltage used in Europe (415 V vs 230 V) may be carried over a greater distance with acceptable power loss. An advantage of the North American system is that failure or maintenance on a single transformer will only affect a few customers. Advantages of the UK system are that the transformers are fewer in number, larger and more efficient, and due to the diversity of many loads there need be less spare capacity in the transformers, reducing waste. In North American city areas with many customers per unit area, network distribution may be used, with multiple transformers interconnected with low voltage distribution buses over several city blocks.

Rural electrification systems, in contrast to urban systems, tend to use higher distribution voltages because of the longer distances covered by distribution lines. 7.2, 12.47, 25, and 34.5 kV distribution is common in the United States; 11 kV and 33 kV are common in the UK, Australia and New Zealand; 11 kV and 22 kV are common in South Africa. Other voltages are occasionally used.

In New Zealand, Australia, Saskatchewan, Canada, and South Africa, single wire earth return systems SWER are used to electrify remote rural areas.

While power electronics now allow for conversion between DC voltage levels, AC is preferred in distribution due to the economy, efficiency and reliability of transformers. High-voltage DC is used for transmission of large blocks of power over long distances, for transmission over submarine cables for medium distances or for interconnecting adjacent AC networks, but not for local distribution to customers. Electric power is normally generated at 11-25kV in a power station. To transmit power over long distances, it is then stepped-up to higher voltages as necessary : 400kV, 330kV, 275kV, 220kV, 132kV 110kV and 66kV are common in UK, Ireland, Australia and New Zealand, while 765kV, 525kV, 360kV, 330kV, 287kV, 230kV, 189kV, 138kV, and 69kV are common in North America. Power is carried through this transmission network of high voltage lines for hundreds of kilometres and delivers the power as an interconnected power pool called the 'electric grid'. This grid is then connected to load centres (cities) through a sub-transmission network of lines at voltages from 33kV up to 230kV or more. These lines terminate at sub-stations, where the voltage is further stepped-down to 25kV or less for power distribution to customers through a distribution network of local lines at these lower voltages. Note that a 'grid' does not actually enable power to flow with no loss from one end to the other - it may be hundreds of kilometers long, but the power flows inside the grid are typically much shorter than that, and it would be very inefficient to treat the 'grid' as a long-distance transmission carrier. The 'grid' really performs a 'balancing' function - enabling local power generators across a country to synchronise their power outputs and thus readily share generated power with their neighbours.

MODERN DISTRIBUTION SYSTEMS

The modern distribution system begins as the primary circuit leaves the sub-station and ends as the secondary service enters the customer's meter socket by way of a service drop. Distribution circuits serve many customers. The voltage used is appropriate for the shorter distance and varies from 2,300 to about 35,000 volts depending on utility standard practice, distance, and load to be served. Distribution circuits are fed from a transformer located in an electrical sub-station, where the voltage is reduced from the high values used for power transmission.

Conductors for distribution may be carried on overhead pole lines, or in densely populated areas, buried underground. Urban and sub-urban distribution is done with three-phase systems to serve both residential, commercial, and industrial loads. Distribution in rural areas may be only single-phase if it is not economical to install three-phase power for relatively few and small customers.

Only large consumers are fed directly from distribution voltages; most utility customers are connected to a transformer, which reduces the distribution voltage to the relatively low voltage used by lighting and interior wiring systems. The transformer may be pole-mounted or set on the ground in a protective enclosure. In rural areas a pole-mount transformer may serve only one customer, but in more built-up areas multiple customers may be connected. In very dense city areas, a secondary network may be formed with many transformers feeding into a common bus at the utilization voltage. Each customer has a service drop connection and a meter for billing. (Some very small loads, such as yard lights, may be too small to meter and so are charged only a monthly rate.)

A ground connection to local earth is normally provided for the customer's system as well as for the equipment owned by the utility. The purpose of connecting the customer's system to ground is to limit the voltage that may develop if high voltage conductors fall down onto lower-voltage conductors which are usually mounted lower to the ground, or if a failure occurs within a distribution transformer. If all conductive objects are bonded to the same earth grounding system, the risk of electric shock is minimized. However, multiple connections between the utility ground and customer ground can lead to stray voltage problems; customer piping, swimming pools or other equipment may develop objectionable voltages. These problems may be difficult to resolve since they often originate from places other than the customer's premises.

International Differences

In many areas, "delta" three phase service is common. Delta service has no distributed neutral wire and is therefore less expensive. In North America and Latin America, three phase service is often a Y (*wye*) in which the neutral is directly connected to the 'electrical center' of the generator stator. The neutral provides a low-resistance metallic return to the distribution transformer. Wye service is recognizable when a line has four conductors, one of which is lightly insulated. Three-phase wye service is ideal for motors and heavy power usage.

Many areas in the world use single-phase 220 V or 230 V residential and light industrial service. In this system, the high voltage distribution network supplies a few sub-stations per area, and the 230 V power from each sub-station is directly distributed. A live (hot) wire and neutral are connected to the building from one phase of three phase service. Single-phase distribution is used where motor loads are light.

Europe

In Europe, electricity is normally distributed for industry and domestic use by the three-phase, four wire system. This gives a three-phase voltage of 400 volts and a single-phase voltage of 230 volts. For industrial customers, 3-phase 690 / 400 volt is also available.

Japan

Japan has a large number of small industrial manufacturers, and therefore supplies standard low-voltage three phase-service in many suburbs. Also, Japan normally supplies residential service as two phases of a three phase service, with a neutral. These work well for both lighting and motors.

Rural Services

Rural services normally try to minimize the number of poles and wires. Single-wire earth return (SWER) is the least expensive, with one wire. It uses higher voltages (than urban distribution), which in turn permits use of galvanized steel wire. The strong steel wire allows for less expensive wide pole spacings. Other areas use higher voltage split-phase or three phase service at higher cost.

Metering

Electricity meters use different *metering equations* depending on the form of electrical service. Since the math differs from service to service, the number of conductors and sensors in the meters also vary.

Terms

Besides referring to the physical wiring, the term *electrical service* also refers in an abstract sense to the provision of electricity to a building.

DISTRIBUTION NETWORK CONFIGURATIONS

Distribution networks are typically of two types, radial or interconnected. A radial network leaves the station and passes through the network area with no normal connection to any other supply. This is typical of long rural lines with isolated load areas. An interconnected network is generally found in more urban areas and will have multiple connections to other points of supply. These points of connection are normally open but allow various configurations by the operating utility by closing and opening switches. Operation of these switches may be by remote control from a control center or by a lineman. The benefit of the interconnected model is that in the event of a fault or required maintenance a small area of network can be isolated and the remainder kept on supply.

Within these networks there may be a mix of overhead line construction utilizing traditional utility poles and wires and, increasingly, underground construction with cables and indoor or cabinet sub-stations. However, underground distribution is significantly more expensive than overhead construction. In part to reduce this cost, underground power lines are sometimes co-located with other utility lines in what are called common utility ducts. Distribution feeders emanating from a sub-station are generally controlled by a circuit breaker which will open when a fault is detected. Automatic circuit reclosers may be installed to further segregate the feeder thus minimizing the impact of faults.

Long feeders experience voltage drop requiring capacitors or voltage regulators to be installed.

Characteristics of the supply given to customers are generally mandated by contract between the supplier and customer. Variables of the supply include :

• AC or DC - Virtually all public electricity supplies are AC today. Users of large amounts of DC power such as some electric railways, telephone exchanges and industrial processes such as aluminium smelting usually either operate their own or have adjacent dedicated generating equipment, or use rectifiers to derive DC from the public AC supply

• Nominal voltage, and tolerance (for example, +/- 5 per cent)

• Frequency, commonly 50 or 60 Hz, 16.7 Hz and 25 Hz for some railways and, in a few older industrial and mining locations, 25 Hz.

- Phase configuration (single-phase, polyphase including two-phase and three-phase)
- Maximum demand (some energy providers measure as the largest mean power delivered within a 15 or 30 minute period during a billing period)
- Load factor, expressed as a ratio of average load to peak load over a period of time. Load factor indicates the degree of effective utilization of equipment (and capital investment) of distribution line or system.
- Power factor of connected load
- Earthing systems - TT, TN-S, TN-C-S or TN-C
- Prospective short circuit current
- Maximum level and frequency of occurrence of transients.

Reconfiguration, by exchanging the functional links between the elements of the system, represents one of the most important measures which can improve the operational performance of a distribution system. The problem of optimization through the reconfiguration of a power distribution system, in terms of its definition, is a historical single objective problem with constraints. Since 1975, when Merlin and Back introduced the idea of distribution system reconfiguration for active power loss reduction, until nowadays, a lot of researchers have proposed diverse methods and algorithms to solve the reconfiguration problem as a single objective problem. Some authors have proposed Pareto optimality based approaches (including active power losses and reliability indices as objectives). For this purpose, different artificial intelligence based methods have been used : microgenetic, branch exchange, particle swarm optimization and non-dominated sorting genetic algorithm.

DISTRIBUTION INDUSTRY

Traditionally the electricity industry has been a publicly owned institution but starting in the 1970s nations began the process of deregulation and privatisation, leading to electricity markets. A major focus of these was the elimination of the former so called *natural monopoly* of generation, transmission, and distribution. As a consequence, electricity has become more of a commodity. The separation has also led to the development of new terminology to describe the business units (*e.g.*, line company, wires business and network company).

ELECTRIC POWER TRANSMISSION

Electric-power transmission is the bulk transfer of electrical energy, from generating power plants to electrical sub-stations located near demand centers. This is distinct from the local wiring between high-voltage sub-stations and customers, which is typically referred to as electric power distribution. Transmission lines, when interconnected with each other, become transmission networks. The combined transmission and distribution network is known as the "power grid" in the United States, or just "the grid". In the United Kingdom, the network is known as the "National Grid".

A wide area synchronous grid, also known as an "interconnection" in North America, directly connects a large number of generators delivering AC power with the same relative phase, to a large number of consumers. For example, there are three major interconnections in North America (the Western Interconnection, the Eastern Interconnection and the Electric Reliability Council of Texas (ERCOT) grid), and one large grid for most of continental Europe.

Historically, transmission and distribution lines were owned by the same company, but starting in the 1990s, many countries have liberalized the regulation of the electricity market in ways that have led to the separation of the electricity transmission business from the distribution business.

System

Most transmission lines are high-voltage three-phase alternating current (AC), although single phase AC is sometimes used in railway electrification systems. High-voltage direct-current (HVDC) technology is used for greater efficiency at very long distances (typically hundreds of miles (kilometres)), or in submarine power cables (typically longer than 30 miles (50 km)). HVDC links are also used to stabilize and control problems in large power distribution networks where sudden new loads or blackouts in one part of a network can otherwise result in synchronization problems and cascading failures.

Electricity is transmitted at high voltages (120 kV or above) to reduce the energy losses in long-distance transmission. Power is usually transmitted through overhead power lines. Underground power transmission has a significantly higher cost and greater operational limitations but is sometimes used in urban areas or sensitive locations.

A key limitation of electric power is that, with minor exceptions, electrical energy cannot be stored, and therefore must be generated as needed. A sophisticated control system is required to ensure electric generation very closely matches the demand. If the demand for power exceeds the supply, generation plant and transmission equipment can shut down, which in the worst case may lead to a major regional blackout, such as occurred in the US Northeast blackout of 1965, 1977, 2003, and other regional blackouts in 1996 and 2011. It is to reduce the risk of such failure that electric transmission networks are interconnected into regional, national or continent wide networks thereby providing multiple redundant alternative routes for power to flow should such equipment failures occur. Much analysis is done by transmission companies to determine the maximum reliable capacity of each line (ordinarily less than its physical or thermal limit) to ensure spare capacity is available should there be any such failure in another part of the network.

Overhead Transmission

High-voltage overhead conductors are not covered by insulation. The conductor material is nearly always an aluminium alloy, made into several strands and possibly reinforced with steel strands. Copper was sometimes used for overhead transmission but aluminium is lighter, yields only marginally reduced performance and costs much less. Overhead conductors are a commodity supplied by several companies worldwide. Improved conductor material and shapes are regularly used to allow increased capacity and modernize transmission circuits. Conductor sizes range from 12 mm² (#6 American wire gauge) to 750 mm² (1,590,000 circular mils area), with varying resistance and current-carrying capacity. Thicker wires would lead to a relatively small increase in capacity due to the skin effect, that causes most of the current to flow close to the surface of the wire. Because of this current limitation, multiple parallel cables (called bundle conductors) are used when higher capacity is needed. Bundle conductors are also used at high voltages to reduce energy loss caused by corona discharge.

Today, transmission-level voltages are usually considered to be 110 kV and above. Lower voltages such as 66 kV and 33 kV are usually considered sub-transmission voltages but are occasionally used on long lines with light loads. Voltages less than 33 kV are usually used for distribution. Voltages above 230 kV are considered extra high voltage and require different designs compared to equipment used at lower voltages.

Since overhead transmission wires depend on air for insulation, design of these lines requires minimum clearances to be observed to maintain safety. Adverse weather conditions of high wind and low temperatures can lead to power outages. Wind speeds as low as 23 knots (43 km/h) can permit conductors to encroach operating clearances, resulting in a flashover and loss of supply. Oscillatory motion of the physical line can be termed gallop or flutter depending on the frequency and amplitude of oscillation.

Underground Transmission

Electric power can also be transmitted by underground power cables instead of overhead power lines. Underground cables take up less right-of-way than overhead lines, have lower visibility, and are less affected by bad weather. However, costs of insulated cable and excavation are much higher than overhead construction. Faults in buried transmission lines take longer to locate and repair. Underground lines are strictly limited by their thermal capacity, which permits less overload or re-rating than overhead lines. Long underground AC cables have significant capacitance, which may reduce their ability to provide useful power to loads beyond 50 miles. Long underground DC cables have no such issue and can run for thousands of miles.

History

In the early days of commercial electric power, transmission of electric power at the same voltage as used by lighting and mechanical loads restricted the distance between generating plant and consumers. In 1882, generation was with direct current (DC), which could not easily be increased in voltage for long-distance transmission. Different classes of loads (for example, lighting, fixed motors, and traction/railway systems) required different voltages, and so used different generators and circuits.

Due to this specialization of lines and because transmission was inefficient for low-voltage high-current circuits, generators needed to be near their loads. It seemed at the time, that the industry would develop into what is now known as a distributed generation system with large numbers of small generators located near their loads.

In 1886, in Great Barrington, Massachusetts, a 1 kV alternating current (AC) distribution system was installed. That same year, AC power at 2 kV, transmitted 30 km, was installed at Cerchi, Italy. At an AIEE meeting on May 16, 1888, Nikola Tesla delivered a lecture entitled *A New System of Alternating Current Motors and Transformers*, describing the equipment which allowed efficient generation and use of polyphase alternating currents. The transformer, and Tesla's polyphase and single-phase induction motors, were essential for a combined AC distribution system for both lighting and machinery. Ownership of the rights to the Tesla patents was a key advantage to the Westing-house Company in offering a complete alternating current power system for both lighting and power.

Regarded as one of the most influential electrical innovations, the *universal system* used transformers to step-up voltage from generators to high-voltage transmission lines, and then to step-down voltage to local distribution circuits or industrial customers. By a suitable choice of utility frequency, both lighting and motor loads could be served. Rotary converters and later mercury-arc valves and other rectifier equipment allowed DC to be provided where needed. Generating stations and loads using different frequencies could be interconnected using rotary converters. By using common generating plants for every type of load, important economies of scale were achieved, lower overall capital investment was required, load factor on each plant was increased allowing for higher efficiency, a lower cost for the consumer and increased overall use of electric power.

By allowing multiple generating plants to be interconnected over a wide area, electricity production cost was reduced. The most efficient available plants could be used to supply the varying loads during the day. Reliability was improved and capital investment cost was reduced, since stand-by generating capacity could be shared over many more customers and a wider geographic area. Remote and low-cost sources of energy, such as hydroelectric power or mine-mouth coal, could be exploited to lower energy production cost.

The first transmission of three-phase alternating current using high voltage took place in 1891 during the international electricity exhibition in Frankfurt. A 25 kV transmission line, approximately 175 km long, connected Lauffen on the Neckar and Frankfurt.

Voltages used for electric power transmission increased throughout the 20th century. By 1914, fifty-five transmission systems each operating at more than 70 kV were in service. The highest voltage then used was 150 kV.

The rapid industrialization in the 20th century made electrical transmission lines and grids a critical infrastructure item in most industrialized nations. Interconnection of local generation plants and small distribution networks was greatly spurred by the requirements of World War I, with large electrical generating plants built by governments to provide power to munitions factories. Later these generating plants were connected to supply civil loads through long-distance transmission.

Bulk Power Transmission

Engineers design transmission networks to transport the energy as efficiently as feasible, while at the same time taking into account economic factors, network safety and redundancy. These networks use components such as power lines, cables, circuit breakers, switches and transformers. The transmission network is usually administered on a regional basis by an entity such as a regional transmission organization or transmission system operator.

Transmission efficiency is greatly improved by devices that increase the voltage, (and thereby proportionately reduce the current) in the line conductors, thus allowing power to be transmitted with acceptable losses. The reduced current flowing through the line reduces the heating losses in the conductors. According to Joule's Law, energy losses are directly proportional to the square of the current. Thus, reducing the current by a factor of 2 will lower the energy lost to conductor resistance by a factor of 4 for any given size of conductor.

The optimum size of a conductor for a given voltage and current can be estimated by Kelvin's law for conductor size which states that the size is at its optimum when the annual cost of energy wasted in the resistance is equal to the annual capital charges of providing the conductor. At times of lower interest rates Kelvin's law indicates that thicker wires are optimal, while when metals are expensive thinner conductors are indicated : however power lines are designed for long-term usage so Kelvin's law has to be used in conjunction with long-term estimates of the price of copper and aluminium as well as interest rates for capital.

The increase in voltage is achieved in AC circuits by using a *step-up transformer*. HVDC systems require relatively costly conversion equipment which may be economically justified for particular projects such as submarine cables and longer distance high capacity point to point transmission. HVDC is necessary for the import and export of energy between grid systems that are not synchronised with each other.

A transmission grid is a network of power stations, transmission lines, and sub-stations. Energy is usually transmitted within a grid with three-phase AC. Single-phase AC is used only for distribution to end users since it is not usable for large polyphase induction motors. In the 19th century, two-phase transmission was used but required either four wires or three wires with unequal currents. Higher order phase systems require more than three wires, but deliver little or no benefit.

The price of electric power station capacity is high, and electric demand is variable, so it is often cheaper to import some portion of the needed power than to generate it locally. Because loads are often regionally correlated (hot weather in the Southwest portion of the US might cause many people to use air conditioners), electric power often comes from distant sources. Because of the economic benefits of load sharing between regions, wide area transmission grids now span countries and even continents. The web of interconnections between power producers and consumers should enable power to flow, even if some links are inoperative.

The unvarying (or slowly varying over many hours) portion of the electric demand is known as the *base load* and is generally served by large facilities (which are more efficient due to economies of scale) with fixed costs for fuel and operation. Such facilities are nuclear, coal-fired or hydroelectric, while other energy sources such as concentrated solar thermal and geothermal power have the potential to provide base load power. Renewable energy sources such as solar photovoltaics, wind, wave, and tidal are, due to their intermittency, not considered as supplying "base load" but will still add power to the grid. The remaining or 'peak' power demand, is supplied by peaking power plants, which are typically smaller, faster-responding, and higher cost sources, such as combined cycle or combustion turbine plants fueled by natural gas.

Long-distance transmission of electricity (thousands of kilometers) is cheap and efficient, with costs of US$0.005–0.02/kWh (compared to annual averaged large producer costs of US$0.01–0.025/kWh, retail rates upwards of US$0.10/kWh, and multiples of retail for instantaneous suppliers at unpredicted highest demand moments). Thus distant suppliers can be cheaper than local sources (*e.g.*, New York often buys over 1000 MW of electricity from Canada). Multiple **local sources** (even if more expensive and infrequently used) can make the transmission grid more fault tolerant to weather and other disasters that can disconnect distant suppliers.

Long-distance transmission allows remote renewable energy resources to be used to displace fossil fuel consumption. Hydro and wind sources cannot be moved closer to populous cities, and solar costs are lowest in remote areas where local power needs are minimal. Connection costs alone can determine whether any particular renewable alternative is economically sensible. Costs can be prohibitive for transmission lines, but various proposals for massive infrastructure investment in high capacity, very long distance super grid transmission networks could be recovered with modest usage fees.

Grid Input

At the power stations the power is produced at a relatively low voltage between about 2.3 kV and 30 kV, depending on the size of the unit. The generator terminal voltage is then stepped up by the power station transformer to a higher voltage (115 kV to 765 kV AC, varying by the transmission system and by country) for transmission over long distances.

Losses

Transmitting electricity at high voltage reduces the fraction of energy lost to resistance, which varies depending on the specific conductors, the current flowing, and the length of the transmission line. For example, a 100 mile 765 kV line carrying 1000 MW of energy can have losses of 1.1% to 0.5%. A 345 kV line carrying the same load across the same distance has losses of 4.2%. For a given amount of power, a higher voltage reduces the current and thus the resistive losses in the conductor. For example, raising the voltage by a factor of 10 reduces the current by a corresponding factor of 10 and therefore the I^2R losses by a factor of 100, provided the same sized conductors are used in both cases. Even if the conductor size (cross-sectional area) is reduced 10-fold to match the lower current the I^2R losses are still reduced 10-fold. Long-distance transmission is typically done with overhead lines at voltages of 115 to 1,200 kV. At extremely high voltages, more than 2,000 kV exists between conductor and ground, corona discharge losses are so large that they can offset the lower resistive losses in the line conductors. Measures to reduce corona losses include conductors having larger diameters; often hollow to save weight, or bundles of two or more conductors.

Transmission and distribution losses in the USA were estimated at 6.6% in 1997 and 6.5% in 2007. By using underground DC transmission these losses can be cut in half. Underground cables can be larger diameter because they do not have the constraint of light weight that overhead cables have, being 100 feet in the air. In general, losses are estimated from the discrepancy between power produced (as reported by power plants) and power sold to end customers; the difference between what is produced and what is consumed constitute transmission and distribution losses, assuming no theft of utility occurs.

As of 1980, the longest cost-effective distance for direct-current transmission was determined to be 7,000 km (4,300 mi). For alternating current it was 4,000 km (2,500 mi), though all transmission lines in use today are substantially shorter than this.

In any alternating current transmission line, the inductance and capacitance of the conductors can be significant. Currents that flow solely in 'reaction' to these properties of the circuit, (which together with the resistance define the impedance) constitute reactive power flow, which transmits no 'real' power to the load. These reactive currents however are **very** real and cause extra-heating losses in the transmission circuit. The ratio of 'real' power (transmitted to the load) to 'apparent' power (sum of 'real' and 'reactive') is the power factor. As reactive current increases, the reactive power increases and the power factor decreases. For transmission systems with low power factor, losses are higher than for systems with high power factor. Utilities add capacitor banks, reactors and other components (such as phase-shifting transformers; static VAR compensators; physical transposition of the phase conductors; and flexible AC transmission systems, FACTS) throughout the system to compensate for the reactive power flow and reduce the losses in power transmission and stabilize system voltages. These measures are collectively called 'reactive support'.

Sub-transmission

Sub-transmission is part of an electric power transmission system that runs at relatively lower voltages. It is uneconomical to connect all distribution sub-stations to the high main transmission voltage, because the equipment is larger and more expensive. Typically, only larger sub-stations connect with this high voltage. It is stepped down and sent to smaller sub-stations in towns and neighborhoods. Sub-transmission circuits are usually arranged in loops so that a single line failure does not cut off service to a large number of customers for more than a short time. While sub-transmission circuits are usually carried on overhead lines, in urban areas buried cable may be used.

There is no fixed cutoff between sub-transmission and transmission, or sub-transmission and distribution. The voltage ranges overlap somewhat. Voltages of 69 kV, 115 kV and 138 kV are often used for sub-transmission in North America. As power systems evolved, voltages formerly used for transmission were used for sub-transmission, and sub-transmission voltages became distribution voltages. Like transmission, sub-transmission moves relatively large amounts of power, and like distribution, sub-transmission covers an area instead of just point to point.

Transmission Grid Exit

At the sub-stations, transformers reduce the voltage to a lower level for distribution to commercial and residential users. This distribution is accomplished with a combination of sub-transmission (33 kV to 132 kV) and distribution (3.3 to 25 kV). Finally, at the point of use, the energy is transformed to low voltage (varying by country and customer requirements).

High-voltage Direct Current

High-voltage direct current (HVDC) is used to transmit large amounts of power over long distances or for interconnections between asynchronous grids. When electrical energy is to be transmitted over very long distances, the power lost in AC transmission becomes appreciable and it is less expensive to use direct current instead of alternating current. For a very long transmission line, these lower losses (and reduced construction cost of a DC line) can offset the additional cost of the required converter stations at each end.

HVDC is also used for submarine cables because over about 30 kilometres (19 mi) lengths AC cannot be supplied. In these cases special high voltage cables for DC are used. Submarine HVDC systems are often used to connect the electricity grids of islands, for example, between Great Britain and mainland Europe, between Great Britain and Ireland, between Tasmania and the Australian mainland, and between the North and South Islands of New Zealand. Submarine connections up to 600 kilometres (370 mi) in length are presently in use.

HVDC links can be used to control problems in the grid with AC electricity flow. The power transmitted by an AC line increases as the phase angle between source end voltage and destination ends increases, but too large a phase angle will allow the systems at either end of the line to fall out of step. Since the power flow in a DC link is controlled independently of the phases of the AC networks at either end of the link, this phase angle limit does not exist, and a DC link is always able to transfer its full rated power. A DC link therefore stabilizes the AC grid at either end, since power flow and phase angle can then be controlled independently.

As an example, to adjust the flow of AC power on a hypothetical line between Seattle and Boston would require adjustment of the relative phase of the two regional electrical grids. This is an everyday occurrence in AC systems, but one that can become disrupted when AC system components fail and place unexpected loads on the remaining working grid system. With an HVDC line instead, such an interconnection would : (1) Convert AC in Seattle into HVDC. (2) Use HVDC for the three thousand miles of cross-country transmission. Then (3) convert the HVDC to locally synchronized AC in Boston, (and possibly in other co-operating cities along the transmission route). Such a system could be less prone to failure if parts of it were suddenly shut down. One example of a long DC transmission line is the Pacific DC Intertie located in the Western United States.

Capacity

The amount of power that can be sent over a transmission line is limited. The origins of the limits vary depending on the length of the line. For a short line, the heating of conductors due to line losses sets a thermal limit. If too much current is drawn, conductors may sag too close to the ground, or conductors and equipment may be damaged by overheating. For intermediate-length lines on the order of 100 km (62 mi), the limit is set by the voltage drop in the line. For longer AC lines, system stability sets the limit to the power that can be transferred. Approximately, the power flowing over an AC line is proportional to the cosine of the phase angle of the voltage and current at the receiving and transmitting ends. Since this angle varies depending on system loading and generation, it is undesirable for the angle to approach 90 degrees. Very approximately, the allowable product of line length and maximum load is proportional to the square of the system voltage. Series capacitors or phase-shifting transformers are used on long lines to improve stability. High-voltage direct current lines are restricted only by thermal and voltage drop limits, since the phase angle is not material to their operation.

Up to now, it has been almost impossible to foresee the temperature distribution along the cable route, so that the maximum applicable current load was usually set as a compromise between understanding of operation conditions and risk minimization. The availability of industrial distributed temperature sensing (DTS) systems that measure in real time temperatures all along the cable is a first step in monitoring the transmission system capacity. This monitoring solution is based on using passive optical fibers as temperature sensors, either integrated directly inside a high voltage cable or mounted externally on the cable insulation. A solution for overhead lines is also available. In this case the optical fiber is integrated into the core of a phase wire of overhead transmission lines (OPPC). The integrated Dynamic Cable Rating (DCR) or also called Real Time Thermal Rating (RTTR) solution enables not only to continuously monitor the temperature of a high voltage cable circuit in real time, but to safely utilize the existing network capacity to its maximum. Furthermore it provides the ability to the operator to predict the behaviour of the transmission system upon major changes made to its initial operating conditions.

Control

To ensure safe and predictable operation the components of the transmission system are controlled with generators, switches, circuit breakers and loads. The voltage, power, frequency, load factor, and reliability capabilities of the transmission system are designed to provide cost effective performance for the customers.

Load Balancing

The transmission system provides for base load and peak load capability, with safety and fault tolerance margins. The peak load times vary by region largely due to the industry mix. In very hot and very cold climates home air conditioning and heating loads have an effect on the overall load. They are typically highest in the late afternoon in the hottest part of the year and in mid-mornings and mid-evenings in the coldest part of the year. This makes the power requirements vary by the season and the time of day. Distribution system designs always take the base load and the peak load into consideration.

The transmission system usually does not have a large buffering capability to match the loads with the generation. Thus generation has to be kept matched to the load, to prevent overloading failures of the generation equipment.

Multiple sources and loads can be connected to the transmission system and they must be controlled to provide orderly transfer of power. In centralized power generation, only local control of generation is necessary, and it involves synchronization of the generation units, to prevent large transients and overload conditions.

In distributed power generation the generators are geographically distributed and the process to bring them online and offline must be carefully controlled. The load control signals can either be sent on separate lines or on the power lines themselves. Voltage and frequency can be used as signalling mechanisms to balance the loads.

In voltage signaling, the variation of voltage is used to increase generation. The power added by any system increases as the line voltage decreases. This arrangement is stable in principle. Voltage-based regulation is complex to use in mesh networks, since the individual components and setpoints would need to be reconfigured every time a new generator is added to the mesh.

In frequency signaling, the generating units match the frequency of the power transmission system. In droop speed control, if the frequency decreases, the power is increased. (The drop in line frequency is an indication that the increased load is causing the generators to slow down.)

Wind turbines, vehicle-to-grid and other distributed storage and generation systems can be connected to the power grid, and interact with it to improve system operation.

Failure Protection

Under excess load conditions, the system can be designed to fail gracefully rather than all at once. Brownouts occur when the supply power drops below the demand. Blackouts occur when the supply fails completely.

Rolling blackouts (also called load shedding) are intentionally engineered electrical power outages, used to distribute insufficient power when the demand for electricity exceeds the supply.

Communications

Operators of long transmission lines require reliable communications for control of the power grid and, often, associated generation and distribution facilities. Fault-sensing protective relays at each end of the line must communicate to monitor the flow of power into and out of the protected line section so that faulted conductors or equipment can be quickly de-energized and the balance of the system restored. Protection of the transmission line from short circuits and other faults is usually so critical that common carrier telecommunications are insufficiently reliable, and in remote areas a common carrier may not be available. Communication systems associated with a transmission project may use :

- Microwaves
- Power line communication
- Optical fibers.

Rarely, and for short distances, a utility will use pilot-wires strung along the transmission line path. Leased circuits from common carriers are not preferred since availability is not under control of the electric power transmission organization.

Transmission lines can also be used to carry data : this is called power-line carrier, or PLC. PLC signals can be easily received with a radio for the long wave range.

Optical fibers can be included in the stranded conductors of a transmission line, in the overhead shield wires. These cables are known as optical ground wire (OPGW). Sometimes a standalone cable is used, all-dielectric self-supporting (ADSS) cable, attached to the transmission line cross arms.

Some jurisdictions, such as Minnesota, prohibit energy transmission companies from selling surplus communication bandwidth or acting as a telecommunications common carrier. Where the regulatory structure permits, the utility can sell capacity in extra dark fibers to a common carrier, providing another revenue stream.

Electricity Market Reform

Some regulators regard electric transmission to be a natural monopoly and there are moves in many countries to separately regulate transmission.

Spain was the first country to establish a regional transmission organization. In that country transmission operations and market operations are controlled by separate companies. The transmission system operator is Red Eléctrica de España (REE) and the wholesale electricity market operator is Operador del Mercado Ibérico de Energía – Polo Español, S.A. (OMEL) [1]. Spain's transmission system is interconnected with those of France, Portugal, and Morocco.

In the United States and parts of Canada, electrical transmission companies operate independently of generation and distribution companies.

Cost of Electric Power Transmission

The cost of high voltage electricity transmission (as opposed to the costs of electric power distribution) is comparatively low, compared to all other costs arising in a consumer's electricity bill. In the UK transmission costs are about 0.2p/kWh compared to a delivered domestic price of around 10 p/kWh.

Research evaluates the level of capital expenditure in the electric power T&D equipment market will be worth $128.9bn in 2011.

Merchant Transmission

Merchant transmission is an arrangement where a third party constructs and operates electric transmission lines through the franchise area of an unrelated utility.

Operating merchant transmission projects in the United States include the Cross Sound Cable from Shoreham, New York to New Haven, Connecticut, Neptune RTS Transmission Line from Sayreville, N.J., to Newbridge, N.Y, ITC Holdings, Inc. transmission system in the midwest, and Path 15 in California. Additional projects are in development or have been proposed throughout the United States.

There is only one unregulated or market interconnector in Australia : Basslink between Tasmania and Victoria. Two DC links originally implemented as market interconnectors Directlink and Murraylink have been converted to regulated interconnectors. NEMMCO

A major barrier to wider adoption of merchant transmission is the difficulty in identifying who benefits from the facility so that the beneficiaries will pay the toll. Also, it is difficult for a merchant transmission line to compete when the alternative transmission lines are subsidized by other utility businesses.

Health Concerns

Some large studies, including a large United States study, have failed to find any link between living near power lines and developing any sickness or diseases such as cancer. One old study from 1997 found that it did not matter how close one was to a power line or a sub-station, there was no increased risk of cancer or illness.

The mainstream scientific evidence suggests that low-power, low-frequency, electromagnetic radiation associated with household currents and high transmission power lines does not constitute a short or long term health hazard. Some studies, however, have found statistical correlations between various diseases and living or working near power lines. No adverse health effects have been substantiated for people not living close to powerlines.

There are established biological effects for acute *high* level exposure to magnetic fields well above 100 μT (1 G). In a residential setting, there is "limited evidence of carcinogenicity in humans and less than sufficient evidence for carcinogenicity in experimental animals", in particular, childhood leukaemia, *associated with* average exposure to residential power-frequency magnetic field above 0.3 μT (3 mG) to 0.4 μT (4 mG). These levels exceed average residential power-frequency magnetic fields in homes which are about 0.07 μT (0.7 mG) in Europe and 0.11 μT (1.1 mG) in North America.

Tree Growth Regulator and Herbicide Control Methods may be used in transmission line right of ways which may have health effects.

United States Government Policy

Historically, local governments have exercised authority over the grid and have significant disincentives to take action that would benefit states other than their own. Localities with cheap electricity have a disincentive to making interstate commerce in electricity trading easier, since other regions will be able to compete for local energy and drive up rates. Some regulators in Maine for example do not wish to address congestion problems because the congestion serves to keep Maine rates low. Further, vocal local constituencies can block or slow permitting by pointing to visual impact, environmental, and perceived health concerns. In the US, generation is growing 4 times faster than transmission, but big transmission upgrades require the co-ordination of multiple states, a multitude of interlocking permits, and co-operation between a significant portion of the 500 companies that own the grid. From a policy perspective, the control of the grid is balkanized, and even former energy secretary Bill Richardson refers to it as a *third world grid*. There have been efforts in the EU and US to confront the problem. The US national security interest in significantly growing transmission capacity drove passage of the 2005 energy act giving the Department of Energy the authority to approve transmission if states refuse to act. However, soon after using its power to designate two National Interest Electric Transmission Corridors, 14 senators signed a letter stating the DOE was being too aggressive.

Special Transmission

Grids for Railways

In some countries where electric locomotives or electric multiple units run on low frequency AC power, there are separate single phase traction power networks operated by the railways. These grids are fed by separate generators in some traction powerstations or by traction sub-stations from the public three phase AC network.

Super-conducting Cables

High-temperature super-conductors (HTS) promise to revolutionize power distribution by providing lossless transmission of electrical power. The development of super-conductors with transition temperatures higher than the boiling point of liquid nitrogen has made the concept of superconducting power lines commercially feasible, at least for high-load applications. It has been estimated that the waste would be halved using this method, since the necessary refrigeration equipment would consume about half the power saved by the elimination of the majority of resistive losses. Some companies such as Consolidated Edison and American Superconductor have already begun commercial production of such systems. In one hypothetical future system called a SuperGrid, the cost of cooling would be eliminated by coupling the transmission line with a liquid hydrogen pipeline.

Superconducting cables are particularly suited to high load density areas such as the business district of large cities, where purchase of an easement for cables would be very costly.

HTS transmission lines				
Location Length (km)		Voltage (kV)	Capacity (GW)	Date
Carrollton, Georgia				2000
Albany, New York	0.35	34.5	0.048	2006
Long Island	0.6	130	0.574	2008
Tres Amigas			5	Proposed 2013
Manhattan : Project Hydra				Proposed 2014
Essen, Germany	1	10	0.04	Proposed

Single Wire Earth Return

Single-wire earth return (SWER) or single wire ground return is a single-wire transmission line for supplying single-phase electrical power for an electrical grid to remote areas at low cost. It is principally used for rural electrification, but also finds use for larger isolated loads such as water pumps. Single wire earth return is also used for HVDC over submarine power cables.

Wireless Power Transmission

Both Nikola Tesla and Hidetsugu Yagi attempted to devise systems for large scale wireless power transmission in the late 1800s and early 1900s, with no commercial success.

In November 2009, Laser Motive won the NASA 2009 Power Beaming Challenge by powering a cable climber 1 km vertically using a ground-based laser transmitter. The system produced up to 1 kW of power at the receiver end. In August 2010, NASA contracted with private companies to pursue the design of laser power beaming systems to power low earth orbit satellites and to launch rockets using laser power beams.

Wireless power transmission has been studied for transmission of power from solar power satellites to the earth. A high power array of microwave or laser transmitters would beam power to a rectenna. Major engineering and economic challenges face any solar power satellite project.

Security of Control Systems

The Federal government of the United States admits that the power grid is susceptible to cyber-warfare. The United States Department of Homeland Security works with industry to identify vulnerabilities and to help industry enhance the security of control system networks, the federal government is also working to ensure that security is built in as the U.S. develops the next generation of 'smart grid' networks.

Records

- Highest capacity system : 6.3 GW HVDC Itaipu (Brazil/Paraguay) (±600 kV DC)
- Highest transmission voltage (AC) :
  o planned : 1.20 MV (Ultra High Voltage) on Wardha-Aurangabad line (India) - under construction. Initially will operate at 400 kV.
  o worldwide : 1.15 MV (Ultra High Voltage) on Ekibastuz-Kokshetau line (Kazakhstan)
  o Europe (under construction) : 750kV (Extra High Voltage) on the Rivne NPP/Khmelnytskyi NPP to Kyivska Sub-station route (Ukraine)
- Largest double-circuit transmission, Kita-Iwaki Powerline (Japan).
- Highest towers : Yangtze River Crossing (China) (height : 345 m or 1,132 ft)
- Longest power line : Inga-Shaba (Democratic Republic of Congo) (length : 1,700 kilometres or 1,056 miles)
- Longest span of power line : 5,376 m (17,638 ft) at Ameralik Span (Greenland, Denmark)
- Longest submarine cables :

- o NorNed, North Sea (Norway/Netherlands) – (length of submarine cable : 580 kilometres or 360 miles)
- o Basslink, Bass Strait, (Australia) – (length of submarine cable : 290 kilometres or 180 miles, total length : 370.1 kilometres or 230 miles)
- o Baltic Cable, Baltic Sea (Germany/Sweden) – (length of submarine cable : 238 kilometres or 148 miles, HVDC length : 250 kilometres or 155 miles, total length : 262 kilometres or 163 miles)
- • Longest underground cables :
- o Murraylink, Riverland/Sunraysia (Australia) – (length of underground cable : 180 kilometres or 112 miles).

POWER-FLOW STUDY

In power engineering, the **power-flow study**, also known as **load-flow study**, is an important tool involving numerical analysis applied to a power system. A power-flow study usually uses simplified notation such as a one-line diagram and per-unit system, and focuses on various forms of AC power (*i.e.* : voltages, voltage angles, real power and reactive power). It analyzes the power systems in normal steady-state operation. A number of software implementations of power-flow studies exist.

In addition to a power-flow study, sometimes called the *base case*, many software implementations perform other types of analysis, such as short-circuit fault analysis, stability studies (transient & steady-state), unit commitment and economic dispatch. In particular, some programs use linear programming to find the *optimal power flow*, the conditions which give the lowest cost per kilowatt hour delivered.

Power-flow or load-flow studies are important for planning future expansion of power systems as well as in determining the best operation of existing systems. The principal information obtained from the power-flow study is the magnitude and phase angle of the voltage at each bus, and the real and reactive power flowing in each line.

Commercial power systems are usually too large to allow for hand solution of the power flow. Special purpose network analyzers were built between 1929 and the early 1960s to provide laboratory models of power systems; large-scale digital computers replaced the analog methods.

Model

An *AC power-flow model* is a model used in electrical engineering to analyze power grids. It provides a non-linear system which describes the energy flow through each transmission line. Due to non-linearity, in many cases the analysis of large network via AC power-flow model is not feasible, and a linear (but less accurate) DC power-flow model is used instead. Both of those models are very crude approximations to reality.

Systems Analysis

Introduction

Prior to purchasing heavy or high-voltage electrical equipment (transformers, breakers, cables, etc.), it is customary for plant owners, engineering firms, design firms, power utilities, etc., to model their electrical system in a digital computer software for the purpose of equipment sizing and performing various "what if" scenarios. These electrical models range from 120V all the way to 765 kV at the transmission level. An electrical software is typically used for performing various analysis from steady state to transient behaviour of the electrical system.

Network Topology Builder

Electrical models can be created by many available methods :

- Dragging and connecting blocks from symbol library to create a logical electrical single-line diagram
- Importing data from other databases and programs like Microsoft Excel, Access, PSS/E, etc.
- Utilizing templates for substations, protection, distribution, switching stations, data center tiers, etc. Templates are groups of pre-built and connected symbols that represent standard electrical power system configurations

Model Verification and Validation

The electrical system one-line diagram must be updated to accurately illustrate the power system. One-line diagram is an important maintenance document in any plant. When any change or addition is made to a power system, the one-line diagram should be updated immediately to show that change. All personals concerned with the maintenance and operation of the electrical system should have access to the latest revised copies on a regular basis. These diagrams should be reviewed and updated periodically. ETAP software provides an integrated facility to create, maintain, track and revise your power system one-line diagrams.

Load Flow

A load flow study is especially valuable for a system with multiple load centers such a refinery complex. The power flow study is an analysis of the system's capability to adequately supply the connected load. The total system losses, as well as individual line losses, also are tabulated. Transformer tap positions are selected to insure the correct voltage at critical locations such as motor control centers. Performing load flow study on an existing system provides insight and recommendations as to the system operation and optimization of control settings to obtain maximum capacity while minimizing the operating costs. By this analysis we get the result of active power, reactive power, magnitude and phase angle.

Short Circuit

As plant expansion and modification occurs, loads may be moved and larger ones added, leading to increased levels of available short-circuit currents. In addition, the power grid supplying the plant may have increased the available fault capacity due to the enlargement of its own system. The possibility of increasing the amount of short-circuit current available into a fault by these changes is the major reason for a periodic system study. If the short-circuit capacity of the system exceeds the capacity of the protective device, a dangerous situation exists for both plant personnel and system equipment. Under fault conditions, the protective devices would attempt to interrupt the fault current, which could cause a catastrophic failure. Therefore, the need and importance of determining the short-circuit capabilities of a system cannot be stressed enough.

Motor Starting

There are many considerations to starting a motor other than effectively connecting it to the line voltage. Nuisance tripping and excessive running currents, as well as dimming of lights, are signs that a power system is not performing properly. The power system should be able to supply inrush to any motor on the system while supplying normal service for the rest of the system. If the system does not have sufficient capacity, there will be excessively low voltage drops and insufficient capacity for motor starting.

Protective Devices

Protective device co-ordination is the ability of the closest upstream protective device to detect and clear the system fault without the operation of another protective device further upstream in the power system. As backup protection, if the closest device fails to operate for some reason, the next set of protective devices should be co-ordinated so they will operate before extensive damage results. Many cases of unexplained outages or heavily damaged equipment are accepted without question or knowledge that the real reason may be improper co-ordination of protective devices. A poorly co-ordinated system will result in nuisance outages during a fault condition. Damage to equipment are more likely under mis-co-ordinated protective devices hence, resulting in unplanned downtime and equipment repair or replacement.

Arc Flash

Arc Flash studies are required and compliance with this standard is mandatory per OSHA. Companies will be cited and fined for not complying with these standards. A facility must provide, and be able to demonstrate, a safety program with defined responsibilities the following :

- Calculations for the degree of arc flash hazard
- Correct personal protective equipment (PPE) for workers

- Training for workers on the hazards of arc flash
- Appropriate tools for safe working
- Warning labels on equipment.

Dynamic Stability

System stability study is essential when adding, upgrading or evaluating existing generators within the facility. This study will evaluate overcurrent relay settings and or modifications to the protection scheme associated with the generators and utility. This time-based analysis will determine relay settings that will allow the generator out-of-step protection and overcurrent protections to operate for a disturbance prior to loss of system stability and damage to equipment's. In general, load shedding (LS) can be defined as the amount of load that must nearly instantly be removed from a power system to keep the remaining portion of the system operational. This load reduction is in response to a system disturbance that results in an unbalanced condition of the amount of system load exceeding the available electric generation. Common disturbances that can cause this condition to occur include faults, loss of generation, switching errors, lightning strikes, etc. Consequences of Improper Load Shed :

- Shedding too much load (Loss of Critical Process)
- Total loss of production
- Unsafe operating condition and environmental concerns
- Costly outages and equipment damage.

Harmonic Analysis

In general, harmonic analysis is commonly done to predict distortion levels for addition of a new harmonic producing load or capacitor bank. Consequences of excessive harmonic distortion include :

- Control / Computer system interference
- Heating of rotating machinery
- Overheating / failure of capacitors
- Costly outages and equipment damage.

Power-flow Problem Formulation

The goal of a power-flow study is to obtain complete voltage angle and magnitude information for each bus in a power system for specified load and generator real power and voltage conditions. Once this information is known, real and reactive power flow on each branch as well as generator reactive power output can be analytically determined. Due to the non-linear nature of this problem, numerical methods are employed to obtain a solution that is within an acceptable tolerance.

The solution to the power-flow problem begins with identifying the known and unknown variables in the system. The known and unknown variables are dependent on the type of bus. A bus without any generators connected to it is called a Load Bus. With one exception, a bus with at least one generator connected to it is called a Generator Bus. The exception is one arbitrarily-selected bus that has a generator. This bus is referred to as the slack bus.

In the power-flow problem, it is assumed that the real power P_D and reactive power Q_D at each Load Bus are known. For this reason, Load Buses are also known as PQ Buses. For Generator Buses, it is assumed that the real power generated P_G and the voltage magnitude $|V|$ is known. For the Slack Bus, it is assumed that the voltage magnitude $|V|$ and voltage phase Θ are known. Therefore, for each Load Bus, both the voltage magnitude and angle are unknown and must be solved for; for each Generator Bus, the voltage angle must be solved for; there are no variables that must be solved for the Slack Bus. In a system with N buses and R generators, there are then $2(N-1)-(R-1)$ unknowns.

In order to solve for the $2(N-1)-(R-1)$ unknowns, there must be $2(N-1)-(R-1)$ equations that do not introduce any new unknown variables. The possible equations to use are power balance equations, which can be written for real and reactive power for each bus. The real power balance equation is :

$$0 = -\ Pi + \sum_{k=1}^{N} |Vi|\,|Vk|(G_{ik}\cos\theta_{ik} + B_{ik}\sin\theta_{ik})$$

where P_i is the net power injected at bus i, G_{ik} is the real part of the element in the bus admittance matrix Y_{BUS} corresponding to the ith row and kth column, B_{ik} is the imaginary part of the element in the Y_{BUS} corresponding to the ith row and kth column and θ_{ik} is the difference in voltage angle between the ith and kth buses ($\theta_{ik} = \delta_i - \delta_k$). The reactive power balance equation is :

$$0 = -\ Q_i + \sum_{k=1}^{N} |V_i|\,|V_k|(G_{ik}\sin\theta_{ik} - B_{ik}\cos\theta_{ik})$$

where Q_i is the net reactive power injected at bus i.

Equations included are the real and reactive power balance equations for each Load Bus and the real power balance equation for each Generator Bus. Only the real power balance equation is written for a Generator Bus because the net reactive power injected is not assumed to be known and therefore including the reactive power balance equation would result in an additional unknown variable. For similar reasons, there are no equations written for the Slack Bus.

In many transmission systems, the voltage angles θ_{ik} are usually relatively small. There is thus a strong coupling between real power and voltage angle, and between reactive power and voltage magnitude, while the coupling between real power and voltage magnitude, as well as reactive power and voltage angle, is weak. As a result, real power is usually transmitted from the bus with higher voltage angle to the bus with lower voltage angle, and reactive power is usually transmitted from the bus with higher voltage magnitude to the bus with lower voltage magnitude. However, this approximation does not hold when the voltage angle is very large.

Newton–Raphson Solution Method

There are several different methods of solving the resulting non-linear system of equations. The most popular is known as the Newton–Raphson method. This method begins with initial guesses of all unknown variables (voltage magnitude and angles at Load Buses and voltage angles at Generator Buses). Next, a Taylor Series is written, with the higher order terms ignored, for each of the power balance equations included in the system of equations . The result is a linear system of equations that can be expressed as :

$$\begin{bmatrix} \Delta\theta \\ \Delta|V| \end{bmatrix} = -J^{-1} \begin{bmatrix} \Delta P \\ \Delta Q \end{bmatrix}$$

where ΔP and ΔQ are called the mismatch equations :

$$\Delta P_i = -P_i + \sum_{k=1}^{N} |V_i| |V_k| (G_{ik} \cos\theta_{ik} + B_{ik} \sin\theta_{ik})$$

$$\Delta Q_i = -Q_i + \sum_{k=1}^{N} |V_i| |V_k| (G_{ik} \sin\theta_{ik} - B_{ik} \cos\theta_{ik})$$

and J is a matrix of partial derivatives known as a Jacobian :

$$J = \begin{bmatrix} \dfrac{\partial \Delta P}{\partial \theta} & \dfrac{\partial \Delta P}{\partial |V|} \\ \dfrac{\partial \Delta \theta}{\partial \theta} & \dfrac{\partial \Delta \theta}{\partial |V|} \end{bmatrix}.$$

The linearized system of equations is solved to determine the next guess $(m + 1)$ of voltage magnitude and angles based on :

$$\theta^{m+1} = \theta^m + \Delta\theta$$

$$|V|^{m+1} = |V|^m + \Delta|V|$$

The process continues until a stopping condition is met. A common stopping condition is to terminate if the norm of the mismatch equations is below a specified tolerance.

A rough outline of solution of the power-flow problem is :

1. Make an initial guess of all unknown voltage magnitudes and angles. It is common to use a "flat start" in which all voltage angles are set to zero and all voltage magnitudes are set to 1.0 p.u.

2. Solve the power balance equations using the most recent voltage angle and magnitude values.

3. Linearize the system around the most recent voltage angle and magnitude values.

4. Solve for the change in voltage angle and magnitude.
5. Update the voltage magnitude and angles.
6. Check the stopping conditions, if met then terminate, else go to step 2.

Power-flow Methodt

Gauss–Seidel method

In numerical linear algebra, the **Gauss–Seidel method**, also known as the **Lieb-mann method** or the **method of successive displacement**, is an iterative method used to solve a linear system of equations. It is named after the German mathematicians Carl Friedrich Gauss and Philipp Ludwig von Seidel, and is similar to the Jacobi method. Though it can be applied to any matrix with non-zero elements on the diagonals, convergence is only guaranteed if the matrix is either diagonally dominant, or symmetric and positive definite. It was only mentioned in a private letter from Gauss to his student Gerling in 1823. A publication was not delivered before 1874 by Seidel.

Description

The Gauss–Seidel method is an iterative technique for solving a square system of n linear equations with unknown $\mathbf{x}$:

$$Ax = b.$$

It is defined by the iteration

$$L_* x^{(k+1)} = \mathbf{b} - U x(k)$$

where the matrix A is decomposed into a lower triangular component L_*, and a strictly upper triangular component $U : A = L_* + U$.

In more detail, write out A, $\mathbf{x}$ and $\mathbf{b}$ in their components :

$$A = \begin{bmatrix} a_{11} & a_{12} & \cdots & a_{1n} \\ a_{21} & a_{22} & \cdots & a_{2n} \\ \vdots & \vdots & \ddots & \vdots \\ a_{n1} & a_{n2} & \cdots & a_{nn} \end{bmatrix}, \quad x = \begin{bmatrix} x_1 \\ x_2 \\ \vdots \\ x_n \end{bmatrix}, \quad b = \begin{bmatrix} b_1 \\ b_2 \\ \vdots \\ b_n \end{bmatrix}$$

Then the decomposition of A into its lower triangular component and its strictly upper triangular component is given by :

$$A = L_* + U \text{ where } \quad L_* = \begin{bmatrix} a_{11} & 0 & \cdots & 0 \\ a_{21} & a_{22} & \cdots & 0 \\ \vdots & \vdots & \ddots & \vdots \\ a_{n1} & a_{n2} & \cdots & a_{nn} \end{bmatrix}, U = \begin{bmatrix} 0 & a_{12} & \cdots & a_{1n} \\ 0 & 0 & \cdots & a_{2n} \\ \vdots & \vdots & \ddots & \vdots \\ 0 & 0 & \cdots & 0 \end{bmatrix}$$

The system of linear equations may be rewritten as :

$$L_*x = b - Ux$$

The Gauss–Seidel method now solves the left hand side of this expression for **x**, using previous value for **x** on the right hand side. Analytically, this may be written as :

$$x^{(k+1)} = L_*^{-1}(b - Ux^{(k)}).$$

However, by taking advantage of the triangular form of L_*, the elements of $x^{(k+1)}$ can be computed sequentially using forward substitution :

$$x_i^{(k+1)} = \frac{1}{a_{ii}}\left(b_i - \sum_{j<i} a_{ij} x_j^{(k+1)} - \sum_{j>i} a_{ij} x_j^{(k)} \right), i, j = 1, 2, ..., n.$$

The procedure is generally continued until the changes made by an iteration are below some tolerance, such as a sufficiently small residual.

Discussion

The element-wise formula for the Gauss–Seidel method is extremely similar to that of the Jacobi method.

The computation of $x_i^{(k+1)}$ uses only the elements of $x^{(k+1)}$ that have already been computed, and only the elements of $x^{(k)}$ that have not yet to be advanced to iteration $k+1$. This means that, unlike the Jacobi method, only one storage vector is required as elements can be overwritten as they are computed, which can be advantageous for very large problems.

However, unlike the Jacobi method, the computations for each element cannot be done in parallel. Furthermore, the values at each iteration are dependent on the order of the original equations.

Gauss-Seidel is the same as SOR (successive over-relaxation) with $\omega = 1$.

Convergence

The convergence properties of the Gauss–Seidel method are dependent on the matrix A. Namely, the procedure is known to converge if either :

- A is symmetric positive-definite, or
- A is strictly or irreducibly diagonally dominant.

The Gauss–Seidel method sometimes converges even if these conditions are not satisfied.

Algorithm

Since elements can be overwritten as they are computed in this algorithm, only one storage vector is needed, and vector indexing is omitted. The algorithm goes as follows :

```
Inputs: A, b
Output: φ
Choose an initial guess φ to the solution
repeat until convergence
      for i from 1 until n do
          σ ← 0
            for j from 1 until n do
                if j ≠ i then
                    σ ← σ + aᵢⱼφⱼ
                end if
            end (j-loop)
```

$$\phi_i \leftarrow \frac{1}{a_{ii}}(b_i - \sigma)$$

```
      end (i-loop)
      check if convergence is reached
end (repeat)
```

Examples

An Example for the Matrix Version

A linear system shown as $Ax = b$ is given by :

$$A = \begin{bmatrix} 16 & 3 \\ 7 & -11 \end{bmatrix} \text{ and } b = \begin{bmatrix} 11 \\ 13 \end{bmatrix}$$

We want to use the equation

$$x^{(k+1)} = L_*^{-1}(b - Ux^{(k)})$$

in the form

$$x^{(k+1)} = Tx^{(k)} + C$$

where :

$$T = -L_*^{-1}U \text{ and } C = L_*^{-1}b.$$

We must decompose A into the sum of a lower triangular component L_* and a strict upper triangular component U :

$$L^* = \begin{bmatrix} 16 & 0 \\ 7 & -11 \end{bmatrix} \quad \text{and } U = \begin{bmatrix} 0 & 3 \\ 0 & 0 \end{bmatrix}.$$

The inverse of $L*$ is :

$$L^{*-1} = \begin{bmatrix} 16 & 0 \\ 7 & -11 \end{bmatrix}^{-1} = \begin{bmatrix} 0.0625 & 0.0000 \\ 0.0398 & -0.0909 \end{bmatrix}.$$

Now we can find :

$$T = - \begin{bmatrix} 0.0625 & 0.0000 \\ 0.0398 & 0.0909 \end{bmatrix} \times \begin{bmatrix} 0 & 3 \\ 0 & 0 \end{bmatrix} = \begin{bmatrix} 0.000 & -0.01875 \\ 0.000 & -0.1193 \end{bmatrix},$$

$$C = - \begin{bmatrix} 0.0625 & 0.0000 \\ 0.0398 & -0.0909 \end{bmatrix} \times \begin{bmatrix} 11 \\ 13 \end{bmatrix} = \begin{bmatrix} 0.6875 \\ -0.7443 \end{bmatrix}.$$

Now we have T and C and we can use them to obtain the vectors x iteratively.

First of all, we have to choose $x^{(0)}$: we can only guess. The better the guess, the quicker the algorithm will perform.

We suppose :

$$x^{(0)} = \begin{bmatrix} 1.0 \\ 1.0 \end{bmatrix}.$$

We can then calculate :

$$x^{(1)} = \begin{bmatrix} 0.000 & -0.1875 \\ 0.000 & -0.1193 \end{bmatrix} \times \begin{bmatrix} 1.0 \\ 1.0 \end{bmatrix} + \begin{bmatrix} 0.6875 \\ -0.7443 \end{bmatrix} = \begin{bmatrix} 0.5000 \\ -0.8636 \end{bmatrix}.$$

$$x^{(2)} = \begin{bmatrix} 0.000 & -0.1875 \\ 0.000 & -0.1193 \end{bmatrix} \times \begin{bmatrix} 0.5000 \\ -0.8636 \end{bmatrix} + \begin{bmatrix} 0.6875 \\ -0.7443 \end{bmatrix} = \begin{bmatrix} 0.8494 \\ -0.6413 \end{bmatrix}.$$

$$x^{(3)} = \begin{bmatrix} 0.000 & -0.1875 \\ 0.000 & -0.1193 \end{bmatrix} \times \begin{bmatrix} 0.8494 \\ -0.6413 \end{bmatrix} + \begin{bmatrix} 0.6875 \\ -0.7443 \end{bmatrix} = \begin{bmatrix} 0.8077 \\ -0.6675 \end{bmatrix}.$$

$$x^{(4)} = \begin{bmatrix} 0.000 & -0.1875 \\ 0.000 & -0.1193 \end{bmatrix} \times \begin{bmatrix} 0.8077 \\ -0.6678 \end{bmatrix} + \begin{bmatrix} 0.6875 \\ -0.7443 \end{bmatrix} = \begin{bmatrix} 0.7127 \\ -0.6646 \end{bmatrix}.$$

$$x^{(5)} = \begin{bmatrix} 0.000 & -0.1875 \\ 0.000 & -0.1193 \end{bmatrix} \times \begin{bmatrix} 0.8127 \\ -0.6646 \end{bmatrix} + \begin{bmatrix} 0.6875 \\ -0.7443 \end{bmatrix} = \begin{bmatrix} 0.8121 \\ -0.6650 \end{bmatrix}$$

$$x^{(6)} = \begin{bmatrix} 0.000 & -0.1875 \\ 0.000 & -0.1193 \end{bmatrix} \times \begin{bmatrix} 0.8121 \\ -06650 \end{bmatrix} + \begin{bmatrix} 0.6875 \\ -0.7443 \end{bmatrix} = \begin{bmatrix} 0.8122 \\ -0.6650 \end{bmatrix}.$$

$$x^{(7)} = \begin{bmatrix} 0.000 & -0.1875 \\ 0.000 & -0.1193 \end{bmatrix} \times \begin{bmatrix} 0.8122 \\ -0.6650 \end{bmatrix} + \begin{bmatrix} 0.6875 \\ -0.7443 \end{bmatrix} + \begin{bmatrix} 0.8122 \\ -0.6650 \end{bmatrix}.$$

As expected, the algorithm converges to the exact solution :

$$x = A^{-1} b = \begin{bmatrix} 0.8122 \\ -0.6650 \end{bmatrix}.$$

In fact, the matrix A is strictly diagonally dominant (but not positive definite).

Another Example for the Matrix Version

Another linear system shown as $Ax = b$ is given by :

$$A = \begin{bmatrix} 2 & 3 \\ 5 & 7 \end{bmatrix} \text{ and } b = \begin{bmatrix} 11 \\ 13 \end{bmatrix}.$$

We want to use the equation

$$x^{(k+1)} = L_*^{-1} (b - Ux^{(k)})$$

in the form

$$x^{(k+1)} = Tx^{(k)} + C$$

where :

$$T = -L_*^{-1}U \text{ and } C = L_*^{-1}b.$$

We must decompose A into the sum of a lower triangular component L_* and a strict upper triangular component U :

$$L_* = \begin{bmatrix} 2 & 0 \\ 5 & 7 \end{bmatrix} \text{ and } U = \begin{bmatrix} 0 & 3 \\ 0 & 0 \end{bmatrix}.$$

The inverse of L_* is :

$$L_*^{-1} = \begin{bmatrix} 2 & 0 \\ 5 & 7 \end{bmatrix}^{-1} = \begin{bmatrix} 0.500 & 0.000 \\ -0.357 & 0.143 \end{bmatrix}.$$

Now we can find :

$$T = - \begin{bmatrix} 0.500 & 0.000 \\ -0.357 & 0.143 \end{bmatrix} \times \begin{bmatrix} 0 & 3 \\ 0 & 0 \end{bmatrix} = \begin{bmatrix} 0.000 & -1.500 \\ 0.000 & 1.071 \end{bmatrix},$$

$$C = \begin{bmatrix} 0.500 & 0.000 \\ -0.357 & 0.143 \end{bmatrix} \times \begin{bmatrix} 11 \\ 13 \end{bmatrix} = \begin{bmatrix} 5.500 \\ -2.071 \end{bmatrix}.$$

Now we have T and C and we can use them to obtain the vectors x iteratively.

First of all, we have to choose $x^{(0)}$: we can only guess. The better the guess, the quicker will perform the algorithm.

We suppose :

$$x^{(0)} = \begin{bmatrix} 1.1 \\ 2.3 \end{bmatrix}.$$

We can then calculate :

$$x^{(1)} = \begin{bmatrix} 0 & -1.500 \\ 0 & 1.071 \end{bmatrix} \times \begin{bmatrix} 1.1 \\ 2.3 \end{bmatrix} + \begin{bmatrix} 5.500 \\ -2.071 \end{bmatrix} = \begin{bmatrix} 2.050 \\ 0.393 \end{bmatrix}.$$

$$x^{(2)} = \begin{bmatrix} 0 & -1.500 \\ 0 & 1.071 \end{bmatrix} \times \begin{bmatrix} 2.050 \\ 0.393 \end{bmatrix} + \begin{bmatrix} 5.500 \\ -2.071 \end{bmatrix} = \begin{bmatrix} 4.911 \\ -1.651 \end{bmatrix}.$$

$$x^{(3)} = \ldots.$$

If we test for convergence we'll find that the algorithm diverges. In fact, the matrix A is neither diagonally dominant nor positive definite. Then, convergence to the exact solution

$$x = A^{-1} b = \begin{bmatrix} -38 \\ 29 \end{bmatrix}$$

is not guaranteed and, in this case, will not occur.

An Example for the Equation Version

Suppose given k equations where x_n are vectors of these equations and starting point x_0. From the first equation solve for x_1 in terms of $x_{n+1}, x_n + 2, ..., x_n$. For the next equations substitute the previous values of xs.

To make it clear let's consider an example.

$$10x_1 - x_2 + 2_{x3} = 6,$$

$$- x_1 + 11x_2 - x_3 + 3x_4 = 25,$$

$$2x_1 - x_2 + 10x_3 - x_4 = - 11,$$

$$3x_2 - x_3 + 8x_4 = 15.$$

Solving for x_1, x_2, x_3 and x_4 gives :

$$x_1 = x_2 / 10 - x_3/5 + 3/5.$$

$$x_2 = x_1/11 + x_3/11 - x_4/11 + 25/11,$$

$$x_3 = - x_1/5 + x_2/10 + x_4/10 - 11/10,$$

$$x_4 = - 3x_2/8 + x_3/8 + 15/8.$$

Suppose we choose $(0, 0, 0, 0)$ as the initial approximation, then the first approximate solution is given by

$$x_1 = 3/5 = 0.6,$$

$$x_2 = (3/5)11 + 25/11 = 3/55 + 25/11 = 2.3272,$$

$$x_3 = -(3/5)5 + (2.3272)/10 - 11/10 = -3/25 + 0.23272 - 1.1 = -0.9873,$$

$$x_4 = -3(2.3272/8 + (-0.9873)/8 + 15/8 = 0.8789.$$

Using the approximations obtained, the iterative procedure is repeated until the desired accuracy has been reached. The following are the approximated solutions after four iterations.

x_1	x_2	x_3	x_4
0.6	2.32727	-0.987273	0.878864
1.03018	2.03694	-1.01446	0.984341
1.00659	2.00356	-1.00253	0.998351
1.00086	2.0003	-1.00031	0.99985

The exact solution of the system is $(1, 2, -1, 1)$.

Newton's Method

In numerical analysis, **Newton's method** (also known as the **Newton–Raphson method**), named after Isaac Newton and Joseph Raphson, is a method for finding successively better approximations to the roots (or zeroes) of a real-valued function.

$$x : f(x) = 0.$$

The Newton–Raphson method in one variable is implemented as follows :

Given a function f defined over the reals x, and its derivative f', we begin with a first guess x_0 for a root of the function f. Provided the function satisfies all the assumptions made in the derivation of the formula, a better approximation x_1 is

$$x_1 = x_0 - \frac{f(x_0)}{f'(x_0)}.$$

Geometrically, $(x_1, 0)$ is the intersection with the x-axis of the tangent to the graph of f at $(x_0, f(x_0))$.

The process is repeated as

$$x_{n+1} = x_n - \frac{f(x_n)}{f'(x_n)}$$

until a sufficiently accurate value is reached.

This algorithm is first in the class of Householder's methods, succeeded by Halley's method. The method can also be extended to complex functions and to systems of equations.

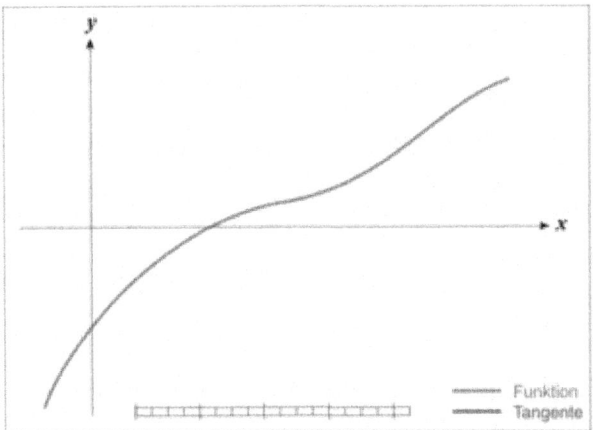

Fig. : The function *f* is shown in blue and the tangent line is in red. We see that x_{n+1} is a better approximation than x_n for the root *x* of the function *f*.

Description

The idea of the method is as follows : one starts with an initial guess which is reasonably close to the true root, then the function is approximated by its tangent line (which can be computed using the tools of calculus), and one computes the *x*-intercept of this tangent line (which is easily done with elementary algebra). This *x*-intercept will typically be a better approximation to the function's root than the original guess, and the method can be iterated.

Suppose $f : [a, b] \rightarrow \mathbf{R}$ is a differentiable function defined on the interval $[a, b]$ with values in the real numbers **R**. The formula for converging on the root can be easily derived. Suppose we have some current approximation x_n. Then we can derive the formula for a better approximation, x_{n+1} by referring to the diagram on the right. The equation of the tangent line to the curve $y = f(x)$ at the point $x = x_n$ is

$$y = f'(x_n)\,(x - x_n) + f(x_n),$$

where, f' denotes the derivative of the function f.

The *x*-intercept of this line (the value of *x* such that *y*=0) is then used as the next approximation to the root, x_{n+1}. In other words, setting *y* to zero and *x* to x_{n+1} gives

$$0 = f'(x_n)\,(x_{n+1} - x_n) + f(x_n).$$

Solving for x_{n+1} gives

$$x_{n+1} = x_n - \frac{f(x_n)}{f'(x_n)}.$$

We start the process off with some arbitrary initial value x_0. (The closer to the zero, the better. But, in the absence of any intuition about where the zero might lie, a "guess and check" method might narrow the possibilities to a reasonably small interval by appealing to the intermediate value theorem.) The method will usually converge, provided this initial guess is close enough to the unknown zero, and that $f'(x_0) \neq 0$. Furthermore, for a zero of multiplicity 1, the convergence is at least quadratic in a neighbourhood of the zero, which intuitively means that the number of correct digits roughly at least doubles in every step.

The Householder's methods are similar but have higher order for even faster convergence. However, the extra computations required for each step can slow down the overall performance relative to Newton's method, particularly if f or its derivatives are computationally expensive to evaluate.

History

The name "Newton's method" is derived from Isaac Newton's description of a special case of the method in *De analysi per aequationes numero terminorum infinitas* (written in 1669, published in 1711 by William Jones) and in *De metodis fluxionum et serierum infinitarum* (written in 1671, translated and published as *Method of Fluxions* in 1736 by John Colson). However, his method differs substantially from the modern method given above : Newton applies the method only to polynomials. He does not compute the successive approximations x_n, but computes a sequence of polynomials, and only at the end arrives at an approximation for the root x. Finally, Newton views the method as purely algebraic and makes no mention of the connection with calculus. Newton may have derived his method from a similar but less precise method by Vieta. The essence of Vieta's method can be found in the work of the Persian mathematician Sharaf al-Din al-Tusi, while his successor Jamshīd al-Kāshī used a form of Newton's method to solve $x^p - N = 0$ to find roots of N (Ypma 1995). A special case of Newton's method for calculating square roots was known much earlier and is often called the Babylonian method.

Newton's method was used by 17th-century Japanese mathematician Seki Kōwa to solve single-variable equations, though the connection with calculus was missing.

Newton's method was first published in 1685 in *A Treatise of Algebra both Historical and Practical* by John Wallis. In 1690, Joseph Raphson published a simplified description in *Analysis aequationum universalis*. Raphson again viewed Newton's method purely as an algebraic method and restricted its use to polynomials, but he describes the method in terms of the successive approximations x_n instead of the more complicated sequence of polynomials used by Newton. Finally, in 1740, Thomas Simpson described Newton's method as an iterative method for solving general non-linear equations using calculus, essentially giving the description above. In the same publication, Simpson also gives the generalization to systems of two equations and notes that Newton's method can be used for solving optimization problems by setting the gradient to zero.

Arthur Cayley in 1879 in *The Newton-Fourier imaginary problem* was the first to notice the difficulties in generalizing Newton's method to complex roots of polynomials with degree greater than 2 and complex initial values. This opened the way to the study of the theory of iterations of rational functions.

Practical Considerations

Newton's method is an extremely powerful technique — in general the convergence is quadratic : as the method converges on the root, the difference between the root and the approximation is squared (the number of accurate digits roughly doubles) at each step. However, there are some difficulties with the method.

Difficulty in Calculating Derivative of a Function

Newton's method requires that the derivative be calculated directly. An analytical expression for the derivative may not be easily obtainable and could be expensive to evaluate. In these situations, it may be appropriate to approximate the derivative by using the slope of a line through two nearby points on the function. Using this approximation would result in something like the secant method whose convergence is slower than that of Newton's method.

Failure of the Method to Converge to the Root

It is important to review the proof of quadratic convergence of Newton's Method before implementing it. Specifically, one should review the assumptions made in the proof. For situations where the method fails to converge, it is because the assumptions made in this proof are not met.

Overshoot

If the first derivative is not well behaved in the neighbourhood of a particular root, the method may overshoot, and diverge from that root. An example of a function with one root, for which the derivative is not well behaved in the neighbourhood of the root is

$$f(x) = |x|^a, 0 < a < \frac{1}{2}$$

for which the root will be overshot and the sequence of x will diverge. For $a = 1/2$, the root will still be overshot, but the sequence will oscillate between two values. For $1/2 < a < 1$, the root will still be overshot but the sequence will converge, and for $a \geq 1$ the root will not be overshot at all.

In some cases, the Newton's method can be stabilized by using successive over-relaxation, or the speed of convergence can be increased by using the same method.

Stationary Point

If a stationary point of the function is encountered, the derivative is zero and the method will terminate due to division by zero.

Poor Initial Estimate

A large error in the initial estimate can contribute to non-convergence of the algorithm.

Mitigation of Non-convergence

In a robust implementation of Newton's method, it is common to place limits on the number of iterations, bound the solution to an interval known to contain the root, and combine the method with a more robust root finding method.

Slow Convergence for Roots of Multiplicity > 1

If the root being sought has multiplicity greater than one, the convergence rate is merely linear (errors reduced by a constant factor at each step) unless special steps are taken. When there are two or more roots that are close together then it may take many iterations before the iterates get close enough to one of them for the quadratic convergence to be apparent. However, if the multiplicity m of the root is known, one can use the following modified algorithm that preserves the quadratic convergence rate :

$$x_{n+1} = x_n - m \frac{f(x_n)}{f'(x_n)}.$$

This is equivalent to using successive over-relaxation. On the other hand, if the multiplicity m of the root is not known, it is possible to estimate m after carrying out one or two iterations, and then use that value to increase the rate of convergence.

Analysis

Suppose that the function f has a zero at α, i.e., $f(\alpha) = 0$.

If f is continuously differentiable and its derivative is non-zero at α, then there exists a neighbourhood of α such that for all starting values x_0 in that neighbourhood, the sequence $\{x_n\}$ will converge to α.

If the function is continuously differentiable and its derivative is not 0 at α and it has a second derivative at α then the convergence is quadratic or faster. If the second derivative is not 0 at α then the convergence is merely quadratic. If the third derivative exists and is bounded in a neighbourhood of α, then :

$$\Delta x_i + 1 = \frac{f''(\alpha)}{2f'(\alpha)}(Dx_i)^2 + O[\Delta x_i]^3,$$

where $\Delta x_i \triangleq x_i - \alpha$.

If the derivative is 0 at α, then the convergence is usually only linear. Specifically, if f is twice continuously differentiable, $f'(a) = 0$ and $f''(a) \neq 0$, then there exists a neighbourhood of α such that for all starting values x_0 in that neighbourhood, the sequence of iterates converges linearly, with rate $\log_{10} 2$ (Süli & Mayers, Exercise 1.6). Alternatively if $f'(a) = 0$ and $f'(x) \neq 0$ for $x \neq \alpha$, x in a neighbourhood U of α, α being a zero of multiplicity r, and if $f \in C'(U)$ then there exists a neighbourhood of α such that for all starting values x_0 in that neighbourhood, the sequence of iterates converges linearly.

However, even linear convergence is not guaranteed in pathological situations.

In practice these results are local, and the neighbourhood of convergence is not known in advance. But there are also some results on global convergence : for instance, given a right neighbourhood U_+ of α, if f is twice differentiable in U_+ and if $f' \neq 0$, $f \cdot f'' > 0$ in U_+, then, for each x_0 in U_+ the sequence x_k is monotonically decreasing to α.

Proof of Quadratic Convergence for Newton's Iterative Method

According to Taylor's theorem, any function $f(x)$ which has a continuous second derivative can be represented by an expansion about a point that is close to a root of f(x). Suppose this root is α. Then the expansion of $f(\alpha)$ about x_n is :

$$f(\alpha) = f(x_n) + f'(x_n)\,(\alpha - x_n) + R_1 \qquad (1)$$

where the Lagrange form of the Taylor series expansion remainder is

$$R_1 = \frac{1}{2} f''(\xi_n)(\alpha - x_n)^2,$$

where ξ_n is in between x_n and α.

Since α is the root, (1) becomes :

$$0 = f(\alpha) = f(x_n) + f'(x_n)\,(\alpha - x_n)\,\frac{1}{2} f''(\xi_n)(\alpha - x_n)^2 \qquad (2)$$

Dividing equation (2) by $f'(x_n)$ and rearranging gives

$$\frac{f(x_n)}{f'(x_n)} + (\alpha - x_n) = \frac{-f''(\xi_n)}{2f'(x_n)}(\alpha - x_n)^2 \qquad (3)$$

Remembering that x_{n+1} is defined by

$$x_{n+1} = x_n - \frac{f(x_n)}{f'(x_n)}, \qquad (4)$$

one finds that

$$\underbrace{\alpha - x_{n+1}}_{\varepsilon_{n+1}} = \frac{-f''(\xi_n)}{2f'(x_n)} \underbrace{(\alpha - x_n)^2}_{\varepsilon_n}.$$

That is,

$$\varepsilon_{n+1} = \frac{-f''(\xi_n)}{2f'(x_n)} \varepsilon_n^2. \qquad (5)$$

Taking absolute value of both sides gives

$$|\varepsilon_{n+1}| = \frac{|f''(\xi_n)|}{2|f'(x_n)|} \varepsilon_n^2 \qquad (6)$$

Equation (6) shows that the rate of convergence is quadratic if following conditions are satisfied :

1. $f'(x) \neq 0; \forall x \in I$, Where I is the interval $[\alpha - r, \alpha + r]$ for some $r \geq |[\alpha - x_0]|$;
2. $f''(x)$ is finie, $\forall x \in I$;
3. x_0 *sufficiently* close to the root α

The term *sufficiently* close in this context means the following :

(a) Taylor approximation is accurate enough such that we can ignore higher order terms,

(b) $\dfrac{1}{2}\left|\dfrac{f''(x_n)}{f'(x_n)}\right| < C\left|\dfrac{f''(\alpha)}{f'(\alpha)}\right|$, for some $C < \infty$,

(c) $C\left|\dfrac{f''(\alpha)}{f'(x_n)}\right| \in_n < 1$, for $n \in Z^+ \cup \{0\}$ and C satisfying condition (b).

Finally, (6) can be expressed in the following way :

$$\in_{n+1} | \leq M\in^2 n$$

where M is the supremum of the variable coefficient of $\in^2 n$ on the interval I defined in the condition 1, that is :

$$M = \sup_{x \in I} \frac{1}{2}\left|\frac{f''(x)}{f'(x)}\right|.$$

The initial point x_0 has to be chosen such that conditions 1 through 3 are satisfied, where the third condition requires that $M|\in_0| < 1$.

Basins of Attraction

The basins of attraction — the regions of the real number line such that within each region iteration from any point leads to one particular root — can be infinite in number and arbitrarily small. For example, for the function $f(x) = x^3 - 2x^2 - 11x + 12$, the following initial conditions are in successive basins of attraction :

     2.35287527 converges to 4;

     2.35284172 converges to −3;

     2.35283735 converges to 4;

     2.352836327 converges to −3;

     2.352836323 converges to 1.

Failure Analysis

Newton's method is only guaranteed to converge if certain conditions are satisfied. If the assumptions made in the proof of quadratic convergence are met, the method will converge. For the following subsections, failure of the method to converge indicates that the assumptions made in the proof were not met.

Bad Starting Points

In some cases the conditions on the function that are necessary for convergence are satisfied, but the point chosen as the initial point is not in the interval where the method converges. This can happen, for example, if the function whose root is sought approaches zero asymptotically as x goes to ∞ or $-\infty$. In such cases a different method, such as bisection, should be used to obtain a better estimate for the zero to use as an initial point.

Iteration Point is Stationary

Consider the function :

$$f(x) = 1 - x^2.$$

It has a maximum at $x = 0$ and solutions of $f(x) = 0$ at $x = \pm 1$. If we start iterating from the stationary point $x_0 = 0$ (where the derivative is zero), x_1 will be undefined, since the tangent at $(0,1)$ is parallel to the x-axis :

$$x_1 = x_0 - \frac{f(x_0)}{f'(x_0)} = 0 - \frac{1}{0}.$$

The same issue occurs if, instead of the starting point, any iteration point is stationary. Even if the derivative is small but not zero, the next iteration will be a far worse approximation.

Starting Point Enters a Cycle

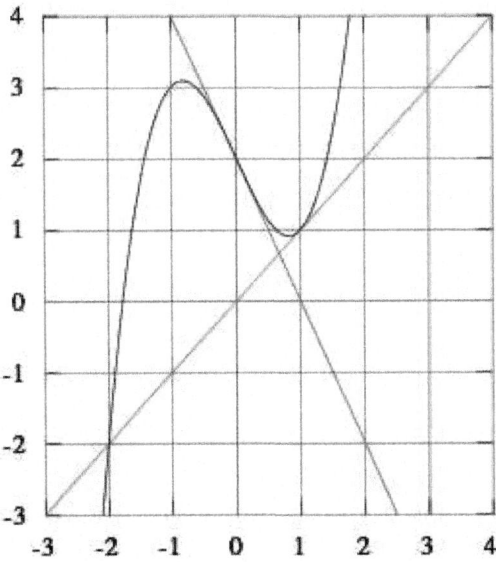

Fig. : The tangent lines of $x^3 - 2x + 2$ at 0 and 1 intersect the x-axis at 1 and 0 respectively, illustrating why Newton's method oscillates between these values for some starting points.

For some functions, some starting points may enter an infinite cycle, prevent-ing convergence. Let

$$f(x) = x^3 - 2x + 2$$

and take 0 as the starting point. The first iteration produces 1 and the second iteration returns to 0 so the sequence will alternate between the two without con-verging to a root. In fact, this 2-cycle is stable : there are neighbourhoods around 0 and around 1 from which all points iterate asymptotically to the 2-cycle (and hence not to the root of the function). In general, the behaviour of the sequence can be very complex.

Derivative Issues

If the function is not continuously differentiable in a neighbourhood of the root then it is possible that Newton's method will always diverge and fail, unless the solution is guessed on the first try.

Derivative Does not Exist at Root

A simple example of a function where Newton's method diverges is the cube root, which is continuous and infinitely differentiable, except for $x = 0$, where its derivative is undefined (this, however, does not affect the algorithm, since it will never require the derivative if the solution is already found) :

$$f(x) = \sqrt[3]{x}.$$

For any iteration point x_n, the next iteration point will be :

$$x_{n+1} = x_n - \frac{f(x_n)}{f'(x_n)} = x_n - \frac{x_n^{\frac{1}{3}}}{\frac{1}{3}x_n^{\frac{1}{3}-1}} = x_n - 3x_n = -2x_n.$$

The algorithm overshoots the solution and lands on the other side of the y-axis, farther away than it initially was; applying Newton's method actually doubles the distances from the solution at each iteration.

In fact, the iterations diverge to infinity for every $f(x) = |x|^\alpha$, where $0 < \alpha < \frac{1}{2}$. In the limiting case of $\alpha = \frac{1}{2}$ (square root), the iterations will alternate indefi-nitely between points x_0 and $-x_0$, so they do not converge in this case either.

Discontinuous Derivative

If the derivative is not continuous at the root, then convergence may fail to occur in any neighbourhood of the root. Consider the function

$$f(x) = \begin{cases} 0 & \text{if } x = 0, \\ x + x^2 \sin\left(\frac{2}{x}\right) & \text{if } x \neq 0. \end{cases}$$

Its derivative is :

$$f'(x) = \begin{cases} 1 & \text{if } x = 0, \\ 1 + 2x \sin\left(\dfrac{2}{x}\right) - 2\cos\left(\dfrac{2}{x}\right) & \text{if } x \neq 0. \end{cases}$$

Within any neighbourhood of the root, this derivative keeps changing sign as x approaches 0 from the right (or from the left) while $f(x) \geq x - x^2 > 0$ for $0 < x < 1$.

So $f(x)/f'(x)$ is unbounded near the root, and Newton's method will diverge almost everywhere in any neighbourhood of it, even though :

- the function is differentiable (and thus continuous) everywhere;
- the derivative at the root is nonzero;
- f is infinitely differentiable except at the root; and
- the derivative is bounded in a neighbourhood of the root (unlike $f(x)/f'(x)$).

Non-quadratic Convergence

In some cases the iterates converge but do not converge as quickly as promised. In these cases simpler methods converge just as quickly as Newton's method.

Zero derivative

If the first derivative is zero at the root, then convergence will not be quadratic. Indeed, let

$$f(x) = x^2$$

then $f'(x) = 2x$ and consequently $x - f(x)/f'(x) = x/2$. So convergence is not quadratic, even though the function is infinitely differentiable everywhere.

Similar problems occur even when the root is only "nearly" double. For example, let

$$f(x) = x^2 (I - 1000) + 1.$$

Then the first few iterates starting at $x_0 = 1$ are 1, 0.500250376, 0.251062828, 0.127507934, 0.067671976, 0.041224176, 0.032741218, 0.031642362; it takes six iterations to reach a point where the convergence appears to be quadratic.

No Second Derivative

If there is no second derivative at the root, then convergence may fail to be quadratic. Indeed, let

$$f(x) = x + x^{\frac{4}{3}}.$$

Then

$$f'(x) = 1 + \frac{4}{3}x^{\frac{1}{3}},$$

And

$$f''(x) = \frac{4}{9}x^{-\frac{2}{3}}$$

except when $x = 0$ where it is undefined. Given x_n,

$$x_{n+1} = x_n - \frac{f(x_n)}{f'(x_n)} = \frac{\frac{1}{3}x_n^{\frac{4}{3}}}{\left(1 + \frac{4}{3}x_n^{\frac{1}{3}}\right)}$$

which has approximately 4/3 times as many bits of precision as x_n has. This is less than the 2 times as many which would be required for quadratic convergence. So the convergence of Newton's method (in this case) is not quadratic, even though : the function is continuously differentiable everywhere; the derivative is not zero at the root; and f is infinitely differentiable except at the desired root.

Generalizations

Complex Functions

Fig. : Basins of attraction for $x^5 - 1 = 0$; darker means more iterations to converge.

When dealing with complex functions, Newton's method can be directly applied to find their zeroes. Each zero has a basin of attraction in the complex plane, the set of all starting values that cause the method to converge to that particular zero. These sets can be mapped as in the image shown. For many complex functions, the boundaries of the basins of attraction are fractals.

In some cases there are regions in the complex plane which are not in any of these basins of attraction, meaning the iterates do not converge. For example, if one uses a real initial condition to seek a root of $x^2 + 1$, all subsequent iterates will be real numbers and so the iterations cannot converge to either root, since both roots are non-real. In this case almost all initial conditions lead to chaotic behaviour, while some initial conditions iterate either to infinity or to repeating cycles of any finite length.

Non-Linear Systems of Equations

k Variables, k Functions

One may also use Newton's method to solve systems of k (non-linear) equations, which amounts to finding the zeroes of continuously differentiable functions $F : \mathbf{R}^k \to \mathbf{R}^k$. In the formulation given above, one then has to left multiply with the inverse of the k-by-k Jacobian matrix $J_F(x_n)$ instead of dividing by $f'(x_n)$.

Rather than actually computing the inverse of this matrix, one can save time by solving the system of linear equations

$$J_F(x_n) (x_{n+1} - x_n = - F(x_n)$$

for the unknown $x_{n+1} - x_n$.

k Variables, m Equations, with m > k

The k-dimensional Newton's method can be used to solve systems of $>k$ (non-linear) equations as well if the algorithm uses the generalized inverse of the non-square Jacobian matrix $J^+ = ((J^T J)^{-1})J^T$ instead of the inverse of J. If the non-linear system has no solution, the method attempts to find a solution in the non-linear least squares sense.

Non-linear Equations in a Banach Space

Another generalization is Newton's method to find a root of a functional F defined in a Banach space. In this case the formulation is

$$X_{n+1} = X_n - [F'(X_n)]{-1} F(X_n),$$

where $F'(X_n)$ is the Fréchet derivative computed at X_n. One needs the Fréchet derivative to be boundedly invertible at each X_n in order for the method to be applicable. A condition for existence of and convergence to a root is given by the Newton–Kantorovich theorem.

Non-linear Equations Over p-adic Numbers

In p-adic analysis, the standard method to show a polynomial equation in one variable has a p-adic root is Hensel's lemma, which uses the recursion from Newton's method on the p-adic numbers. Because of the more stable behaviour of addition and multiplication in the p-adic numbers compared to the real numbers (specifically, the unit ball in the p-adics is a ring), convergence in Hensel's lemma can be guaranteed under much simpler hypotheses than in the classical Newton's method on the real line.

Newton-Fourier Method

Assume that $f(x)$ is twice continuously differentiable on $[a, b]$ and that f contains a root in this interval. Assume that $f'(x) f''(x) \neq 0$ on this interval (this is the case for instance if $f(a) < 0$, $f(b) > 0$, and $f'(x) > 0$, and $f''(x) > 0$ on this interval). This guarantees that there is a unique root on this interval, call it α. If it is concave down instead of concave up then replace $f(x)$ by $-f(i)$ since they have the same roots.

Let $x_0 = b$ be the right endpoint of the interval and let $z_0 = a$ be the left end point of the interval. Given x_n, define $x_{n+1} = x_n - \dfrac{f(x_n)}{f'(x_n)}$, which is just Newton's

method as before. Then define $z_{n+1} = z_n - \dfrac{f(z_n)}{f'(x_n)}$ and note that the denominator

has $f'(x_n)$ and not $f'(z_n)$. The iterates x_n will be strictly decreasing to the root while

the iterates z_n will be strictly increasing to the root. Also, $\displaystyle\lim_{n \to \infty} \frac{x_{n+1} - z_{n+1}}{(x_n - z_n)^2} = \frac{f''(\alpha)}{2f'(\alpha)}$

so that distance between x_n and z_n decreases quadratically.

Applications

Minimization and Maximization Problems

Newton's method can be used to find a minimum or maximum of a function. The derivative is zero at a minimum or maximum, so minima and maxima can be found by applying Newton's method to the derivative. The iteration becomes :

$$x_{n+1} = x_n - \frac{f'(x_n)}{f''(x_n)}.$$

Multiplicative Inverses of Numbers and Power Series

An important application is Newton–Raphson division, which can be used to quickly find the reciprocal of a number, using only multiplication and subtraction.

Finding the reciprocal of a amounts to finding the root of the function

$$f(x) = a - \frac{1}{x}$$

Newton's iteration is

$$x_{n+1} = x_n - \frac{f(x_n)}{f'(x_n)}$$

$$= x_n - \frac{a - \dfrac{1}{x_n}}{\dfrac{1}{x_n^2}}$$

$$= x_n (2 - a x_n)$$

Therefore, Newton's iteration needs only two multiplications and one subtraction.

This method is also very efficient to compute the multiplicative inverse of a power series.

Solving Transcendental Equations

Many transcendental equations can be solved using Newton's method. Given the equation

$$g(x) = h(x),$$

with $g(x)$ and/or $h(x)$ a transcendental function, one writes

$$f(x) = g(x) - h(x).$$

The values of x that solves the original equation are then the roots of $f(x)$, which may be found via Newton's method.

Examples

Square Root of a Number

Consider the problem of finding the square root of a number. Newton's method is one of many methods of computing square roots.

For example, if one wishes to find the square root of 612, this is equivalent to finding the solution to

$$x^2 = 612$$

The function to use in Newton's method is then,

$$f(x) = x^2 - 612$$

with derivative,

$$f'(x) = 2x.$$

With an initial guess of 10, the sequence given by Newton's method is

$$x_1 = \quad x_0 - \frac{f(x_0)}{f'(x_0)} \quad = 10 - \frac{10^2 - 612}{2 \cdot 10} \quad = 35.6$$

$$x_2 = \quad x_1 - \frac{f(x_1)}{f'(x_1)} \quad = 35.6 - \frac{35.6^2 - 612}{2 \cdot 35.6} \quad = 26.395505617978...$$

$$x_3 = \quad \vdots \quad = \vdots \quad = 24.790635492455...$$

$$x_4 = \quad \vdots \quad = \vdots \quad = 24.738688294075...$$

$$x_5 = \quad \vdots \quad = \vdots \quad = 24.738633753767...$$

Where the correct digits are underlined. With only a few iterations one can obtain a solution accurate to many decimal places.

Solution of cos(x) = x³

Consider the problem of finding the positive number x with $\cos(x) = x^3$. We can rephrase that as finding the zero of $f(x) = \cos(x) - x^3$. We have $f'(x) = -\sin(x) - 3x^2$. Since $\cos(x) \leq 1$ for all x and $x^3 > 1$ for $x > 1$, we know that our zero lies between 0 and 1. We try a starting value of $x_0 = 0.5$. (Note that a starting value of 0 will lead to an undefined result, showing the importance of using a starting point that is close to the zero.)

$$x_1 = x_0 - \frac{f(x_0)}{f'(x_0)} \qquad = 0.5 - \frac{\cos(0.5) - (0.5)^3}{-\sin(0.5) - 3(0.5)^2} = 1.112141637097$$

$$x_2 = x_1 - \frac{f(x_1)}{f'(x_1)} \qquad = \vdots \qquad\qquad = \underline{0.}909672693736$$

$$x_3 = \quad \vdots \qquad\qquad = \vdots \qquad\qquad = \underline{0.867263}818209$$

$$x_4 = \quad \vdots \qquad\qquad = \vdots \qquad\qquad = \underline{0.8654}77135298$$

$$x_5 = \quad \vdots \qquad\qquad = \vdots \qquad\qquad = \underline{0.865474033}111$$

$$x_6 = \quad \vdots \qquad\qquad = \vdots \qquad\qquad = \underline{0.865474033102}$$

The correct digits are underlined in the above example. In particular, x_6 is correct to the number of decimal places given. We see that the number of correct digits after the decimal point increases from 2 (for x_3) to 5 and 10, illustrating the quadratic convergence.

Holomorphic Embedding Load Flow Method

The **Holomorphic Embedding Load-flow Method (HELM)** is a solution method for the power flow equations of electrical power systems. Its main features are that it is direct (that is, non-iterative) and that it mathematically guarantees a consistent selection of the correct operative branch of the multi-valued problem, also signalling the condition of voltage collapse when there is no solution. These properties are relevant not only for the reliability of existing off-line and real-time applications, but also because they enable new types of analytical tools that would be impossible to build with existing iterative load flows (due to their convergence problems). An example of this would be decision-support tools providing validated action plans in real time.

The HELM load flow algorithm was invented by Antonio Trias and has been granted two US Patents. A detailed description was presented at the 2012 IEEE PES General Meeting, and published in. The method is founded on advanced concepts and results from complex analysis, such as holomorphicity, the theory of algebraic curves, and analytic continuation. However, the numerical implementation is rather straightforward as it uses standard linear algebra and Padé approximation. Additionally, since the limiting part of the computation is the factorization of the admittance matrix and this is done only once, its performance is competitive with established fast-decoupled loadflows. The method is currently implemented into industrial-strength real-time and off-line packaged EMS applications.

Background

The load-flow calculation is one of the most fundamental components in the analysis of power systems and is the cornerstone for almost all other tools used in power system simulation and management. The load-flow equations can be written in the following general form :

$$\sum_k Y_{ik} V_k + Y_i^{sh} V_i = \frac{S_i^*}{V_i^*} \tag{1}$$

where the given (complex) parameters are the admittance matrix Y_{ik}, the bus shunt admittances Y_i^{sh}, and the bus power injections S_i representing constant-power loads and generators.

To solve this non-linear system of algebraic equations, traditional load-flow algorithms were developed based on three iterative techniques : the Gauss-Seidel method, which has poor convergence properties but very little memory requirements and is straightforward to implement; the full Newton-Raphson method , which has fast (quadratic) iterative convergence properties, but it is computationally costly; and the Fast Decoupled Load-Flow (FDLF) method, which is based on Newton-Raphson, but greatly reduces its computational cost by means of a decoupling approximation that is valid in most transmission networks. Many other incremental improvements exist; however, the underlying technique in all of them is still an iterative solver, either of Gauss-Seidel or of Newton type. There are two fundamental problems with all iterative schemes of this type. On the one hand, there is no guarantee that the iteration will always converge to a solution; on the other, since the system has multiple solutions, it is not possible to control which solution will be selected. As the power system approaches the point of voltage collapse, spurious solutions get closer to the correct one, and the iterative scheme may be easily attracted to one of them because of the phenomenon of Newton fractals : when the Newton method is applied to complex functions, the basins of attraction for the various solutions show fractal behaviour. As a result, no matter how close the chosen initial point of the iterations (seed) is to the correct solution, there is always some non-zero chance of straying off to a different solution. These fundamental problems of iterative load flows have been extensively documented. A simple illustration for the two-bus model is provided in Although there exist homotopic continuation techniques that alleviate the problem to some degree, the fractal nature of the basins of attraction precludes a 100% reliable method for all electrical scenarios.

The key differential advantage of the HELM is that it is fully deterministic and unambiguous : it guarantees that the solution always corresponds to the correct operative solution, when it exists; and it signals the non-existence of the solution when the conditions are such that there is no solution (voltage collapse). Additionally, the method is competitive with the FDNR method in terms of computational cost. It brings a solid mathematical treatment of the load-flow problem that provides new insights not previously available with the iterative numerical methods.

Methodology and Applications

HELM is grounded on a rigorous mathematical theory, and in practical terms it could be summarized as follows :

1. Define a specific (holomorphic) embedding for the equations in terms of a complex parameter s, such that for $s=0$ the system has an obvious correct solution, and for $s=1$ one recovers the original problem.
2. Given this holomorphic embedding, it is now possible to compute univocally power series for voltages as analytic functions of s. The correct load-flow solution at $s=1$ will be obtained by analytic continuation of the known correct solution at $s=0$.
3. Perform the analytic continuation using algebraic approximants, which in this case are guaranteed to either converge to the solution if it exists, or not converge if the solution does not exist (voltage collapse).

HELM provides a solution to a long-standing problem of all iterative load-flow methods, namely the unreliability of the iterations in finding the correct solution (or any solution at all).

This makes HELM particularly suited for real-time applications, and mandatory for any EMS software based on exploratory algorithms, such as contingency analysis, and under alert and emergency conditions solving operational limits violations and restoration providing guidance through action plans.

Holomorphic Embedding

For the purposes of the discussion, we will omit the treatment of controls, but the method can accommodate all types of controls. For the constraint equations imposed by these controls, an appropriate holomorphic embedding must be also defined.

The method uses an embedding technique by means of a complex parameter s. The first key ingredient in the method lies in requiring the embedding to be holomorphic, that is, that the system of equations for voltages V is turned into a system of equations for functions $V(s)$ in such a way that the new system defines $V(s)$ as holomorphic functions (*i.e.* complex analytic) of the new complex variable s. The aim is to be able to use the process of analytic continuation which will allow the calculation of $V(s)$ at $s=1$. Looking at equations (1), a necessary condition for the embedding to be holomorphic is that V^* is replaced under the embedding with $V^*(s^*)$, not $V^*(s)$. This is because complex conjugation itself is not a holomorphic function. On the other hand, it is easy to see that the replacement $V^*(s^*)$ does allow the equations to define a holomorphic function $V(s)$. However, for a given arbitrary embedding, it remains to be proven that $V(s)$ is indeed holomorphic. Taking into account all these considerations, an embedding of this type is proposed :

$$\sum_k Y_{ik} V_k(s) + Y_i^{sh} V_i(s) = s \frac{S_i^*}{V_i^*(s*)} \tag{1}$$

With this choice, at $s=0$ the right hand side terms become zero, (provided that the denominator is not zero), this corresponds to the case where all the injections are zero and this case has a well known and simple operational solution : all voltages are equal and all flow intensiti es are zero. Therefore this choice for the embedding provides at $s=0$ a well known operational solution.

Now using classical techniques for variable elimination in polynomial systems (results from the theory of Resultants and Gröbner basis it can be proven that equations (1) do in fact define $V(s)$ as holomorphic functions. More significantly, they define $V(s)$ as algebraic curves. It is this specific fact, which becomes true because the embedding is holomorphic that guarantees the uniqueness of the result. The solution at $s=0$ determines uniquely the solution everywhere (except on a finite number of branch cuts), thus getting rid of the multi-valuedness of the load-flow problem.

The technique to obtain the coefficients for the power series expansion (on $s=0$) of voltages V is quite straightforward, once one realizes that equations (2) can be used to obtain them order after order. Consider the power series expansion for $V(s) = \sum_{n=0}^{\infty} a[n] s^n$ and $1/V(s) = \sum_{n=0}^{\infty} b[n] s^n$. By substitution into equations (1) and identifying terms at each order in sn, one obtains :

$$\sum_k Y_{ik} a_k[n] + Y_i^{sh} a_i[n] = S_i^* b_i^*[n-1] \quad (n = 0, ..., \infty) \tag{2}$$

It is then straightforward to solve the sequence of linear systems (2) successively order after order, starting from $n=0$. Note that the coefficients of the expansions for V and $1/V$ are related by the simple convolution formulas derived from the following identity :

$$1 = V(s)V^{-1}(s)$$

$$= \left(\sum_{n=0}^{\infty} a_n s^n \right) \left(\sum_{n=0}^{\infty} b_n s^n \right) \tag{3}$$

$$= a_0 b_0 + \left(\sum_{k+0}^{\infty} a_{1-k} b_k \right) s + \left(\sum_{k=0}^{2} a_{2-k} b_k \right) s^2 + ... + \left(\sum_{k=0}^{n} a_{n-k} b_k \right) s^n + ...$$

so that the right-hand side in (2) can always be calculated from the solution of the system at the previous order. Note also how the procedure works by solving just linear systems, in which the matrix remains constant. A more detailed discussion about this procedure is offered in Ref.

Analytic Continuation

Once the power series at $s=0$ are calculated to the desired order, the problem of calculating them at $s=1$ becomes one of analytic continuation. It should be strongly remarked that this does not have anything in common with the techniques of homotopic continuation. Homotopy is powerful since it only makes use of the concept of continuity and thus it is applicable to general smooth non-linear systems, but on the other hand it does not always provide a reliable method to approximate the functions (as it relies on iterative schemes such as Newton-Raphson).

It can be proven that algebraic curves are complete global analytic functions, that is, knowledge of the power series expansion at one point (the so-called germ of the function) uniquely determines the function everywhere on the complex plane, except on a finite number of branch cuts. Stahl's extremal domain theorem further asserts that there exists a maximal domain for the analytic continuation of the function, which corresponds to the choice of branch cuts with minimal logarithmic capacity measure. In the case of algebraic curves the number of cuts is finite, therefore it would be feasible to find maximal continuations by finding the combination of cuts with minimal capacity. For further improvements, Stahl's theorem on the convergence of Padé Approximants states that the diagonal and supra-diagonal Padé (or equivalently, the continued fraction approximants to the power series) converge to the maximal analytic continuation. The zeros and poles of the approximants remarkably accumulate on the set of branch cuts having minimal capacity.

These properties confer the load-flow method with the ability to unequivocally detect the condition of voltage collapse : the algebraic approximations are guaranteed to either converge to the solution if it exists, or not converge if the solution does not exist.

Chapter 9

ELECTRICAL RESISTIVITY AND CONDUCTIVITY

Electrical resistivity (also known as **resistivity, specific electrical resistance,** or **volume resistivity**) quantifies how strongly a given material opposes the flow of electric current. A low resistivity indicates a material that readily allows the movement of electric charge. Resistivity is commonly represented by the Greek letter ρ (rho). The SI unit of electrical resistivity is the ohm·metre (Ω·m) although other units like ohm·centimetre (Ω·cm) are also in use. As an example, if a 1 × 1 × 1 m solid cube of material has sheet contacts on two opposite faces, and the resistance between these contacts is 1 Ω, then the resistivity of the material is 1 Ω·m.

Electrical conductivity or **specific conductance** is the reciprocal of electrical resistivity, and measures a material's ability to conduct an electric current. It is commonly represented by the Greek letter σ (sigma), but κ (kappa) (especially in electrical engineering) or γ (gamma) are also occasionally used. Its SI unit is siemens per metre (S/m) and CGSE unit is reciprocal second (s^{-1}).

DEFINITION

Resistors or Conductors with Uniform Cross-section

Many resistors and conductors have a uniform cross-section with a uniform flow of electric current, and are made of one material. In this case, the electrical resistivity ρ (Greek : rho) is defined as :

$$\rho = R\frac{A}{\ell}$$

where :

R is the electrical resistance of a uniform specimen of the material (measured in ohms, Ω)

ℓ is the length of the piece of material (measured in metres, m)

A is the cross-sectional area of the specimen (measured in square metres, m²).

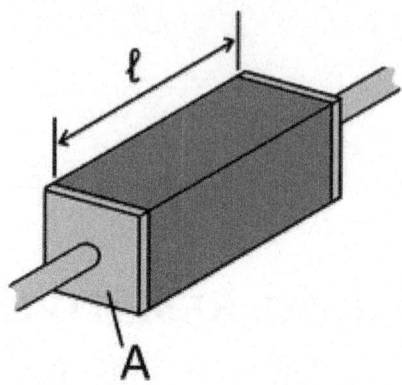

Fig. : A piece of resistive material with electrical contacts on both ends.

The reason resistivity is defined this way is that it makes resistivity an *intrinsic property*, unlike resistance. All copper wires, irrespective of their shape and size, have approximately the same *resistivity*, but a long, thin copper wire has a much larger *resistance* than a thick, short copper wire. Every material has its own characteristic resistivity – for example, rubber's resistivity is far larger than copper's.

In a hydraulic analogy, passing current through a high-resistivity material is like pushing water through a pipe full of sand, while passing current through a low-resistivity material is like pushing water through an empty pipe. If the pipes are the same size and shape, the pipe full of sand has higher resistance to flow. But resistance is not *solely* determined by the presence or absence of sand; it also depends on the length and width of the pipe : short or wide pipes will have lower resistance than narrow or long pipes.

The above equation can be transposed to get **Pouillet's law** (named after Claude Pouillet) :

$$R = \rho\frac{\ell}{A}.$$

The resistance of a given material will increase with the length, but decrease with increasing cross-sectional area. From the above equations, resistivity has SI units of ohm·metre. Other units like ohm·cm or ohm·inch are also sometimes used.

The formula $R = \rho\ell/A$ can be used to intuitively understand the meaning of a resistivity value. For example, if $A = 1m^2$ and $\ell = 1m$ (forming a cube with perfectly-conductive contacts on opposite faces), then the resistance of this element in ohms is numerically equal to the resistivity of the material it is made of in ohm-meters. Likewise, a 1 ohm·cm material would have a resistance of 1 ohm if contacted on opposite faces of a 1 cm×1 cm×1 cm cube.

Conductivity σ (Greek : sigma) is defined as the inverse of resistivity :

$$r = \frac{1}{\rho}.$$

Conductivity has SI units of siemens per meter (S/m).

General Definition

The above definition was specific to resistors or conductors with a uniform cross-section, where current flows uniformly through them. A more basic and general definition starts from the fact that if there is electric field inside a material, it will cause electric current to flow. The electrical resistivity ρ is defined as the ratio of the electric field to the density of the current it creates :

$$\rho = \frac{E}{J},$$

where :

ρ is the resistivity of the conductor material (measured in ohm·metres, Ω·m),

E is the magnitude of the electric field (in volts per metre, V·m^{-1}),

J is the magnitude of the current density (in amperes per square metre, A·m^{-2}),

in which E and J are inside the conductor.

Conductivity is the inverse :

$$\rho = \frac{1}{\rho} = \frac{J}{E}.$$

For example, rubber is a material with large ρ and small σ, because even a very large electric field in rubber will cause almost no current to flow through it. On the other hand, copper is a material with small ρ and large σ, because even a small electric field pulls a lot of current through it.

CAUSES OF CONDUCTIVITY

Band Theory Simplified

Quantum mechanics states that electrons in an atom cannot take on any arbitrary energy value. Rather, there are fixed energy levels which the electrons can occupy, and values in between these levels are impossible. When a large number of such allowed energy levels are spaced close together (in energy-space) i.e. have similar (minutely differing) energies then we can talk about these energy levels together as an "energy band." There can be many such energy bands in a material, depending on the atomic number (number of electrons) and their distribution (besides external factors like environment modifying the energy bands).

The material's electrons seek to minimize the total energy in the material by going to low energy states, however the Pauli exclusion principle means that they cannot all go to the lowest state. The electrons instead "fill up" the band structure starting from the bottom. The characteristic energy level up to which the electrons have filled is called the Fermi level. The position of the Fermi level with respect to the band structure is very important for electrical conduction : only electrons in energy levels near the Fermi level are free to move around since the electrons can easily jump among the partially occupied states in that region. In contrast, the low energy states are rigidly filled with a fixed number of electrons at all times, and the high energy states are empty of electrons at all times.

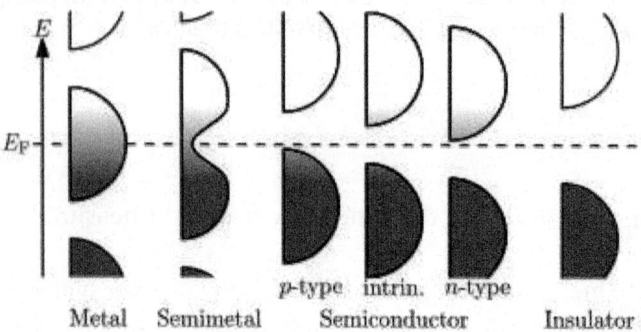

Fig. : Filling of electronic band structure in various types of material at equilibrium. In metals and semimetals the Fermi level E_F lies inside at least one band. In insulators and semi-conductors the Fermi level is inside a band gap, however in semi-conductors the bands are near enough to the Fermi level to be thermally populated with electrons or holes.

In metals there are many energy levels near the Fermi level, meaning that there are many electrons available to move. This is what causes the high electronic conductivity in metals.

An important part of band theory is that there may be forbidden bands in energy : energy intervals which do not contain any energy levels. In insulators and semi-conductors, the number of electrons happens to be just the right amount to fill a certain integer number of low energy bands, exactly to the boundary. In this case, the Fermi level falls within a band gap. Since there are no available states near the Fermi level, and the electrons are not freely movable, the electronic conductivity is very low.

In Metals

A metal consists of a lattice of atoms, each with an outer shell of electrons which freely dissociate from their parent atoms and travel through the lattice. This is also known as a positive ionic lattice. This 'sea' of dissociable electrons allows the metal to conduct electric current. When an electrical potential difference (a voltage) is applied across the metal, the resulting electric field causes electrons to move from one end of the conductor to the other.

Near room temperatures, metals have resistance. The primary cause of this resistance is the thermal motion of ions. This acts to scatter electrons (due to destructive interference of free electron waves on non-correlating potentials of ions) . Also contributing to resistance in metals with impurities are the resulting imperfections in the lattice. In pure metals this source is negligible .

The larger the cross-sectional area of the conductor, the more electrons per unit length are available to carry the current. As a result, the resistance is lower in larger cross-section conductors. The number of scattering events encountered by an electron passing through a material is proportional to the length of the conductor. The longer the conductor, therefore, the higher the resistance. Different materials also affect the resistance.

In Semi-conductors and Insulators

In metals, the Fermi level lies in the conduction band giving rise to free conduction electrons. However, in semi-conductors the position of the Fermi level is within the band gap, approximately half-way between the conduction band minimum and valence band maximum for intrinsic (undoped) semi-conductors. This means that at 0 kelvin, there are no free conduction electrons and the resistance is infinite. However, the resistance will continue to decrease as the charge carrier density in the conduction band increases. In extrinsic (doped) semi-conductors, dopant atoms increase the majority charge carrier concentration by donating electrons to the conduction band or accepting holes in the valence band. For both types of donor or acceptor atoms, increasing the dopant density leads to a reduction in the resistance, hence highly doped semi-conductors behave metallically. At very high temperatures, the contribution of thermally generated carriers will dominate over the contribution from dopant atoms and the resistance will decrease exponentially with temperature.

In Ionic Liquids/Electrolytes

In electrolytes, electrical conduction happens not by band electrons or holes, but by full atomic species (ions) travelling, each carrying an electrical charge. The resistivity of ionic liquids varies tremendously by the concentration – while distilled water is almost an insulator, salt water is a very efficient electrical conductor. In biological membranes, currents are carried by ionic salts. Small holes in the membranes, called ion channels, are selective to specific ions and determine the membrane resistance.

Superconductivity

The electrical resistivity of a metallic conductor decreases gradually as temperature is lowered. In ordinary conductors, such as copper or silver, this decrease is limited by impurities and other defects. Even near absolute zero, a real sample of a normal conductor shows some resistance. In a superconductor, the resistance drops abruptly to zero when the material is cooled below its critical temperature.

An electric current flowing in a loop of superconducting wire can persist indefinitely with no power source.

In 1986, it was discovered that some cuprate-perovskite ceramic materials have a critical temperature above 90 K (−183 °C). Such a high transition temperature is theoretically impossible for a conventional superconductor, leading the materials to be termed high-temperature superconductors. Liquid nitrogen boils at 77 K, facilitating many experiments and applications that are less practical at lower temperatures. In conventional superconductors, electrons are held together in pairs by an attraction mediated by lattice phonons. The best available model of high-temperature superconductivity is still somewhat crude. There is a hypothesis that electron pairing in high-temperature superconductors is mediated by short-range spin waves known as paramagnons.

Plasma

Plasmas are very good electrical conductors and electric potentials play an important role. The potential as it exists on average in the space between charged particles, independent of the question of how it can be measured, is called the "plasma potential", or the "space potential". If an electrode is inserted into a plasma, its potential will generally lie considerably below the plasma potential due to what is termed a Debye sheath. The good electrical conductivity of plasmas makes their electric fields very small. This results in the important concept of "quasineutrality", which says the density of negative charges is approximately equal to the density of positive charges over large volumes of the plasma ($n_e = <Z>n_i$), but on the scale of the Debye length there can be charge imbalance. In the special case that *double layers* are formed, the charge separation can extend some tens of Debye lengths.

The magnitude of the potentials and electric fields must be determined by means other than simply finding the net charge density. A common example is to assume that the electrons satisfy the Boltzmann relation :

$$n_e \propto e^{\varepsilon\Phi}/k_B T_e.$$

Differentiating this relation provides a means to calculate the electric field from the density :

$$\vec{E} = (k_B Te/e)\,(\nabla n_e/n_e)$$

It is possible to produce a plasma that is not quasineutral. An electron beam, for example, has only negative charges. The density of a non-neutral plasma must generally be very low, or it must be very small, otherwise it will be dissipated by the repulsive electrostatic force.

In astrophysical plasmas, Debye screening prevents electric fields from directly affecting the plasma over large distances, *i.e.*, greater than the Debye length. However, the existence of charged particles causes the plasma to generate, and be affected by, magnetic fields. This can and does cause extremely complex behaviour, such as the generation of plasma double layers, an object that separates charge over a few tens of Debye lengths. The dynamics of plasmas interacting with external and self-generated magnetic fields are studied in the academic discipline of magnetohydrodynamics.

Plasma is often called the *fourth state of matter* after solid, liquids and gases. It is distinct from these and other lower-energy states of matter. Although it is closely related to the gas phase in that it also has no definite form or volume, it differs in a number of ways, including the following :

Property	Gas	Plasma
Electrical conductivity	**Very low** : Air is an excellent insulator until it breaks down into plasma at electric field strengths above 30 kilovolts per centimeter.	**Usually very high** : For many purposes, the conductivity of a plasma may be treated as infinite.
Independently acting species	**One** : All gas particles behave in a similar way, influenced by gravity and by collisions with one another.	**Two or three** : Electrons, ions, protons and neutrons can be distinguished by the sign and value of their charge so that they behave independently in many circumstances, with different bulk velocities and temperatures, allowing phenomena such as new types of waves and instabilities.
Velocity distribution	**Maxwellian** : Collisions usually lead to a Maxwellian velocity distribution of all gas particles, with very few relatively fast particles.	**Often non-Maxwellian** : Collisional interactions are often weak in hot plasmas and external forcing can drive the plasma far from local equilibrium and lead to a significant population of unusually fast particles.
Interactions	**Binary** : Two-particle collisions are the rule, three-body collisions extremely rare.	**Collective** : Waves, or organized motion of plasma, are very important because the particles can interact at long ranges through the electric and magnetic forces.

RESISTIVITY OF VARIOUS MATERIALS

- A conductor such as a metal has high conductivity and a low resistivity.

- An insulator like glass has low conductivity and a high resistivity.

- The conductivity of a semi-conductor is generally intermediate, but varies widely under different conditions, such as exposure of the material to electric fields or specific frequencies of light, and, most important, with temperature and composition of the semi-conductor material.

The degree of doping in semi-conductors makes a large difference in conductivity. To a point, more doping leads to higher conductivity. The conductivity of a solution of water is highly dependent on its concentration of dissolved salts, and other chemical species that ionize in the solution. Electrical conductivity of water samples is used as an indicator of how salt-free, ion-free, or impurity-free the sample is; the purer the water, the lower the conductivity (the higher the resistivity). Conductivity measurements in water are often reported as *specific conductance*, relative to the conductivity of pure water at 25 °C. An EC meter is normally used to measure conductivity in a solution. A rough summary is as follows :

Material	Resistivity ρ (Ω m)
Superconductors	0
Metals	10^{-8}
Semi-conductors	variable
Electrolytes	variable
Insulators	10^{16}

This table shows the resistivity, conductivity and temperature coefficient of various materials at 20 °C (68 °F, 293 K)

Material	ρ (Ω m) at 20 °C	σ (S/m) at 20 °C	Temperature coefficient (K^{-1})
Carbon (graphene)	1×10^{-8}	-	-0.0002
Silver	1.59×10^{-8}	6.30×10^{7}	0.0038
Copper	1.68×10^{-8}	5.96×10^{7}	0.003862
Annealed copper[note 2]	1.72×10^{-8}	5.80×10^{7}	0.00393
Gold[note 3]	2.44×10^{-8}	4.10×10^{7}	0.0034
Aluminium[note 4]	2.82×10^{-8}	3.5×10^{7}	0.0039
Calcium	3.36×10^{-8}	2.98×10^{7}	0.0041
Tungsten	5.60×10^{-8}	1.79×10^{7}	0.0045
Zinc	5.90×10^{-8}	1.69×10^{7}	0.0037
Nickel	6.99×10^{-8}	1.43×10^{7}	0.006
Lithium	9.28×10^{-8}	1.08×10^{7}	0.006
Iron	1.0×10^{-7}	1.00×10^{7}	0.005
Platinum	1.06×10^{-7}	9.43×10^{6}	0.00392
Tin	1.09×10^{-7}	9.17×10^{6}	0.0045
Carbon steel (1010)	1.43×10^{-7}	6.99×10^{6}	
Lead	2.2×10^{-7}	4.55×10^{6}	0.0039
Titanium	4.20×10^{-7}	2.38×10^{6}	X
Grain oriented electrical steel	4.60×10^{-7}	2.17×10^{6}	
Manganin	4.82×10^{-7}	2.07×10^{6}	0.000002
Constantan	4.9×10^{-7}	2.04×10^{6}	0.000008

Material	ρ (Ω m) at 20 °C	σ (S/m) at 20 °C	Temperature coefficient (K⁻¹)
Stainless steel	6.9×10^{-7}	1.45×10^{6}	
Mercury	9.8×10^{-7}	1.02×10^{6}	0.0009
Nichrome	1.10×10^{-6}	9.09×10^{5}	0.0004
GaAs	1×10^{-3} to 1×10^{8}	1×10^{-8} to 10^{3}	
Carbon (amorphous)	5×10^{-4} to 8×10^{-4}	1.25×10^{3} to 2×10^{3}	−0.0005
Carbon (graphite)	2.5×10^{-6} to 5.0×10^{-6} // basal plane 3.0×10^{-3}⊥basal plane	2×105 to 3×105 //basal plane 3.3×10^{2} ⊥basal plane	
Carbon (diamond)	1×10^{12}	$\sim 10^{-13}$	
Germanium	4.6×10^{-1}	2.17	−0.048
Sea water	2×10^{-1}	4.8	
Drinking water	2×10^{1} to 2×10^{3}	5×10^{-4} to 5×10^{-2}	
Silicon	6.40×10^{2}	1.56×10^{-3}	−0.075
Wood (damp)	1×10^{3} to 1×10^{4}	10^{-4} to 10^{-3}	
Deionized water	1.8×10^{5}	5.5×10^{-6}	
Glass	10×10^{10} to 10×10^{14}	10^{-11} to 10^{-15}	?
Hard rubber	1×10^{13}	10^{-14}	?
Wood (oven dry)	1×10^{14} to 1×10^{16}	10^{-16} to 10^{-14}	
Sulfur	1×10^{15}	10^{-16}	?
Air	1.3×10^{16} to 3.3×10^{16}	3×10^{-15} to 8×10^{-15}	
PEDOT:PSS	1×10^{-3} to 1×10^{-1}	1×10^{1} to 1×10^{3}	?
Fused quartz	7.5×10^{17}	1.3×10^{-18}	?
PET	10×10^{20}	10^{-21}	?
Teflon	10×10^{22} to 10×10^{24}	10^{-25} to 10^{-23}	?

The effective temperature coefficient varies with temperature and purity level of the material. The 20 °C value is only an approximation when used at other temperatures. For example, the coefficient becomes lower at higher temperatures for copper, and the value 0.00427 is commonly specified at 0 °C.

The extremely low resistivity (high conductivity) of silver is characteristic of metals. George Gamow tidily summed up the nature of the metals' dealings with electrons in his science-popularizing book, *One, Two, Three...Infinity* (1947) : "The metallic substances differ from all other materials by the fact that the outer shells of their atoms are bound rather loosely, and often let one of their electrons go free. Thus the interior of a metal is filled up with a large number of unattached electrons that travel aimlessly around like a crowd of displaced persons. When a metal wire is subjected to electric force applied on its opposite ends, these free electrons rush in the direction of the force, thus forming what we call an electric current." More technically, the free electron model gives a basic description of electron flow in metals.

Wood is widely regarded as an extremely good insulator, but its resistivity is sensitively dependent on moisture content, with damp wood being a factor of at least 1010 worse insulator than oven-dry. In any case, a sufficiently high voltage – such as that in lightning strikes or some high-tension powerlines – can lead to insulation breakdown and electrocution risk even with apparently dry wood.

TEMPERATURE DEPENDENCE

Linear Approximation

The electrical resistivity of most materials changes with temperature. If the temperature T does not vary too much, a linear approximation is typically used :

$$r(T) = \rho_0[1 + \alpha(T - T_0)]$$

where α is called the *temperature coefficient of resistivity*, T_0 is a fixed reference temperature (usually room temperature), and ρ_0 is the resistivity at temperature T_0. The parameter α is an empirical parameter fitted from measurement data. Because the linear approximation is only an approximation, α is different for different reference temperatures. For this reason it is usual to specify the temperature that α was measured at with a suffix, such as α_{15}, and the relationship only holds in a range of temperatures around the reference. When the temperature varies over a large temperature range, the linear approximation is inadequate and a more detailed analysis and understanding should be used.

Metals

In general, electrical resistivity of metals increases with temperature. Electron–phonon interactions can play a key role. At high temperatures, the resistance of a metal increases linearly with temperature. As the temperature of a metal is reduced, the temperature dependence of resistivity follows a power law function of temperature. Mathematically the temperature dependence of the resistivity ρ of a metal is given by the Bloch–Grüneisen formula :

$$\rho(T) = \rho(0) + A\left(\frac{T}{\Theta_R}\right)^n \int_0^{\frac{\Theta_R}{T}} \frac{x^n}{(e^x - 1)(1 - e^{-x})}\,dx$$

where $\rho(0)$ is the residual resistivity due to defect scattering, A is a constant that depends on the velocity of electrons at the Fermi surface, the Debye radius and the number density of electrons in the metal. Φ_R is the Debye temperature as obtained from resistivity measurements and matches very closely with the values of Debye temperature obtained from specific heat measurements. n is an integer that depends upon the nature of interaction :

1. n=5 implies that the resistance is due to scattering of electrons by phonons (as it is for simple metals)
2. n=3 implies that the resistance is due to s-d electron scattering (as is the case for transition metals)
3. n=2 implies that the resistance is due to electron–electron interaction.

If more than one source of scattering is simultaneously present, Matthiessen's Rule (first formulated by Augustus Matthiessen in the 1860s) says that the total resistance can be approximated by adding up several different terms, each with the appropriate value of n.

As the temperature of the metal is sufficiently reduced (so as to 'freeze' all the phonons), the resistivity usually reaches a constant value, known as the **residual resistivity**. This value depends not only on the type of metal, but on its purity and thermal history. The value of the residual resistivity of a metal is decided by its impurity concentration. Some materials lose all electrical resistivity at sufficiently low temperatures, due to an effect known as superconductivity.

An investigation of the low-temperature resistivity of metals was the motivation to Heike Kamerlingh Onnes's experiments that led in 1911 to discovery of superconductivity.

Semi-conductors

In general, resistivity of intrinsic semi-conductors decreases with increasing temperature. The electrons are bumped to the conduction energy band by thermal energy, where they flow freely and in doing so leave behind holes in the valence band which also flow freely. The electric resistance of a typical intrinsic (non doped) semi-conductor decreases exponentially with the temperature :

$$\rho = \rho_0 e^{-aT}$$

An even better approximation of the temperature dependence of the resistivity of a semi-conductor is given by the Steinhart–Hart equation :

$$1/T = A + B \ln(\rho) + C(\ln(\rho))^3$$

where A, B and C are the so-called **Steinhart–Hart coefficients**.

This equation is used to calibrate thermistors.

Extrinsic (doped) semi-conductors have a far more complicated temperature profile. As temperature increases starting from absolute zero they first decrease steeply in resistance as the carriers leave the donors or acceptors. After most of the donors or acceptors have lost their carriers the resistance starts to increase again slightly due to the reducing mobility of carriers (much as in a metal). At higher temperatures it will behave like intrinsic semi-conductors as the carriers from the donors/acceptors become insignificant compared to the thermally generated carriers.

In non-crystalline semi-conductors, conduction can occur by charges quantum tunnelling from one localised site to another. This is known as variable range hopping and has the characteristic form of

$$\rho = A \, \exp(T^{-1/n}),$$

where $n = 2, 3, 4$, depending on the dimensionality of the system.

COMPLEX RESISTIVITY AND CONDUCTIVITY

When analyzing the response of materials to alternating electric fields, in applications such as electrical impedance tomography, it is necessary to replace resistivity with a complex quantity called **impeditivity** (in analogy to electrical impedance). Impeditivity is the sum of a real component, the resistivity, and an imaginary component, the **reactivity** (in analogy to reactance). The magnitude of Impeditivity is the square root of sum of squares of magnitudes of resistivity and reactivity.

Conversely, in such cases the conductivity must be expressed as a complex number (or even as a matrix of complex numbers, in the case of anisotropic materials) called the *admittivity*. Admittivity is the sum of a real component called the conductivity and an imaginary component called the susceptivity.

An alternative description of the response to alternating currents uses a real (but frequency-dependent) conductivity, along with a real permittivity. The larger the conductivity is, the more quickly the alternating-current signal is absorbed by the material (*i.e.*, the more opaque the material is).

TENSOR EQUATIONS FOR ANISOTROPIC MATERIALS

Some materials are anisotropic, meaning they have different properties in different directions. For example, a crystal of graphite consists microscopically of a stack of sheets, and current flows very easily through each sheet, but moves much less easily from one sheet to the next.

For an anisotropic material, it is not generally valid to use the scalar equations

$$J = \sigma E \rightleftharpoons E \rightleftharpoons \rho J.$$

For example, the current may not flow in exactly the same direction as the electric field. Instead, the equations are generalized to the 3D tensor form

$$\mathbf{J} = \sigma\mathbf{E} \rightleftharpoons \mathbf{E} \rightleftharpoons \rho\mathbf{J}.$$

where the conductivity σ and resistivity ρ are rank-2 tensors (in other words, 3×3 matrices). The equations are compactly illustrated in component form (using index notation and the summation convention) :

$$J_i = \sigma_{ij} E_j \rightleftharpoons E_i = \rho_{ij}J_j.$$

The σ and ρ tensors are inverses (in the sense of a matrix inverse). The individual components are not necessarily inverses; for example, σ_{xx} may not be equal to $1/\rho_{xx}$.

RESISTANCE VERSUS RESISTIVITY IN COMPLICATED GEOMETRIES

If the material's resistivity is known, calculating the resistance of something made from it may, in some cases, be much more complicated than the formula $R = \rho\ell/A$ above. One example is Spreading Resistance Profiling, where the material is inhomogeneous (different resistivity in different places), and the exact paths of current flow are not obvious.

In cases like this, the formulas

$$J = \sigma E \rightleftharpoons E \rightleftharpoons \rho J.$$

need to be replaced with

$$J(r) = \sigma(r)E(r) \rightleftharpoons E(r) = \rho(r)J(r),$$

where **E** and **J** are now vector fields. This equation, along with the continuity equation for **J** and the Poisson's equation for **E**, form a set of partial differential equations. In special cases, an exact or approximate solution to these equations can be worked out by hand, but for very accurate answers in complex cases, computer methods like finite element analysis may be required.

RESISTIVITY DENSITY PRODUCTS

In some applications where the weight of an item is very important resistivity density products are more important than absolute low resistivity – it is often possible to make the conductor thicker to make up for a higher resistivity; and then a low resistivity density product material (or equivalently a high conductance to density ratio) is desirable. For example, for long distance overhead power lines, aluminium is frequently used rather than copper because it is lighter for the same conductance.

Material	Resistivity (nΩ m)	Density (g/cm³)	Resistivity-density product (nΩ m g/cm³)
Sodium	47.7	0.97	46
Lithium	92.8	0.53	49
Calcium	33.6	1.55	52
Potassium	72.0	0.89	64
Beryllium	35.6	1.85	66
Aluminium	26.50	2.70	72
Magnesium	43.90	1.74	76.3
Copper	16.78	8.96	150
Silver	15.87	10.49	166
Gold	22.14	19.30	427
Iron	96.1	7.874	757

Silver, although it is the least resistive metal known, has a high density and does poorly by this measure. Calcium and the alkali metals have the best resistivity-density products, but are rarely used for conductors due to their high reactivity with water and oxygen. Aluminium is far more stable. Two other important attributes, price and toxicity, exclude the (otherwise) best choice : Beryllium. Thus, aluminium is usually the metal of choice when the weight of some required conduction (and/or the cost of conduction) is the driving consideration.

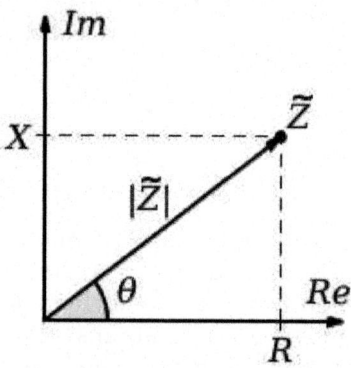

Fig. : A graphical representation of the complex impedance plane.

ELECTRICAL IMPEDANCE

Electrical impedance is the measure of the opposition that a circuit presents to a current when a voltage is applied.

In quantitative terms, it is the complex ratio of the voltage to the current in an alternating current (AC) circuit. Impedance extends the concept of resistance to AC circuits, and possesses both magnitude and phase, unlike resistance, which has only magnitude. When a circuit is driven with direct current (DC), there is no distinction between impedance and resistance; the latter can be thought of as impedance with zero phase angle.

It is necessary to introduce the concept of impedance in AC circuits because there are two additional impeding mechanisms to be taken into account besides the normal resistance of DC circuits : the induction of voltages in conductors self-induced by the magnetic fields of currents (inductance), and the electrostatic storage of charge induced by voltages between conductors (capacitance). The impedance caused by these two effects is collectively referred to as reactance and forms the imaginary part of complex impedance whereas resistance forms the real part.

The symbol for impedance is usually Z and it may be represented by writing its magnitude and phase in the form $|Z| \angle \theta$. However, complex number representation is often more powerful for circuit analysis purposes. The term *impedance* was coined by Oliver Heaviside in July 1886. Arthur Kennelly was the first to represent impedance with complex numbers in 1893.

Impedance is defined as the frequency domain ratio of the voltage to the current. In other words, it is the voltage–current ratio for a single complex exponential at a particular frequency ω. In general, impedance will be a complex number, with the same units as resistance, for which the SI unit is the ohm (Ω). For a sinusoidal current or voltage input, the polar form of the complex impedance relates the amplitude and phase of the voltage and current. In particular,

- The magnitude of the complex impedance is the ratio of the voltage amplitude to the current amplitude.
- The phase of the complex impedance is the phase shift by which the current lags the voltage.

The reciprocal of impedance is admittance (*i.e.*, admittance is the current-to-voltage ratio, and it conventionally carries units of siemens, formerly called mhos).

Complex Impedance

Impedance is represented as a complex quantity z and the term *complex impedance* may be used interchangeably; the polar form conveniently captures both magnitude and phase characteristics,

$$Z = |Z| e^{j \arg(Z)}$$

where the magnitude $|Z|$ represents the ratio of the voltage difference amplitude to the current amplitude, while the argument $\arg(Z)$ (commonly given the symbol θ) gives the phase difference between voltage and current. j is the imaginary unit, and is used instead of i in this context to avoid confusion with the symbol for electric current. In Cartesian form,

$$Z = R + jX$$

where the real part of impedance is the resistance R and the imaginary part is the reactance X.

Where it is required to add or subtract impedances the cartesian form is more convenient, but when quantities are multiplied or divided the calculation becomes simpler if the polar form is used. A circuit calculation, such as finding the total impedance of two impedances in parallel, may require conversion between forms several times during the calculation. Conversion between the forms follows the normal conversion rules of complex numbers.

Ohm's Law

The meaning of electrical impedance can be understood by substituting it into Ohm's law.

$$V = IZ = I |Z| e^{j \arg(Z)}$$

The magnitude of the impedance $|Z|$ acts just like resistance, giving the drop in voltage amplitude across an impedance Z for a given current I. The phase factor tells us that the current lags the voltage by a phase of $\theta = \arg(Z)$ (*i.e.*, in the time domain, the current signal is shifted $\frac{\theta}{2\pi}T$ later with respect to the voltage signal).

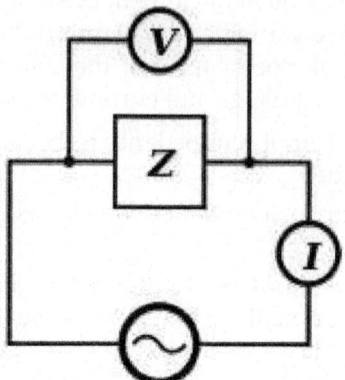

Fig. : An AC supply applying a voltage V, across a load Z, driving a current I.

Just as impedance extends Ohm's law to cover AC circuits, other results from DC circuit analysis such as voltage division, current division, Thévenin's theorem, and Norton's theorem can also be extended to AC circuits by replacing resistance with impedance.

Complex Voltage and Current

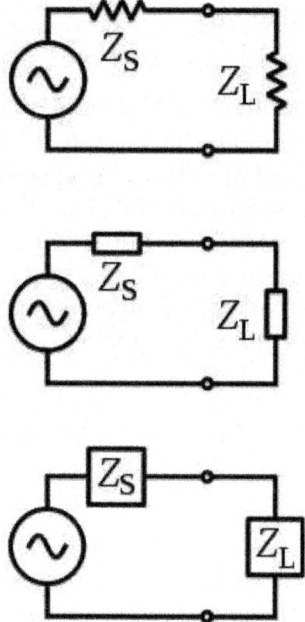

Fig. : Generalized impedances in a circuit can be drawn with the same symbol as a resistor (US ANSI or DIN Euro) or with a labelled box.

In order to simplify calculations, sinusoidal voltage and current waves are commonly represented as complex-valued functions of time denoted as V and I.

$$V = |V|e^{j(\omega t + \phi_V)}$$

$$I = |I|\,e^{j(\omega t + \phi_I)}$$

Impedance is defined as the ratio of these quantities.

$$Z = \frac{V}{I}$$

Substituting these into Ohm's law we have

$$|V|e^{j(\omega t + \phi_V)} = |I|e^{j(\omega t + \phi_I)}\,|Z|e^{j\theta}$$

$$= |I|\,|Z|e^{j(\omega t + \phi_I + \theta)}$$

Noting that this must hold for all t, we may equate the magnitudes and phases to obtain

$$|V| = |I|\,|Z|$$

$$\phi_V = \phi_I + \theta$$

The magnitude equation is the familiar Ohm's law applied to the voltage and current amplitudes, while the second equation defines the phase relationship.

Validity of Complex Representation

This representation using complex exponentials may be justified by noting that (by Euler's formula) :

$$\cos(\omega t + \phi) = \frac{1}{2}\left[e^{j(\omega t + \phi)} + e^{-j(\omega t + \phi)}\right]$$

The real-valued sinusoidal function representing either voltage or current may be broken into two complex-valued functions. By the principle of superposition, we may analyse the behaviour of the sinusoid on the left-hand side by analysing the behaviour of the two complex terms on the right-hand side. Given the symmetry, we only need to perform the analysis for one right-hand term; the results will be identical for the other. At the end of any calculation, we may return to real-valued sinusoids by further noting that

$$\cos(\omega t + \phi) = \Re\{e^{j(\omega t + \phi)}\}$$

Phasors

A phasor is a constant complex number, usually expressed in exponential form, representing the complex amplitude (magnitude and phase) of a sinusoidal function of time. Phasors are used by electrical engineers to simplify computations involving sinusoids, where they can often reduce a differential equation problem to an algebraic one.

The impedance of a circuit element can be defined as the ratio of the phasor voltage across the element to the phasor current through the element, as determined by the relative amplitudes and phases of the voltage and current. This is identical to the definition from Ohm's law given above, recognising that the factors of $e^{j\omega t}$ cancel.

Device Examples

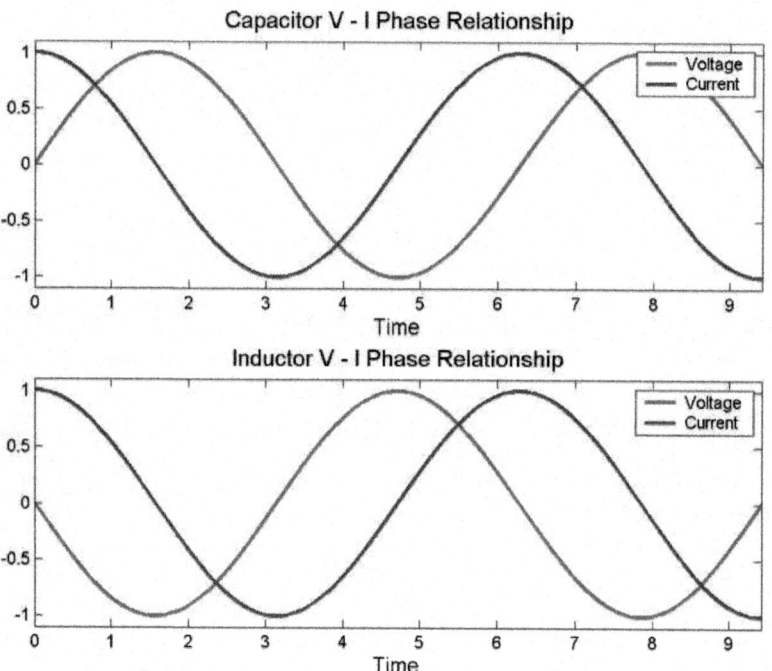

Fig. : The phase angles in the equations for the impedance of inductors and capacitors indicate that the voltage across a capacitor *lags* the current through it by a phase of $\pi/2$, while the voltage across an inductor *leads* the current through it by $\pi/2$. The identical voltage and current amplitudes indicate that the magnitude of the impedance is equal to one.

The impedance of an ideal resistor is purely real and is referred to as a *resistive impedance* :

$$Z_R = R$$

In this case, the voltage and current waveforms are proportional and in phase.

Ideal inductors and capacitors have a purely imaginary *reactive impedance* :

the impedance of inductors increases as frequency increases;

$$Z_L = j\omega L$$

the impedance of capacitors decreases as frequency increases;

$$Z_C = \frac{1}{j\omega C}$$

In both cases, for an applied sinusoidal voltage, the resulting current is also sinusoidal, but in quadrature, 90 degrees out of phase with the voltage. However, the phases have opposite signs : in an inductor, the current is *lagging*; in a capacitor the current is *leading*.

Note the following identities for the imaginary unit and its reciprocal :

$$j \equiv \cos \, \cos \left(\frac{\pi}{2}\right) + j \sin \left(\frac{\pi}{2}\right) \equiv e^{j\frac{\pi}{2}}$$

$$\frac{1}{j} \equiv -j \equiv \cos \left(-\frac{\pi}{2}\right) + j \sin \left(-\frac{\pi}{2}\right) \equiv e^{j\left(-\frac{\pi}{2}\right)}$$

Thus the inductor and capacitor impedance equations can be rewritten in polar form :

$$Z_L = \omega L e^{j\frac{\pi}{2}}$$

$$Z_C = \frac{1}{\omega C} e^{j\left(-\frac{\pi}{2}\right)}$$

The magnitude gives the change in voltage amplitude for a given current amplitude through the impedance, while the exponential factors give the phase relationship.

Deriving the Device-specific Impedances

What follows below is a derivation of impedance for each of the three basic circuit elements : the resistor, the capacitor, and the inductor. Although the idea can be extended to define the relationship between the voltage and current of any arbitrary signal, these derivations will assume sinusoidal signals, since any arbitrary signal can be approximated as a sum of sinusoids through Fourier analysis.

Resistor

For a resistor, there is the relation :

$$v_R(t) = i_R(t)R$$

This is Ohm's law.

Considering the voltage signal to be

$$v_R(t) = V_p \sin(\omega t)$$

it follows that

$$\frac{v_R(t)}{i_R(t)} = \frac{V_p \sin(\omega t)}{I_p \sin(\omega t)} = R$$

This says that the ratio of AC voltage amplitude to alternating current (AC) amplitude across a resistor is R, and that the AC voltage leads the current across a resistor by 0 degrees.

This result is commonly expressed as

$$Z_{resistor} = R$$

Capacitor

For a capacitor, there is the relation :

$$i_C(t) = C \frac{d\, v_C(t)}{dt}$$

Considering the voltage signal to be

$$v_C(t) = V_p \sin(\omega t)$$

it follows that

$$\frac{d\, v_C(t)}{dt} = \omega V_p \cos(\omega t)$$

And thus

$$\frac{v_C(t)}{i_C(t)} = \frac{V_p \; \sin(\omega t)}{\omega V_p C \cos(\omega t)} = \frac{\sin(\omega t)}{\omega C \sin\left(\omega t + \frac{\pi}{2}\right)}$$

This says that the ratio of AC voltage amplitude to AC current amplitude across a capacitor is $\dfrac{1}{\omega C}$, and that the AC voltage lags the AC current across a capacitor by 90 degrees (or the AC current leads the AC voltage across a capacitor by 90 degrees).

This result is commonly expressed in polar form, as

$$Z_{capacitor} = \frac{1}{\omega C} e^{-j\frac{\pi}{2}}$$

or, by applying Euler's formula, as

$$Z_{capacitor} = j \frac{1}{\omega C} = \frac{1}{j\omega C}$$

Inductor

For the inductor, we have the relation :

$$v_L(t) = L \frac{d\, i_L(t)}{dt}$$

This time, considering the current signal to be

$$i_L(t) = I_p \sin(\omega t)$$

it follows that

$$\frac{d\, i_L(t)}{dt} = \omega i_p \cos(\omega t)$$

And thus

$$\frac{v_L(t)}{i_L(t)} = \frac{\omega I_p L \cos(\omega t)}{I_p \sin(\omega t)} = \frac{\omega_L \sin\left(\omega t + \dfrac{\pi}{2}\right)}{\sin(\omega t)}$$

This says that the ratio of AC voltage amplitude to AC current amplitude across an inductor is ω_L, and that the AC voltage leads the AC current across an inductor by 90 degrees.

This result is commonly expressed in polar form, as

$$Z_{inductor} = \omega L e^{j\frac{\pi}{2}}$$

or, using Euler's formula, as

$$Z_{inductor} = j\omega L$$

Generalised s-plane Impedance

Impedance defined in terms of $j\omega$ can strictly only be applied to circuits which are driven with a steady-state AC signal. The concept of impedance can be extended to a circuit energised with any arbitrary signal by using complex frequency instead of $j\omega$. Complex frequency is given the symbol s and is, in general, a complex number. Signals are expressed in terms of complex frequency by taking the Laplace transform of the time domain expression of the signal. The impedance of the basic circuit elements in this more general notation is as follows :

Element	Impedance expression
Resistor	R
Inductor	sL
Capacitor	$\dfrac{1}{sC}$

For a DC circuit this simplifies to $s = 0$. For a steady-state sinusoidal AC signal $s = j\omega$.

Resistance vs Reactance

Resistance and reactance together determine the magnitude and phase of the impedance through the following relations :

$$|Z| = \sqrt{ZZ^*} = \sqrt{R^2 + X^2}$$

$$\theta = \arctan\left(\frac{X}{R}\right)$$

In many applications the relative phase of the voltage and current is not critical so only the magnitude of the impedance is significant.

Resistance

Resistance R is the real part of impedance; a device with a purely resistive impedance exhibits no phase shift between the voltage and current.

$$R = |Z|\cos\theta$$

Reactance

Reactance x is the imaginary part of the impedance; a component with a finite reactance induces a phase shift θ between the voltage across it and the current through it.

$$X = |Z|\sin\theta$$

A purely reactive component is distinguished by the sinusoidal voltage across the component being in quadrature with the sinusoidal current through the component. This implies that the component alternately absorbs energy from the circuit and then returns energy to the circuit. A pure reactance will not dissipate any power.

Capacitive Reactance

A capacitor has a purely reactive impedance which is inversely proportional to the signal frequency. A capacitor consists of two conductors separated by an insulator, also known as a dielectric.

$$X_C = (\omega C)^{-1} = (2\pi f C)^{-1}$$

At low frequencies a capacitor is open circuit, as no charge flows in the dielectric. A DC voltage applied across a capacitor causes charge to accumulate on one side; the electric field due to the accumulated charge is the source of the opposition to the current. When the potential associated with the charge exactly balances the applied voltage, the current goes to zero.

Driven by an AC supply, a capacitor will only accumulate a limited amount of charge before the potential difference changes sign and the charge dissipates. The higher the frequency, the less charge will accumulate and the smaller the opposition to the current.

Inductive Reactance

Inductive reactance X_L is proportional to the signal frequency f and the inductance L.

$$X_L = \omega L = 2\pi f L$$

An inductor consists of a coiled conductor. Faraday's law of electromagnetic induction gives the back emf ε(voltage opposing current) due to a rate-of-change of magnetic flux density B through a current loop.

$$\varepsilon = - \frac{d\Phi_B}{dt}$$

For an inductor consisting of a coil with N loops this gives.

$$\varepsilon = - N\frac{d\Phi_B}{dt}$$

The back-emf is the source of the opposition to current flow. A constant direct current has a zero rate-of-change, and sees an inductor as a short-circuit (it is typically made from a material with a low resistivity). An alternating current has a time-averaged rate-of-change that is proportional to frequency, this causes the increase in inductive reactance with frequency.

Total Reactance

The total reactance is given by

$$X = X_L - X_C$$

so that the total impedance is

$$Z = R + jX$$

Combining Impedances

The total impedance of many simple networks of components can be calculated using the rules for combining impedances in series and parallel. The rules are identical to those used for combining resistances, except that the numbers in general will be complex numbers. In the general case however, equivalent impedance transforms in addition to series and parallel will be required.

Series Combination

For components connected in series, the current through each circuit element is the same; the total impedance is the sum of the component impedances.

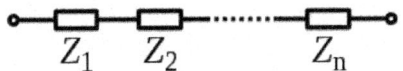

$$Z_{eq} = Z_1 + Z_2 + \ldots + Z_n$$

Or explicitly in real and imaginary terms :

$$Z_{eq} = R + X = (R_1 + R_2 + \ldots + R_n) + j(X_1 + X_2 + \ldots + X_n)$$

Parallel Combination

For components connected in parallel, the voltage across each circuit element is the same; the ratio of currents through any two elements is the inverse ratio of their impedances.

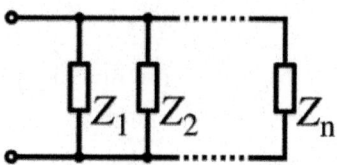

Hence the inverse total impedance is the sum of the inverses of the component impedances :

$$\frac{1}{Z_{eq}} = \frac{1}{Z_1} + \frac{1}{Z_2} + \cdots + \frac{1}{Z_n}$$

or, when $n = 2$:

$$\frac{1}{Z_{eq}} = \frac{1}{Z_1} + \frac{1}{Z_2} = \frac{Z_1 + Z_2}{Z_1 Z_2}$$

$$Z_{eq} = \frac{Z_1 Z_2}{Z_1 + Z_2}$$

The equivalent impedance Z_{eq} can be calculated in terms of the equivalent series resistance R_{eq} and reactance X_{eq}.

$$Z_{eq} = R_{eq} + j X_{eq}$$

$$R_{eq} = \frac{(X_1 R_2 + X_2 R_1)(X_1 + X_2) + (R_1 R_2 - X_1 X_2)(R_1 + R_2)}{(R_1 + R_2) + (X_1 + X_2)}$$

$$X_{eq} = \frac{(X_1 R_2 + \qquad (R_1 + R_2)^2 + (X_1 + X_2)^2 \qquad (X_1 + X_2)}{(R_1 + R_2)^2 + (X_1 + X_2)^2}$$

Measurement

The measurement of the impedance of devices and transmission lines is a practical problem in radio technology and others. Measurements of impedance may be carried out at one frequency, or the variation of device impedance over a range of frequencies may be of interest. The impedance may be measured or displayed directly in ohms, or other values related to impedance may be displayed; for example, in a radio antenna the standing wave ratio or reflection coefficient may be more useful than the impedance alone. Measurement of impedance requires measurement of the magnitude of voltage and current, and the phase difference between them. Impedance is often measured by "bridge" methods, similar to the direct-current Wheatstone bridge; a calibrated reference impedance is adjusted to balance off the effect of the impedance of the device under test. Impedance measurement in power electronic devices may require simultaneous measurement and provision of power to the operating device.

The impedance of a device can be calculated by complex division of the voltage and current. The impedance of the device can be calculated by applying a sinusoidal voltage to the device in series with a resistor, and measuring the voltage across the resistor and across the device. Performing this measurement by sweeping the frequencies of the applied signal provides the impedance phase and magnitude.

The use of an impulse response may be used in combination with the fast Fourier transform (FFT) to rapidly measure the electrical impedance of various electrical devices.

The LCR meter (Inductance (L), Capacitance (C), and Resistance (R)) is a device commonly used to measure the inductance, resistance and capacitance of a component; from these values the impedance at any frequency can be calculated.

Variable Impedance

In general, neither impedance nor admittance can be time varying as they are defined for complex exponentials for $-\infty < t < +\infty$. If the complex exponential voltage–current ratio changes over time or amplitude, the circuit element cannot be described using the frequency domain. However, many systems (*e.g.*, varicaps that are used in radio tuners) may exhibit non-linear or time-varying voltage–current ratios that appear to be linear time-invariant (LTI) for small signals over small observation windows; hence, they can be roughly described as having a time-varying impedance. That is, this description is an approximation; over large signal swings or observation windows, the voltage–current relationship is non-LTI and cannot be described by impedance.

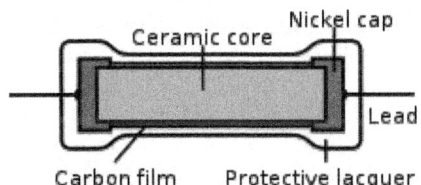

Fig. : Resistor based on the sheet resistance of carbon film.

SHEET RESISTANCE

Sheet resistance is a measure of resistance of thin films that are nominally uniform in thickness. It is commonly used to characterize materials made by semi-conductor doping, metal deposition, resistive paste printing, and glass coating. Examples of these processes are : doped semi-conductor regions (*e.g.*, silicon or polysilicon), and the resistors that are screen printed onto the substrates of thick-film hybrid microcircuits.

The utility of sheet resistance as opposed to resistance or resistivity is that it is directly measured using a four-terminal sensing measurement (also known as a four-point probe measurement).

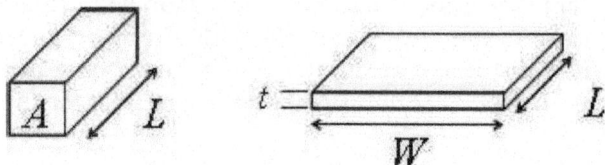

Fig. : Geometry for defining resistivity (left) and sheet resistance (right). In both cases, the current is parallel to the direction of the double-arrow near the letter "L".

Calculations

Sheet resistance is applicable to two-dimensional systems in which thin films are considered as two-dimensional entities. When the term sheet resistance is used, it is implied that the current is along the plane of the sheet, not perpendicular to it.

In a regular three-dimensional conductor, the resistance can be written as :

$R = \rho \dfrac{L}{A} = \rho \dfrac{L}{Wt}$ where ρ is the resistivity, A is the cross-sectional area and L is the

length. The cross-sectional area can be split into the width W and the sheet thickness t.

Upon combining the resistivity with the thickness, the resistance can then be written as :

$$R = \frac{\rho}{t} \frac{L}{W} = R_s \frac{L}{W}$$

where R_s is the sheet resistance. If the film thickness is known, the bulk resistivity ρ (in ohm cm) can be calculated by multiplying the sheet resistance by the film thickness in cm.

$$\rho = R_s \times t$$

Units

Sheet Resistance is a special case of resistivity for a uniform sheet thickness. Commonly, resistivity (also known as bulk resistance, specific electrical resistance, or volume resistivity) is in units of $\Omega \cdot cm$, which is more completely stated in units of $\Omega \cdot cm^2 / cm$ ($\Omega \cdot$Area/Length). When divided by the sheet thickness, $\cdot 1/cm$, the units are $\Omega \cdot cm \cdot (cm/cm) \cdot 1/cm = \Omega$. The term "(cm/cm)" cancels, but represents a special "square" situation yielding an answer in ohms. An alternate, common unit is "ohms per square" (denoted "Ω/sq" or "Ω /$\square$"), which is dimensionally equal to an ohm, but is exclusively used for sheet resistance. This is an advantage, because sheet resistance of 1 Ω could be taken out of context and mis-interpreted as bulk resistance of 1 ohm, whereas sheet resistance of 1 Ω/sq cannot thusly be mis-interpreted.

The reason for the name "ohms per square" is that a square sheet with sheet resistance 10 ohm/square has an actual resistance of 10 ohm, regardless of the size of the square. (For a square, $L = W$, so $R_s = R$.) The unit can be thought of as, loosely, "ohms · aspect ratio". Example : A 3 unit wide by 1 unit tall (aspect ratio = 3) sheet made of material having a Sheet Resistance of 7 Ω/sq would measure 21 Ω, if the 1 unit edges were attached to an ohmmeter that made contact entirely over each edge.

For Semi-conductors

For semi-conductors doped through diffusion or surface peaked ion implantation we define the sheet resistance using the average resistivity $\bar{\rho} = \dfrac{1}{\bar{\sigma}}$ of the material:

$$R_s = \bar{\rho} / x_j = \left(\bar{\sigma} \, x_j\right)^{-1} = \frac{1}{\int_0^{x_j} \sigma(x)\, dx}$$

which in materials with majority-carrier properties can be approximated by (neglecting intrinsic charge carriers) :

$$R_s = \frac{1}{\int_0^{x_j} \mu q \, N\,(x) dx}$$

where x_j is the junction depth, μ is the majority-carrier mobility, q is the carrier charge and $N(x)$ is the net impurity concentration in terms of depth. Knowing the background carrier concentration N_B and the surface impurity concentration the *sheet resistance-junction depth* product $R_s x_j$ can be found using Irvin's curves, which are numerical solutions to the above equation.

Measurement

A four point probe is used to avoid contact resistance, which can often be the same magnitude as the sheet resistance. Typically a constant current is applied to two probes and the potential on the other two probes is measured with a high imped-ance voltmeter. A geometry factor needs to be applied according to the shape of the four point array. Two common arrays are square and in-line.

Measurement may also be made by applying high conductivity buss bars to opposite edges of a square (or rectangular) sample. Resistance across a square area will equal Ω/sq. For a rectangle an appropriate geometric factor is added. Buss bars must make ohmic contact.

Inductive measurement is used as well. This method measures the shield-ing effect created by eddy currents. In one version of this technique a conductive sheet under test is placed between two coils. This non-contact sheet resistance measurement method also allows to characterize encapsulated thin-films or films with rough surfaces.

A very crude two point probe method is to measure resistance with the probes close together and the resistance with the probes far apart. The difference between these two resistances will be the order of magnitude of the sheet resistance.

Chapter 10

SEMI-CONDUCTOR

A **semi-conductor** is a material which has electrical conductivity between that of a conductor such as copper and that of an insulator such as glass. Semi-conductors are the foundation of modern electronics, including transistors, solar cells, light-emitting diodes (LEDs), quantum dots and digital and analog integrated circuits. The modern understanding of the properties of a semi-conductor relies on quantum physics to explain the movement of electrons inside a lattice of atoms. The increasing understanding of semi-conductor materials and fabrication processes has made possible continuing increases in the complexity and speed of semi-conductor devices, an effect known as Moore's Law.

The conductivity of a semi-conductor material increases with increasing temperature, behaviour opposite to that of a metal. Semi-conductors can display a range of useful properties such as passing current more easily in one direction than the other, variable resistance, and sensitivity to light or heat. Because the conductive properties of a semi-conductor material can be modified by controlled addition of impurities or by the application of electrical fields or light, devices made with semi-conductors are very useful for amplification of signals, switching, and energy conversion.

Current conduction in a semi-conductor occurs through the movement of free electrons and "holes", collectively known as charge carriers. Adding impurity atoms to a semiconducting material, known as "doping", greatly increases the number of charge carriers within it. When a doped semi-conductor contains excess holes it is called "p-type", and when it contains excess free electrons it is known as "n-type". The semi-conductor material used in devices is doped under highly controlled conditions to precisely control the location and concentration of p- and n-type dopants. A single semi-conductor crystal can have multiple p- and n-type regions; the p-n junctions between these regions have many useful electronic properties.

Some of the properties of semi-conductor materials were observed throughout the mid 19th and first decades of the 20th century. Development of quantum physics in turn allowed the development of the transistor in 1948. Although some pure elements and many compounds display semi-conductor properties, silicon, germanium, and compounds of gallium are the most wideley used in electronic devices.

PROPERTIES

Variable conductivity : A pure semi-conductor is a poor electrical conductor as a consequence of having just the right number of electrons to completely fill its valence bonds. Through various techniques (*e.g.*, doping or gating), the semi-conductor can be modified to have an excess of electrons (becoming an *n*-type **semi-conductor**) or a deficiency of electrons (becoming a *p*-**type semi-conductor**). In both cases, the semi-conductor becomes much more conductive (the conductivity can be increased by one million fold or more). Semi-conductor devices exploit this effect to shape electrical current.

Depletion :When doped semi-conductors are joined to metals, to different semi-conductors, and to the same semi-conductor with different doping, the resulting junction often strips the electron excess or deficiency out from the semi-conductor near the junction. This depletion region is rectifying (only allowing current to flow in one direction), and used to further shape electrical currents in semi-conductor devices.

Energetic electrons travel far : Electrons can be excited across the energy band gap of a semi-conductor by various means. These electrons can carry their excess energy over distance scales of micrometers before dissipating their energy into heat – a significantly longer distance than is possible in metals. This property is essential to the operation of, e. g., bipolar junction transistors and solar cells.

Light emission : In certain semi-conductors, excited electrons can relax by emitting light instead of producing heat. These semi-conductors are used in the construction of light emitting diodes and fluorescent quantum dots.

Thermal energy conversion : Semi-conductors have large thermoelectric power factors making them useful in thermoelectric generators, as well as high thermoelectric figures of merit making them useful in thermoelectric coolers.

MATERIALS

A large number of elements and compounds have semiconducting properties, including :

- Certain pure elements are found in Group XIV of the periodic table; the most commercially important of these elements are silicon and germanium. Silicon and germanium are used here effectively because they have 4 valence electrons in their outermost shell which gives them the ability to gain or lose electrons equally at the same time.
- Binary compounds, particularly between elements in Groups III and V, such as gallium arsenide, Groups II and VI, groups IV and VI, and between different group IV elements, *e.g.* silicon carbide.

- Certain ternary compounds, oxides and alloys.
- Organic semi-conductors, made of organic compounds.

Most common semiconducting materials are crystalline solids, but amorphous and liquid semi-conductors are also known. These include hydrogenated amorphous silicon and mixtures of arsenic, selenium and tellurium in a variety of proportions. These compounds share with better known semi-conductors the properties of intermediate conductivity and a rapid variation of conductivity with temperature, as well as occasional negative resistance. Such disordered materials lack the rigid crystalline structure of conventional semi-conductors such as silicon. They are generally used in thin film structures, which do not require material of higher electronic quality, being relatively insensitive to impurities and radiation damage.

Preparation of Semi-conductor Materials

Semi-conductors with predictable, reliable electronic properties are necessary for mass production. The level of chemical purity needed is extremely high because the presence of impurities even in very small proportions can have large effects on the properties of the material. A high degree of crystalline perfection is also required, since faults in crystal structure (such as dislocations, twins, and stacking faults) interfere with the semiconducting properties of the material. Crystalline faults are a major cause of defective semi-conductor devices. The larger the crystal, the more difficult it is to achieve the necessary perfection. Current mass production processes use crystal ingots between 100 and 300 mm (4 and 10 in) in diameter which are grown as cylinders and sliced into wafers.

Because of the required level of chemical purity and the perfection of the crystal structure which are needed to make semi-conductor devices, special methods have been developed to produce the initial semi-conductor material. A technique for achieving high purity includes growing the crystal using the Czochralski process. An additional step that can be used to further increase purity is known as zone refining. In zone refining, part of a solid crystal is melted. The impurities tend to concentrate in the melted region, while the desired material recrystalizes leaving the solid material more pure and with fewer crystalline faults.

In manufacturing semi-conductor devices involving heterojunctions between different semi-conductor materials, it is often important to align the crystal lattices of the two materials by using epitaxial techniques. The lattice constant, which is the length of the repeating element of the crystal structure, is important for determining the compatibility of materials.

PHYSICS OF SEMI-CONDUCTORS

Energy Bands and Electrical Conduction

Semi-conductors are defined by their unique electric conductive behaviour, somewhere between that of a metal and an insulator. The differences between these materials can be understood in terms of the quantum states for electrons, each of which may contain zero or one electron (by the Pauli exclusion principle). These states are associated with the electronic band structure of the material. Electrical conductivity arises due to the presence of electrons in states that are delocalized (extending through the material), however in order to transport electrons a state must be *partially filled*, containing an electron only part of the time. If the state is always occupied with an electron, then it is inert, blocking the passage of other electrons via that state. The energies of these quantum states are critical, since a state is partially filled only if its energy is near to the Fermi level.

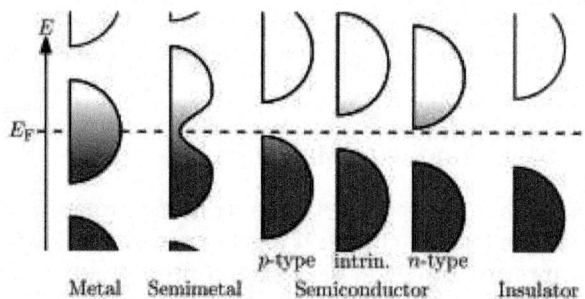

Fig. : Filling of electronic band structure in various types of material at equilibrium. In metals and semi-metals the Fermi level E_F lies inside at least one band. In insulators and semiconductors the Fermi level is inside a band gap, however in semi-conductors the bands are near enough to the Fermi level to be thermally populated with electrons or holes.

High conductivity in a material comes from it having many partially filled states and much state delocalization. Metals are good electrical conductors and have many partially filled states with energies near their Fermi level. Insulators, by contrast, have few partially filled states, their Fermi levels sit within band gaps with few energy states to occupy. Importantly, an insulator can be made to conduct by increasing its temperature : heating provides energy to promote some electrons across the band gap, inducing partially filled states in both the band of states beneath the band gap (valence band) and the band of states above the band gap (conduction band). An (intrinsic) semi-conductor has a band gap that it is smaller than that of an insulator and at room temperature significant numbers of electrons can be excited to cross the band gap.

A pure semi-conductor, however, is not very useful, as it is neither a very good insulator nor a very good conductor. However, one important feature of semi-conductors (and some insulators, known as *semi-insulators*) is that their conductivity can be increased and controlled by doping with impurities and gating with electric fields. Doping and gating move either the conduction or valence band much closer to the Fermi level, and greatly increase the number of partially filled states.

Some wider-band gap semi-conductor materials are sometimes referred to as **semi-insulators**. When undoped, these have electrical conductivity nearer to that of electrical insulators, however they can be doped (making them as useful as semi-conductors). Semi-insulators find niche applications in micro-electronics, such as substrates for HEMT. An example of a common semi-insulator is gallium arsenide. Some materials, such as titanium dioxide, can even be used as insulating materials for some applications, while being treated as wide-gap semi-conductors for other applications.

Charge Carriers (Electrons and Holes)

The partial filling of the states at the bottom of the conduction band can be understood as adding electrons to that band. The electrons do not stay indefinitely (due to the natural thermal recombination) but they can move around for some time. The actual concentration of electrons is typically very dilute, and so (unlike in metals) it is possible to think of the electrons in the conduction band of a semi-conductor as a sort of classical ideal gas, where the electrons fly around freely without being subject to the Pauli exclusion principle. In most semi-conductors the conduction bands have a parabolic dispersion relation, and so these electrons respond to forces (electric field, magnetic field, etc.) much like they would in a vacuum, though with a different effective mass. Because the electrons behave like an ideal gas, one may also think about conduction in very simplistic terms such as the Drude model, and introduce concepts such as electron mobility.

For partial filling at the top of the valence band, it is helpful to introduce the concept of an electron hole. Although the electrons in the valence band are always moving around, a completely full valence band is inert, not conducting any current. If an electron is taken out of the valence band, then the trajectory that the electron would normally have taken is now missing its charge. For the purposes of electric current, this combination of the full valence band, minus the electron, can be converted into a picture of a completely empty band containing a positively charged particle that moves in the same way as the electron. Combined with the *negative* effective mass of the electrons at the top of the valence band, we arrive at a picture of a positively charged particle that responds to electric and magnetic fields just as a normal positively charged particle would do in vacuum, again with some positive effective mass. This particle is called a hole, and the collection of holes in the valence can again be understood in simple classical terms (as with the electrons in the conduction band).

Carrier Generation and Recombination

When ionizing radiation strikes a semi-conductor, it may excite an electron out of its energy level and consequently leave a hole. This process is known as *electron–hole pair generation*. Electron-hole pairs are constantly generated from thermal energy as well, in the absence of any external energy source.

Electron-hole pairs are also apt to recombine. Conservation of energy demands that these recombination events, in which an electron loses an amount of energy larger than the band gap, be accompanied by the emission of thermal energy (in the form of phonons) or radiation (in the form of photons).

In some states, the generation and recombination of electron–hole pairs are in equipoise. The number of electron-hole pairs in the steady state at a given temperature is determined by quantum statistical mechanics. The precise quantum mechanical mechanisms of generation and recombination are governed by conservation of energy and conservation of momentum.

As the probability that electrons and holes meet together is proportional to the product of their amounts, the product is in steady state nearly constant at a given temperature, providing that there is no significant electric field (which might "flush" carriers of both types, or move them from neighbour regions containing more of them to meet together) or externally driven pair generation. The product is a function of the temperature, as the probability of getting enough thermal energy to produce a pair increases with temperature, being approximately $\exp(-E_G/kT)$, where k is Boltzmann's constant, T is absolute temperature and E_G is band gap.

The probability of meeting is increased by carrier traps — impurities or dislocations which can trap an electron or hole and hold it until a pair is completed. Such carrier traps are sometimes purposely added to reduce the time needed to reach the steady state.

Doping

The conductivity of semi-conductors may easily be modified by introducing impurities into their crystal lattice. The process of adding controlled impurities to a semi-conductor is known as *doping*. The amount of impurity, or dopant, added to an *intrinsic* (pure) semi-conductor varies its level of conductivity. Doped semi-conductors are referred to as *extrinsic*. By adding impurity to the pure semi-conductors, the electrical conductivity may be varied by factors of thousands or millions.

A 1 cm^3 specimen of a metal or semi-conductor has of the order of 10^{22} atoms. In a metal, every atom donates at least one free electron for conduction, thus 1 cm^3 of metal contains on the order of 10^{22} free electrons, whereas a 1 cm^3 sample of pure germanium at 20 °C contains about 4.2×10^{22} atoms, but only 2.5×10^{13} free electrons and 2.5×10^{13} holes. The addition of 0.001 % of arsenic (an impurity) donates an extra 10^{17} free electrons in the same volume and the electrical conductivity is increased by a factor of 10,000.

The materials chosen as suitable dopants depend on the atomic properties of both the dopant and the material to be doped. In general, dopants that produce the desired controlled changes are classified as either electron acceptors or donors. Semi-conductors doped with *donor* impurities are called *n-type*, while those doped with *acceptor* impurities are known as *p-type*. The n and p type designations indicate which charge carrier acts as the material's majority carrier. The opposite carrier is called the minority carrier, which exists due to thermal excitation at a much lower concentration compared to the majority carrier.

For example, the pure semi-conductor silicon has four valence electrons which bond each silicon atom to its neighbours. In silicon, the most common dopants are *group III* and *group V* elements. Group III elements all contain three valence electrons, causing them to function as acceptors when used to dope silicon. When an acceptor atom replaces a silicon atom in the crystal, a vacant state (an electron "hole") is created, which can move around the lattice and functions as a charge carrier. Group V elements have five valence electrons, which allows them to act as a donor; substitution of these atoms for silicon creates an extra free electron. Therefore, a silicon crystal doped with boron creates a p-type semi-conductor whereas one doped with phosphorus results in an n-type material.

During manufacture, dopants can be diffused into the semi-conductor body by contact with gaseous compounds of the desired element, or ion implantation can be used to accurately position the doped regions.

EARLY HISTORY OF SEMI-CONDUCTORS

The history of the understanding of semi-conductors begins with experiments on the electrical properties of materials. The properties of negative temperature coefficient of resistance, rectification, and light-sensitivity were observed starting in the early 19th century.

In 1833, Michael Faraday reported that the resistance of specimens of silver sulfide decreases when they are heated. This is contrary to the behaviour of metallic substances such as copper. In 1839, A. E. Becquerel reported observation of a voltage between a solid and a liquid electrolyte when struck by light, the photovoltaic effect. In 1873 Willoughby Smith observed that selenium resistors exhibit decreasing resistance when light falls on them. In 1874 Karl Ferdinand Braun observed conduction and rectification in metallic sulphides, and Arthur Schuster found that a copper oxide layer on wires has rectification properties that ceases when the wires are cleaned. Adams and Day observed the photovoltaic effect in selenium in 1876.

A unified explanation of these phenomena required a theory of solid-state physics which developed greatly in the first half of the 20th Century. In 1878 Edwin Herbert Hall demonstrated the deflection of flowing charge carriers by an applied magnetic field, the Hall effect. The discovery of the electron by J.J. Thomson in 1897 prompted theories of electron-based conduction in solids. Karl Baedeker, by observing a Hall effect with the reverse sign to that in metals, theorized that copper iodide had positive charge carriers. Johan Koenigsberger classified solid materials as metals, insulators and "variable conductors" in 1914. Felix Bloch published a theory of the movement of electrons through atomic lattices in 1928. In 1930, B. Gudden stated that conductivity in semi-conductors was due to minor concentrations of impurities. By 1931, the band theory of conduction had been established by Alan Herries Wilson and the concept of band gaps had been developed. Walter H. Schottky and Nevill Francis Mott developed models of the potential barrier and of the characteristics of a metal-semi-conductor junction. By 1938, Boris Davydov had developed a theory of the copper-oxide rectifer, identifying the effect of the p–n junction and the importance of minority carriers and surface states.

Agreement between theoretical predictions (based on developing quantum mechanics) and experimental results was sometimes poor. This was later explained by John Bardeen as due to the extreme "structure sensitive" behaviour of semi-conductors, whose properties change dramatically based on tiny amounts of impurities. Commercially pure materials of the 1920s containing varying proportions of trace contaminants produced differing experimental results. This spurred the development of improved material refining techniques, culminating in modern semi-conductor refineries producing materials with parts-per-trillion purity.

Devices using semi-conductors were at first constructed based on empirical knowledge, before semi-conductor theory provided a guide to construction of more capable and reliable devices.

Alexander Graham Bell used the light-sensitive property of selenium to Photophone transmit sound over a beam of light in 1880. A working solar cell, of low efficiency, was constructed by Charles Fritts in 1883 using a metal plate coated with selenium and a thin layer of gold; the device became commercially useful in photographic light meters in the 1930s. Point-contact microwave detector rectifiers made of lead sulfide were used by Jagadish Chandra Bose in 1904; the cat's-whisker detector using natural galena or other materials became a common device in the development of radio. However, it was somewhat unpredictable in operation and required manual adjustment for best performance. In 1906 H.J. Round observed light emission when electric current passed through silicon carbide crystals, the principle behind the light emitting diode. Oleg Losev observed similar light emission in 1922 but at the time the effect had no practical use. Power rectifiers, using copper oxide and selenium, were developed in the 1920s and became commercially important as an alternative to vacuum tube rectifiers.

In the years preceding World War II, infra-red detection and communications devices prompted research into lead-sulfide and lead-selenide materials. These devices were used for detecting ships and aircraft, for infrared rangefinders, and for voice communication systems. The point-contact crystal detector became vital for microwave radio systems, since available vacuum tube devices could not serve as detectors above about 4000 MHz; advanced radar systems relied on the fast response of crystal detectors. Considerable research and development of silicon materials occurred during the war to develop detectors of consistent quality.

Detector and power rectifiers could not amplify a signal. Many efforts were made to develop a solid-state amplifier, but these were unsuccessful because of limited theoretical understanding of semi-conductor materials. In 1922 Oleg Losev developed two-terminal, negative resistance amplifiers for radio; however, he perished in the Siege of Leningrad. In 1926 Julius Edgard Lilenfeld patented a device resembling a modern field-effect transistor, but it was not practical. R. Hilsch and R. W. Pohl in 1938 demonstrated a solid-state amplifier using a structure resembling the control grid of a vacuum tube; although the device displayed power gain, it had a cut-off frequency of one cycle per second, too low for any practical applications, but an effective application of the available theory. At Bell Labs, William Shockley and A. Holden started investigating solid-state amplifiers in 1938. The first p–n junction in silicon was observed by Russell Ohl about 1941, when a specimen was found to be light-sensitive, with a sharp boundary between p-type impurity at one end and n-type at the other. A slice cut from the specimen at the p–n boundary developed a voltage when exposed to light.

In France, during the war, Herbert Mataré had observed amplification between adjacent point contacts on a germanium base. After the war, Mataré's group announced their "Transistron" amplifier only shortly after Bell Labs announced the "transistor".

Chapter 11

RELATIVE PERMITTIVITY

Relative permittivities of some materials at room temperature under 1 kHz
(corresponds to an electromagnetic wave with wavelength of 300 km)

Material	ε_r
Vacuum	1 (by definition)
Air	1.00058986 ± 0.00000050 (at STP, for 0.9 MHz),
PTFE/Teflon	2.1
Polyethylene	2.25
Polyimide	3.4
Polypropylene	2.2–2.36
Polystyrene	2.4–2.7
Carbon disulfide	2.6
Paper	3.85
Electroactive polymers	2–12
Silicon dioxide	3.9
Concrete	4.5
Pyrex (Glass)	4.7 (3.7–10)
Rubber	7
Diamond	5.5–10
Salt	3–15
Graphite	10–15
Silicon	11.68
Ammonia	26, 22, 20, 17 (−80, −40, 0, 20 °C)
Methanol	30
Ethylene Glycol	37
Furfural	42.0

Material	ε_r
Glycerol	41.2, 47, 42.5 (0, 20, 25 °C)
Water	88, 80.1, 55.3, 34.5 (0, 20, 100, 200 °C) for visible light : 1.77
Hydrofluoric acid	83.6 (0 °C)
Formamide	84.0 (20 °C)
Sulfuric acid	84–100 (20–25 °C)
Hydrogen peroxide	128 aq–60 (–30–25 °C)
Hydrocyanic acid	158.0–2.3 (0–21 °C)
Titanium dioxide	86–173
Strontium titanate	310
Barium strontium titanate	500
Barium titanate	1250–10,000 (20–120 °C)
Lead zirconium titanate	500–6000
Conjugated polymers	1.8–6 up to 100,000
Calcium copper titanate	>250,000

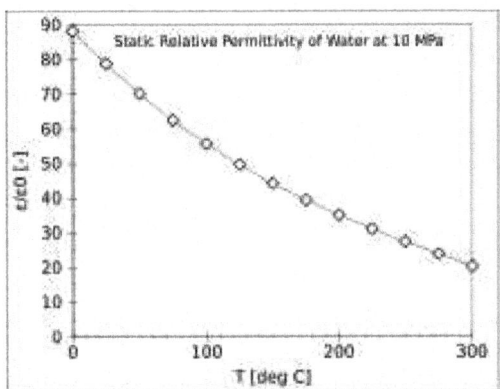

Fig. : Temperature dependence of the relative static permittivity of water.

The **relative permittivity** of a material under given conditions reflects the extent to which it concentrates electrostatic lines of flux. In technical terms, it is the ratio of the amount of electrical energy stored in a material by an applied voltage, relative to that stored in a vacuum. Likewise, it is also the ratio of the capacitance of a capacitor using that material as a dielectric, compared to a similar capacitor that has a vacuum as its dielectric.

DEFINITION

Relative permittivity is typically denoted as $\varepsilon_r(\omega)$ (sometimes κ or K) and is defined as

$$\varepsilon_r(\omega) = \frac{\varepsilon(\omega)}{\varepsilon_0}$$

where $\varepsilon(\omega)$ is the complex frequency-dependent absolute permittivity of the material, and ε_0 is the vacuum permittivity.

Relative permittivity is a dimensionless number that is in general complex-valued; its real and imaginary parts are denoted as :

$$\varepsilon_r(\omega) = \varepsilon'_r(\omega) + \varepsilon''_r(\omega).$$

The relative permittivity of a medium is related to its electric susceptibility, χ_e, as $\varepsilon_r(\omega) = 1 + \chi_e$.

In anisotropic media (such as non cubic crystals) the relative permittivity is a second rank tensor.

The relative permittivity of a material for a frequency of zero is known as its **static relative permittivity**.

Terminology

Dielectric constant is the historical term which, although still very common, has been deprecated by the relevant standards organizations. There is potential ambiguity in this predecessor name, as some older authors used it for the absolute permittivity ε while in most modern usage it refers to a relative permittivity ε_r, which in its turn may be either its static or the frequency-dependent variant, depending on context. It has also been used to refer to only the real component ε'_r of the complex-valued relative permittivity.

Physics

The imaginary portion of the permittivity corresponds to a phase shift of the polarization P relative to E and leads to the attenuation of electromagnetic waves passing through the medium. By definition, the linear relative permittivity of vacuum is equal to 1, that is $\varepsilon = \varepsilon_0$, although there are theoretical non-linear quantum effects in vacuum that exist at high field strengths.

MEASUREMENT

The relative static permittivity, ε_r, can be measured for static electric fields as follows : first the capacitance of a test capacitor, C_0, is measured with vacuum between its plates. Then, using the same capacitor and distance between its plates the capacitance C_x with a dielectric between the plates is measured. The relative dielectric constant can be then calculated as

$$\varepsilon_r = \frac{C_x}{C_0}.$$

For time-variant electromagnetic fields, this quantity becomes frequency-dependent. An indirect technique to calculate ε_r is conversion of radio frequency S-parameter measurement results. A description of frequently used S-parameter conversions for determination of the frequency-dependent ε_r of dielectrics can be found in this bibliographic source. Alternatively, resonance based effects may be employed at fixed frequencies.

APPLICATIONS

Energy

The dielectric constant is an essential piece of information when designing capacitors, and in other circumstances where a material might be expected to introduce capacitance into a circuit. If a material with a high dielectric constant is placed in an electric field, the magnitude of that field will be measurably reduced within the volume of the dielectric. This fact is commonly used to increase the capacitance of a particular capacitor design. The layers beneath etched conductors in printed circuit boards (PCBs) also act as dielectrics.

Communication

Dielectrics are used in RF transmission lines. In a coaxial cable, polyethylene can be used between the center conductor and outside shield. It can also be placed inside waveguides to form filters. Optical fibers are examples of *dielectric waveguides*. They consist of dielectric materials that are purposely doped with impurities so as to control the precise value of ε_r within the cross-section. This controls the refractive index of the material and therefore also the optical modes of transmission. However, in these cases it is technically the relative permittivity that matters, as they are not operated in the electrostatic limit.

Environmental

The relative permittivity of air changes with temperature, humidity, and barometric pressure. Sensors can be constructed to detect changes in capacitance caused by changes in the relative permittivity. Most of this change is due to effects of temperature and humidity as the barometric pressure is fairly stable. Using the capacitance change, along with the measured temperature, the relative humidity can be obtained using engineering formulas.

Chemical

The relative static permittivity of a solvent is a relative measure of its polarity. For example, water (very polar) has a dielectric constant of 80.10 at 20 °C while *n*-hexane (very non-polar) has a dielectric constant of 1.89 at 20 °C. This information is of great value when designing separation, sample preparation and chromatography techniques in analytical chemistry.

The correlation should, however, be treated with caution. For instance, dichloromethane has a value of ε_r of 9.08 (20 °C) and is rather poorly soluble in water (13 g/L or 9.8 mL/L at 20 °C); at the same time, tetrahydrofuran has its $\varepsilon_r = 7.52$ at 22 °C, but it is completely miscible with water.

This is even more apparent when considering the ε_r of acetic acid (6.2528) and that of iodoethane (7.6177). The large numerical value of ε_r is not surprising in the second case, as the iodine atom is easily polarizable; nevertheless, this does not imply that it is polar, too (electronic polarizability prevails over the orientational one in this case).

LOSSY MEDIUM

Again, similar as for absolute permittivity, relative permittivity for lossy materials can be formulated as :

$$\varepsilon_r = \varepsilon'_r + \frac{i\sigma}{\omega\varepsilon_0},$$

in terms of a "dielectric conductivity" σ (units S/m, siemens per meter), which "sums over all the dissipative effects of the material; it may represent an actual [electrical] conductivity caused by migrating charge carriers and it may also refer to an energy loss associated with the dispersion of ε' [the real-valued permittivity]" (, p. 8). Expanding the angular frequency $\omega = 2\pi c/\lambda$ and the electric constant $\varepsilon_0 = 1/(\mu_0 c^2)$, it reduces to :

$$\varepsilon_r = \varepsilon'_r + i\sigma\lambda k,$$

where λ is the wavelength, c is the speed of light in vacuum and $\kappa = \mu_0 c/2\pi \approx 60.0$ S^{-1} is a newly introduced constant (units reciprocal of siemens, such that $\sigma\lambda\kappa = \varepsilon_r''$ remains unitless).

METALS

Although permittivity is typically associated with dielectric materials, we may still speak of an effective permittivity of a metal, with real relative permittivity equal to one (eq.(4.6). In the low-frequency region (which extends from radio frequencies to the far infrared and terahertz region), the plasma frequency of the electron gas is much greater than the electromagnetic propagation frequency, so the complex index n of a metal is practically a purely imaginary number, expressed in terms of effective relative permittivity it has a low imaginary value (loss) and a negative real-value (high conductivity). (eq.(4.8)–(4.9).

DIELECTRIC STRENGTH

In physics, the term **dielectric strength** has the following meanings :

- Of an insulating material, the maximum electric field that a pure material can withstand under ideal conditions without breaking down (*i.e.,* without experiencing failure of its insulating properties).

- For a specific configuration of dielectric material and electrodes, the minimum applied electric field (*i.e.,* the applied voltage divided by electrode separation distance) that results in breakdown.

The theoretical dielectric strength of a material is an intrinsic property of the bulk material and is dependent on the configuration of the material or the electrodes with which the field is applied. The "intrinsic dielectric strength" is measured using pure materials under ideal laboratory conditions. At breakdown, the electric field frees bound electrons. If the applied electric field is sufficiently high, free electrons from background radiation may become accelerated to velocities that can liberate additional electrons during collisions with neutral atoms or molecules in a process called avalanche breakdown. Breakdown occurs quite abruptly (typically in nano-seconds), resulting in the formation of an electrically conductive path and a disruptive discharge through the material. For solid materials, a breakdown event severely degrades, or even destroys, its insulating capability.

Factors affecting apparent dielectric strength :

- it increases slightly with increased sample thickness.
- it decreases with increased operating temperature.
- it decreases with increased frequency.
- for gases (*e.g.* nitrogen, sulfur hexafluoride) it normally decreases with increased humidity.
- for air, dielectric strength increases slightly as humidity increases.

Breakdown Field Strength

The field strength at which breakdown occurs depends on the respective geometries of the dielectric (insulator) and the electrodes with which the electric field is applied, as well as the rate of increase at which the electric field is applied. Because dielectric materials usually contain minute defects, the practical dielectric strength will be a fraction of the intrinsic dielectric strength of an ideal, defect-free, material. Dielectric films tend to exhibit greater dielectric strength than thicker samples of the same material. For instance, the dielectric strength of silicon dioxide films of a few hundred nm to a few μm thick is approximately 0.5GV/m. However very thin layers (below, say, 100 nm) become partially conductive because of electron tunneling. Multiple layers of thin dielectric films are used where maximum practical dielectric strength is required, such as high voltage capacitors and pulse transformers. Since the dielectric strength of gases varies depending on the shape and configuration of the electrodes, it is usually measured as a fraction of the dielectric strength of Nitrogen gas.

Dielectric strength (in MV/m, or 10^6 Volt/meter) of various common materials :

Substance	Dielectric Strength (MV/m)
Helium (relative to nitrogen)	0.15
Air	3.0
Alumina	13.4
Window glass	9.8 - 13.8
Silicone oil, Mineral oil	10 - 15
Benzene	163
Polystyrene	19.7
Polyethylene	18.9 - 21.7
Neoprene rubber	15.7 - 26.7
Distilled Water	65 - 70
High Vacuum (field emission limited)	20 - 40 (depends on electrode shape)
Fused silica	25–40 at 20 °C
Waxed paper	40 - 60
PTFE (Teflon, Extruded)	19.7
PTFE (Teflon, Insulating Film)	60 - 173
Mica	118
Diamond	2000
Vacuum	10

Units

In SI, the unit of dielectric strength is volts per meter (V/m). It is also common to see related units such as volts per centimeter (V/cm), megavolts per meter (MV/m), and so on.

In United States customary units, dielectric strength is often specified in volts per mil (a mil is 1/1000 inch). The conversion is :

$$1 \text{ V/m} = 2.54 \text{ 5–5 V/mil}$$
$$1 \text{ V/mil} = 3.94 \times 104 \text{ V/m}$$

FERROELECTRICITY

Ferroelectricity is a property of certain materials that have a spontaneous electric polarization that can be reversed by the application of an external electric field. The term is used in analogy to ferromagnetism, in which a material exhibits a permanent magnetic moment. Ferromagnetism was already known when ferroelectricity was discovered in 1920 in Rochelle salt by Valasek. Thus, the prefix *ferro*, meaning iron, was used to describe the property despite the fact that most ferroelectric materials do not contain iron.

Polarization

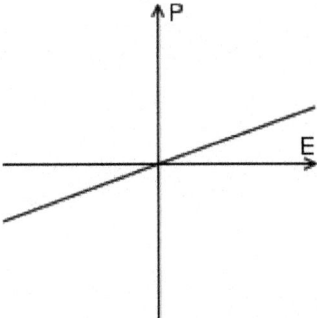

Fig. : Dielectric polarization.

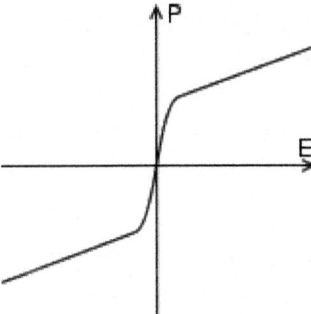

Fig. : Paraelectric polarization.

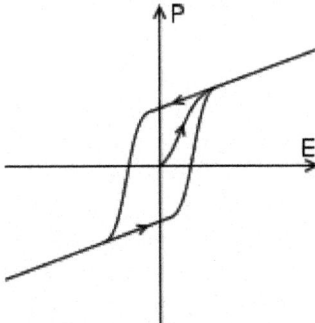

Fig. : Ferroelectric polarization.

When most materials are polarized, the polarization induced, P, is almost exactly proportional to the applied external electric field E; so the polarization is a linear function. This is called dielectric polarization. Some materials, known as paraelectric materials, show a more enhanced non-linear polarization. The electric permittivity, corresponding to the slope of the polarization curve, is not constant as in dielectrics but is a function of the external electric field.

In addition to being non-linear, ferroelectric materials demonstrate a spontaneous non-zero polarization when the applied field E is zero. The distinguishing feature of ferroelectrics is that the spontaneous polarization can be reversed by an applied electric field; the polarization is dependent not only on the current electric field but also on its history, yielding a hysteresis loop. They are called ferroelectrics by analogy to ferromagnetic materials, which have spontaneous magnetization and also exhibit hysteresis loops.

Typically, materials demonstrate ferroelectricity only below a certain phase transition temperature, called the Curie temperature, T_c, and are paraelectric above this temperature.

Applications

The non-linear nature of ferroelectric materials can be used to make capacitors with tunable capacitance. Typically, a ferroelectric capacitor simply consists of a pair of electrodes sandwiching a layer of ferroelectric material. The permittivity of ferroelectrics is not only tunable but commonly also very high in absolute value, especially when close to the phase transition temperature. Because of this, ferroelectric capacitors are small in physical size compared to dielectric (non-tunable) capacitors of similar capacitance.

The spontaneous polarization of ferroelectric materials implies a hysteresis effect which can be used as a memory function, and ferroelectric capacitors are indeed used to make ferroelectric RAM for computers and RFID cards. In these applications thin films of ferroelectric materials are typically used, as this allows the field required to switch the polarization to be achieved with a moderate voltage. However, when using thin films a great deal of attention needs to be paid to the interfaces, electrodes and sample quality for devices to work reliably.

Ferroelectric materials are required by symmetry considerations to be also piezoelectric and pyroelectric. The combined properties of memory, piezoelectricity, and pyroelectricity make ferroelectric capacitors very useful, *e.g.* for sensor applications. Ferroelectric capacitors are used in medical ultrasound machines (the capacitors generate and then listen for the ultrasound ping used to image the internal organs of a body), high quality infrared cameras (the infrared image is projected onto a two dimensional array of ferroelectric capacitors capable of detecting temperature differences as small as millionths of a degree Celsius), fire sensors, sonar, vibration sensors, and even fuel injectors on diesel engines.

Another idea of recent interest is the *ferroelectric tunnel junction (FTJ)* in which a contact made up by nanometer-thick ferroelectric film placed between metal electrodes. The thickness of the ferroelectric layer is small enough to allow tunneling of electrons. The piezoelectric and interface effects as well as the depolarization field may lead to a giant electroresistance (GER) switching effect.

Yet another hot topic is multiferroics, where researchers are looking for ways to couple magnetic and ferroelectric ordering within a material or heterostructure; there are several recent reviews on this topic.

Materials

The internal electric dipoles of a ferroelectric material are coupled to the material lattice so anything that changes the lattice will change the strength of the dipoles (in other words, a change in the spontaneous polarization). The change in the spontaneous polarization results in a change in the surface charge. This can cause current flow in the case of a ferroelectric capacitor even without the presence of an external voltage across the capacitor. Two stimuli that will change the lattice dimensions of a material are force and temperature. The generation of a surface charge in response to the application of an external stress to a material is called piezoelectricity. A change in the spontaneous polarization of a material in response to a change in temperature is called pyroelectricity.

Generally, there are 230 space groups among which 32 crystalline classes can be found in crystals. There are 21 non-centrosymmetric classes, within which 20 are piezoelectric. Among the piezoelectric classes, 10 have a spontaneous electric polarization, that varies with the temperature, therefore they are pyroelectric. Among pyroelectric materials, some of them are ferroelectric.

32 Crystalline classes			
20 classes piezoelectric			non piezoelectric
10 classes pyroelectric		non-pyroelectric	
ferroelectric	non-ferroelectric		
eg : $BaTiO_3$, $PbTiO_3$	eg : Tourmaline	eg : Quartz	

Ferroelectric phase transitions are often characterized as either displacive (such as $BaTiO_3$) or order-disorder (such as $NaNO_2$), though often phase transitions will demonstrate elements of both behaviours. In barium titanate, a typical ferroelectric of the displacive type, the transition can be understood in terms of a polarization catastrophe, in which, if an ion is displaced from equilibrium slightly, the force from the local electric fields due to the ions in the crystal increases faster than the elastic-restoring forces. This leads to an asymmetrical shift in the equilibrium ion positions and hence to a permanent dipole moment. The ionic displacement in barium titanate concerns the relative position of the titanium ion within the oxygen octahedral cage. In lead titanate, another key ferroelectric material, although the structure is rather similar to barium titanate the driving force for ferroelectricity is more complex with interactions between the lead and oxygen ions also playing an important role. In an order-disorder ferroelectric, there is a dipole moment in each unit cell, but at high temperatures they are pointing in random directions. Upon lowering the temperature and going through the phase transition, the dipoles order, all pointing in the same direction within a domain.

An important ferroelectric material for applications is lead zirconate titanate (PZT), which is part of the solid solution formed between ferroelectric lead titanate and anti-ferroelectric lead zirconate. Different compositions are used for different applications; for memory applications, PZT closer in composition to lead titanate is preferred, whereas piezoelectric applications make use of the diverging piezoelectric coefficients associated with the morphotropic phase boundary that is found close to the 50/50 composition.

Ferroelectric crystals often show several transition temperatures and domain structure hysteresis, much as do ferromagnetic crystals. The nature of the phase transition in some ferroelectric crystals is still not well understood.

In 1974 R.B. Meyer used symmetry arguments to predict ferroelectric liquid crystals, and the prediction could immediately be verified by several observations of behaviour connected to ferroelectricity in smectic liquid-crystal phases that are chiral and tilted. The technology allows the building of flat-screen monitors. Mass production between 1994 and 1999 was carried out by Canon. Ferroelectric liquid crystal are used in production of reflective LCoS.

In 2010 David Field found that prosaic films of chemicals such as nitrous oxide or propane exhibited ferroelectric properties. This new class of ferroelectric materials exhibit "spontelectric" properties, and may have wide ranging applications in device and nano-technology and also influence the electrical nature of dust in the interstellar medium.

Other ferroelectric materials used include triglycine sulfate, polyvinylidene fluoride (PVDF) and lithium tantalate.

Theory

An introduction to Landau theory can be found here. Based on Ginzburg–Landau theory, the free energy of a ferroelectric material, in the absence of an electric field and applied stress may be written as a Taylor expansion in terms of the order parameter, P. If a sixth order expansion is used (*i.e.* 8th order and higher terms truncated), the free energy is given by :

$$\Delta E = \frac{1}{2}\alpha_0 (T - T_0)(P_x^2 + P_y^2 + P_z^2) + \frac{1}{4}\alpha_{11}(P_x^4 + P_y^4 + P_z^4)$$

$$+ \frac{1}{2}\alpha_{12}(P_x^2 P_y^2 + P_y^2 P_z^2 + P_z^2 P_x^2)$$

$$+ \frac{1}{6}\alpha_{111}(P_x^6 P_y^2 + P_y^2 P_z^2 + P_z^2 P_x^2)$$

$$+ \frac{1}{2}\alpha_{112}[P_x^4(P_y^2 + P_z^2) + P_y^4(P_x^2 + P_z^2) + P_z^2(P_x^2 + P_y^2)]$$

$$+ \frac{1}{2} \alpha_{123} \; P_x^2 P_y^2 P_z^2$$

where P_x, P_y, and P_z are the components of the polarization vector in the x, y, and z directions respectively, and the coefficients, α_i, α_{ij}, α_{ijk} must be consistent with the crystal symmetry. To investigate domain formation and other phenomena in ferroelectrics, these equations are often used in the context of a phase field model. Typically, this involves adding a gradient term, an electrostatic term and an elastic term to the free energy. The equations are then discretized onto a grid using the finite difference method and solved subject to the constraints of Gauss's law and Linear elasticity.

In all known ferroelectrics, $\alpha_0 > 0$ and $\alpha_{111} > 0$. These coefficients may be obtained experimentally or from ab-initio simulations. For ferroelectrics with a first order phase transition, $\alpha_{11} < 0$ and $\alpha_{11} > 0$ for a second order phase transition.

The spontaneous polarization, P_s of a ferroelectric for a cubic to tetragonal phase transition may be obtained by considering the 1D expression of the free energy which is :

$$\Delta E = \frac{1}{2} \alpha_0 \, (T - T_0) \, P_x^2 + \frac{1}{4} \alpha_{11} \, P_x^4 + \frac{1}{6} \alpha_{111} \, P_x^6$$

This free energy has the shape of a double well potential with two free energy minima at $P = \pm P_s$, where P_s is the spontaneous polarization. At these two minima, the derivative of the free energy is zero, i.e. :

$$\frac{\partial \Delta E}{\partial P_x} = \alpha_0 \, (T - T_0) \, P_x + a_{11} \, P_x^3 + \alpha_{111} \, P_x^5 = 0$$

$$P_x \, [\alpha_0 \, (T - T_0) + \alpha_{11} \, P_x^2 + \alpha_{111} \, P_x^4] = 0$$

Since $P_x = 0$ corresponds to a free energy maxima in the ferroelectric phase, the spontaneous polarization, P_s, is obtained from the solution of the equation :

$$P_x \, [\alpha_0 \, (T - T_0) + \alpha_{11} \, P_x^2 + \alpha_{111} \, P_x^4] = 0$$

which is :

$$P_s^2 = \frac{1}{2\alpha_{111}} \left[-\alpha_{11} \pm \sqrt{a_{11}^2 - 4\alpha_0 \, \alpha_{111} (T - T_0)} \right]$$

and elimination of solutions yielding a negative square root (for either the first or second order phase transitions) gives :

$$P_s = \frac{1}{2\alpha_{111}} \left[-\alpha_{11} \pm \sqrt{a_{11}^2 - 4\alpha_0 \, \alpha_{111} (T - T_0)} \right]$$

If $\alpha_{11} = 0$, using the same approach as above, the spontaneous polarization may be obtained as :

$$P_s = \sqrt{-\frac{\alpha_0 \, (T - T_0)}{\alpha_{111}}}$$

The hysteresis loop (P_x versus E_x) may be obtained from the free energy expansion by adding an additional electrostatic term, $E_x P_x$, as follows :

$$\Delta E = \frac{1}{2}\alpha_0 (T - T_0)\, P_x^2 + \frac{1}{4}\alpha_{11}\, P_x^4 + \frac{1}{6}\alpha_{111} P_x^6 - E_x P_x$$

$$\frac{\partial \Delta E}{\partial P_x} = \alpha_0 (T - T_0)P_x + \alpha_{11}\, P_3^x + a_{111}\, P_x^5 - E_x = 0$$

$$E_x = \alpha_0 (T - T_0)\, P_x + \alpha_{11}\, P_x^3 + \alpha_{111}\, P_x^5$$

Plotting E_x as a function of P_x and reflecting the graph about the 45 degree line gives an 'S' shaped curve. The central part of the 'S' corresponds to a free energy local maximum (since $\dfrac{\partial^2 \Delta E}{\partial P_x^2} < 0$). Elimination of this region, and connection of the top and bottom portions of the 'S' curve by vertical lines at the discontinuities gives the hysteresis loop.

LOW-K DIELECTRIC

In semi-conductor manufacturing, a **low-κ dielectric** is a material with a small dielectric constant relative to silicon dioxide. Although the proper symbol for the dielectric constant is the Greek letter κ (kappa), in conversation such materials are referred to as being "low-k" (low-kay) rather than "low-κ" (low-kappa). Low-κ dielectric material implementation is one of several strategies used to allow continued scaling of microelectronic devices, colloquially referred to as extending Moore's law. In digital circuits, insulating dielectrics separate the conducting parts (wire interconnects and transistors) from one another. As components have scaled and transistors have got closer together, the insulating dielectrics have thinned to the point where charge build up and crosstalk adversely affect the performance of the device. Replacing the silicon dioxide with a low-κ dielectric of the same thickness reduces parasitic capacitance, enabling faster switching speeds and lower heat dissipation.

Low-κ Materials

The dielectric constant of SiO_2, the insulating material used in silicon chips, is 3.9. This number is the ratio of the permittivity of SiO_2 divided by permittivity of vacuum, $\varepsilon_{SiO_2}/\varepsilon_0$, where $\varepsilon_0 = 8.854\times10\ pF/\mu m$. There are many materials with lower dielectric constants but few of them can be suitably integrated into a manufacturing process. Development efforts have focused primarily on three classes of materials :

Fluorine-doped Silicon Dioxide

By doping SiO_2 with fluorine to produce fluorinated silica glass, the dielectric constant is lowered from 3.9 to 3.5.

Carbon-doped Silicon Dioxide

By doping SiO_2 with carbon, one can lower the dielectric constant to 3.0,density 1400 and thermal conductivity of 0.39w/(m.k).

Porous Silicon Dioxide

Various methods may be employed to create large voids or pores in a silicon dioxide dielectric. Voids can have a dielectric constant of nearly 1, thus the dielectric constant of the porous material may be reduced by increasing the porosity of the film. Dielectric constants lower than 2.0 have been reported. Integration difficulties related to porous silicon dioxide implementation include low mechanical strength and difficult integration with etch and polish processes.

Porous Carbon-doped Silicon Dioxide

By UV curing, floating methyl groups in carbon doped silicon dioxide can be eliminated and pores can be introduced to the carbon doped silicon dioxide low-κ materials.

Spin-on Organic Polymeric Dielectrics

Polymeric dielectrics are generally deposited by a spin-on approach, such as those traditionally used to deposit photoresist, rather than chemical vapour deposition. Integration difficulties include low mechanical strength and thermal stability. Some examples of spin-on organic low-κ polymers are polyimide, polynorbornenes, benzocyclobutene, and PTFE.

Spin-on silicon based polymeric dielectric

There are two kinds of silicon based polymeric dielectric materials, hydrogen silsesquioxane (HSQ) and methylsilsesquioxane (MSQ).

HIGH-K DIELECTRIC

The term **high-κ dielectric** refers to a material with a high dielectric constant κ (as compared to silicon dioxide). High-κ dielectrics are used in semi-conductor manufacturing processes where they are usually used to replace a silicon dioxide gate dielectric or another dielectric layer of a device. The implementation of high-κ gate dielectrics is one of several strategies developed to allow further miniaturization of microelectronic components, colloquially referred to as extending Moore's Law.

Need for High-κ Materials

Silicon dioxide has been used as a gate oxide material for decades. As transistors have decreased in size, the thickness of the silicon dioxide gate dielectric has steadily decreased to increase the gate capacitance and thereby drive current, raising device performance. As the thickness scales below 2 nm, leakage currents due to tunneling increase drastically, leading to high power consumption and reduced device reliability. Replacing the silicon dioxide gate dielectric with a high-κ material allows increased gate capacitance without the associated leakage effects.

First Principles

The gate oxide in a MOSFET can be modelled as a parallel plate capacitor. Ignoring quantum mechanical and depletion effects from the Si substrate and gate, the capacitance C of this parallel plate capacitor is given by

$$C = \frac{k\varepsilon_0 A}{t}$$

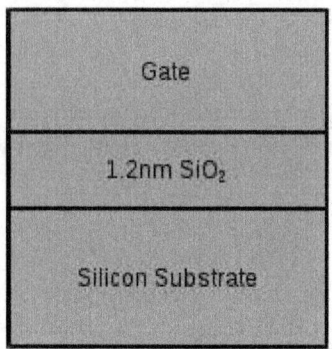

Existing 90nm Process
Capacitance = 1x
Leakage Current = 1x

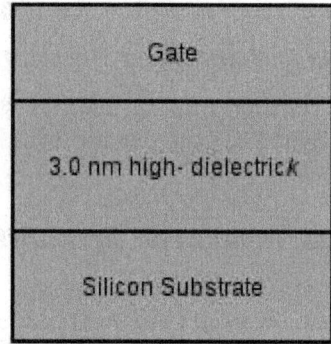

A potential high- process k
Capacitance = 1.6x
Leakage Current = 0.01x

Conventional silicon dioxide gate dielectric structure compared to a potential high-k dielectric structure

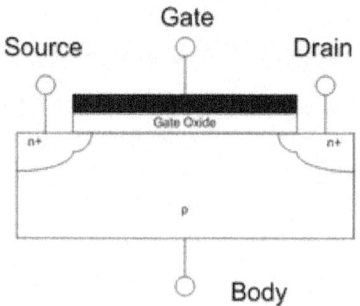

Fig. : Cross-section of an N channel MOSFET transistor showing the gate oxide dielectric.

Where :

- A is the capacitor area
- κ is the relative dielectric constant of the material (3.9 for silicon dioxide)

- ε_0 is the permittivity of free space
- t is the thickness of the capacitor oxide insulator

Since leakage limitation constrains further reduction of t, an alternative method to increase gate capacitance is alter κ by replacing silicon dioxide with a high-κ material. In such a scenario, a thicker gate oxide layer might be used which can reduce the leakage current flowing through the structure as well as improving the gate dielectric reliability.

Gate Capacitance Impact on Drive Current

The drain current I_D for a MOSFET can be written (using the gradual channel approximation) as

$$I_{D,\,Sat} = \frac{W}{L}\,\mu C_{inv}\,\frac{(V_G - V_{th})^2}{2}$$

Where :

- W is the width of the transistor channel
- L is the channel length
- μ is the channel carrier mobility (assumed constant here)
- C_{inv} is the capacitance density associated with the gate dielectric when the underlying channel is in the inverted state
- V_G is the voltage applied to the transistor gate
- V_D is the voltage applied to the transistor drain
- V_{th} is the threshold voltage.

The term $V_G - V_{th}$ is limited in range due to reliability and room temperature operation constraints, since a too large V_G would create an undesirable, high electric field across the oxide. Furthermore, V_{th} cannot easily be reduced below about 200 mV, because leakage currents due to increased oxide leakage (that is, assuming high-κ dielectrics are not available) and subthreshold conduction raise stand-by power consumption to unacceptable levels. Thus, according to this simplified list of factors, an increased $I_{D,sat}$ requires a reduction in the channel length or an increase in the gate dielectric capacitance.

Materials and Considerations

Replacing the silicon dioxide gate dielectric with another material adds complexity to the manufacturing process. Silicon dioxide can be formed by oxidizing the underlying silicon, ensuring a uniform, conformal oxide and high interface quality. As a consequence, development efforts have focused on finding a material with a requisitely high dielectric constant that can be easily integrated into a manufacturing process. Other key considerations include band alignment to silicon (which may alter leakage current), film morphology, thermal stability, maintenance of a high mobility of charge carriers in the channel and minimization of electrical defects in the film/interface. Materials which have received considerable attention are hafnium silicate, zirconium silicate, hafnium dioxide and zirconium dioxide, typically deposited using atomic layer deposition.

It is expected that defect states in the high-k dielectric can influence its electrical properties. Defect states can be measured for example by using zero-bias thermally stimulated current, zero-temperature-gradient zero-bias thermally stimulated current spectroscopy, or inelastic electron tunneling spectroscopy (IETS).

Use in Industry

The industry has employed oxynitride gate dielectrics since the 1990s, wherein a conventionally formed silicon oxide dielectric is infused with a small amount of nitrogen. The nitride content subtly raises the dielectric constant and is thought to offer other advantages, such as resistance against dopant diffusion through the gate dielectric.

In early 2007, Intel announced the deployment of hafnium-based high-k dielectrics in conjunction with a metallic gate for components built on 45 nanometer technologies, and has shipped it in the 2007 processor series codenamed Penryn. At the same time, IBM announced plans to transition to high-k materials, also hafnium-based, for some products in 2008. While not identified, the most likely dielectric used in such applications are some form of nitrided hafnium silicates (HfSiON). HfO_2 and HfSiO are susceptible to crystallization during dopant activation annealing. NEC Electronics has also announced the use of a HfSiON dielectric in their 55 nm *UltimateLowPower* technology. However, even HfSiON is susceptible to trap-related leakage currents, which tend to increase with stress over device lifetime. This leakage effect becomes more severe as hafnium concentration increases. There is no guarantee however, that hafnium will serve as a de facto basis for future high-k dielectrics. The 2006 ITRS roadmap predicted the implementation of high-k materials to be commonplace in the industry by 2010.

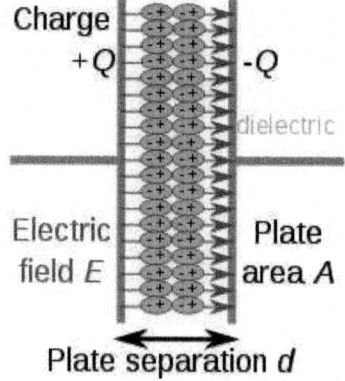

Fig. : A polarized dielectric material.

DIELECTRIC

A **dielectric material** (**dielectric** for short) is an electrical insulator that can be polarized by an applied electric field. When a dielectric is placed in an electric field, electric charges do not flow through the material as they do in a conductor, but only slightly shift from their average equilibrium positions causing **dielectric polarization**. Because of dielectric polarization, positive charges are displaced toward the field and negative charges shift in the opposite direction. This creates an internal electric field that reduces the overall field within the dielectric itself. If a dielectric is composed of weakly bonded molecules, those molecules not only become polarized, but also reorient so that their symmetry axis aligns to the field.

The study of dielectric properties concerns storage and dissipation of electric and magnetic energy in materials. It is important to explain various phenomena in electronics, optics, and solid-state physics.

Terminology

While the term *insulator* implies low electrical conduction, *dielectric* typically means materials with a high polarizability. The latter is expressed by a number called the relative permittivity (also known in older texts as dielectric constant). The term insulator is generally used to indicate electrical obstruction while the term dielectric is used to indicate the energy storing capacity of the material (by means of polarization). A common example of a dielectric is the electrically insulating material between the metallic plates of a capacitor. The polarization of the dielectric by the applied electric field increases the capacitor's surface charge.

The term "dielectric" was coined by William Whewell (from "dia-electric") in response to a request from Michael Faraday. A *perfect dielectric* is a material with zero electrical conductivity. (cf. perfect conductor), thus exhibiting only a displacement current; therefore it stores and returns electrical energy as if it were an ideal capacitor.

Electric Susceptibility

The **electric susceptibility** χ_e of a dielectric material is a measure of how easily it polarizes in response to an electric field. This, in turn, determines the electric permittivity of the material and thus influences many other phenomena in that medium, from the capacitance of capacitors to the speed of light.

It is defined as the constant of proportionality (which may be a tensor) relating an electric field **E** to the induced dielectric polarization density **P** such that

$$\mathbf{P} = \varepsilon_0 Xe\mathbf{E},$$

where ε_0 is the electric permittivity of free space.

The susceptibility of a medium is related to its relative permittivity ε_r by

$$X_e = \varepsilon_r - 1.$$

So in the case of a vacuum,

$$X_e = 0.$$

The electric displacement $\mathbf{D}$ is related to the polarization density $\mathbf{P}$ by

$$\mathbf{D} = \varepsilon_0\, \mathbf{E} + \mathbf{P} = \varepsilon_0(1 + X_e)\mathbf{E} = \varepsilon_r\varepsilon_0\, \mathbf{E}.$$

Dispersion and Causality

In general, a material cannot polarize instantaneously in response to an applied field. The more general formulation as a function of time is

$$\mathbf{P}(t) = \varepsilon_0\int_{-\infty}^{t} \chi_e(t - t')\, \mathbf{E}(t')\, dt'.$$

That is, the polarization is a convolution of the electric field at previous times with time-dependent susceptibility given by $\chi_e(\Delta t)$. The upper limit of this integral can be extended to infinity as well if one defines $\chi_e(\Delta t) = 0$ for $\Delta t < 0$. An instantaneous response corresponds to Dirac delta function susceptibility $\chi_e(\Delta t) = \chi_e\delta(\Delta t)$.

It is more convenient in a linear system to take the Fourier transform and write this relationship as a function of frequency. Due to the convolution theorem, the integral becomes a simple product,

$$\mathbf{P}(\omega) = \varepsilon_0\chi_e\mathbf{E}(\omega).$$

Note the simple frequency dependence of the susceptibility, or equivalently the permittivity. The shape of the susceptibility with respect to frequency characterizes the dispersion properties of the material.

Moreover, the fact that the polarization can only depend on the electric field at previous times (*i.e.*, $\chi_e(\Delta t) = 0$ for $\Delta t < 0$), a consequence of causality, imposes Kramers–Kronig constraints on the susceptibility $\chi_e(0)$.

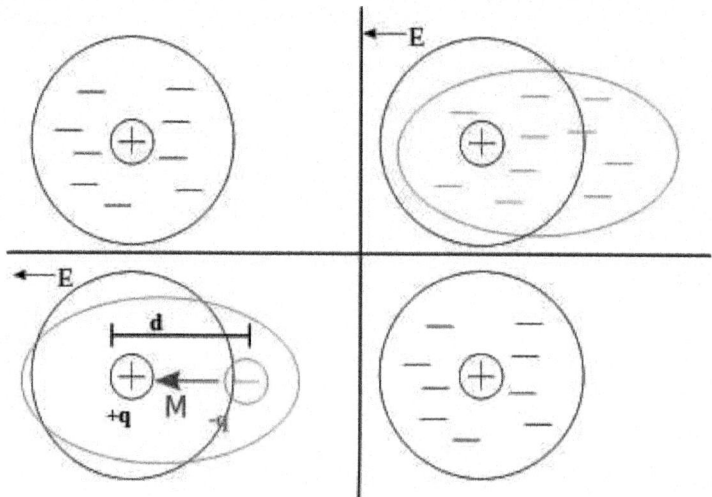

Fig. : Electric field interaction with an atom under the classical dielectric model.

Dielectric Polarization

Basic Atomic Model

In the classical approach to the dielectric model, a material is made up of atoms. Each atom consists of a cloud of negative charge (Electrons) bound to and surrounding a positive point charge at its center. In the presence of an electric field the charge cloud is distorted.

This can be reduced to a simple dipole using the superposition principle. A dipole is characterized by its dipole moment. It is the relationship between the electric field and the dipole moment that gives rise to the behaviour of the dielectric. (Note that the dipole moment points in the same direction as the electric field. This isn't always correct, and is a major simplification, but is suitable for many materials.)

When the electric field is removed the atom returns to its original state. The time required to do so is the so-called relaxation time; an exponential decay.

This is the essence of the model in physics. The behaviour of the dielectric now depends on the situation. The more complicated the situation, the richer the model must be to accurately describe the behaviour. Important questions are :

- Is the electric field constant or does it vary with time?
  - o If the electric field does vary, at what rate?
- What are the characteristics of the material?
  - o Is the direction of the field important (isotropy)?
  - o Is the material the same all the way through (homogeneous)?
  - o Do any boundaries/interfaces have to be taken into account?
- Is the system linear, or do non-linearities have to be taken into account?

The relationship between the electric field **E** and the dipole moment **M** gives rise to the behaviour of the dielectric, which, for a given material, can be characterized by the function **F** defined by the equation :

$$\mathbf{M} = \mathbf{F(E)}.$$

When both the type of electric field and the type of material have been defined, one then chooses the simplest function *F* that correctly predicts the phenomena of interest. Examples of phenomena that can be so modeled include :

- Refractive index
- Group velocity dispersion
- Birefringence
- Self-focusing
- Harmonic generation.

Dipolar Polarization

Dipolar polarization is a polarization that is either inherent to polar molecules (**orientation polarization**), or can be induced in any molecule in which the asymmetric distortion of the nuclei is possible (**distortion polarization**). Orientation polarization results from a permanent dipole, *e.g.*, that arising from the 104.45° angle between the asymmetric bonds between oxygen and hydrogen atoms in the water molecule, which retains polarization in the absence of an external electric field. The assembly of these dipoles forms a macroscopic polarization.

When an external electric field is applied, the distance between charges, which is related to chemical bonding, remains constant in orientation polarization; however, the polarization itself rotates. This rotation occurs on a timescale that depends on the torque and surrounding local viscosity of the molecules. Because the rotation is not instantaneous, dipolar polarizations lose the response to electric fields at the lowest frequency in polarizations. A molecule rotates about 1ps per radian in a fluid, thus this loss occurs at about 10^{11} Hz (in the microwave region). The delay of the response to the change of the electric field causes friction and heat.

When an external electric field is applied in the infrared, a molecule is bent and stretched by the field and the molecular moment changes in response. The molecular vibration frequency is approximately the inverse of the time taken for the molecule to bend, and the **distortion polarization** disappears above the infrared.

Ionic Polarization

Ionic polarization is polarization caused by relative displacements between positive and negative ions in ionic crystals (for example, NaCl).

If crystals or molecules do not consist of only atoms of the same kind, the distribution of charges around an atom in the crystals or molecules leans to positive or negative. As a result, when lattice vibrations or molecular vibrations induce relative displacements of the atoms, the centers of positive and negative charges might be in different locations. These center positions are affected by the symmetry of the displacements. When the centers don't correspond, polarizations arise in molecules or crystals. This polarization is called **ionic polarization**.

Ionic polarization causes ferroelectric transition as well as dipolar polarization. The transition, which is caused by the order of the directional orientations of permanent dipoles along a particular direction, is called **order-disorder phase transition**. Transition caused by ionic polarizations in crystals is called **displacive phase transition**.

Dielectric Dispersion

In physics, **dielectric dispersion** is the dependence of the permittivity of a dielectric material on the frequency of an applied electric field. Because there is always a lag between changes in polarization and changes in an electric field, the permittivity of the dielectric is a complicated, complex-valued function of frequency of the electric field. It is very important for the application of dielectric materials and the analysis of polarization systems.

This is one instance of a general phenomenon known as material dispersion : a frequency-dependent response of a medium for wave propagation.

When the frequency becomes higher :

1. It becomes impossible for dipolar polarization to follow the electric field in the microwave region around 10^{10} Hz;

2. In the infrared or far-infrared region around 10^{13} Hz, ionic polarization and molecular distortion polarization lose the response to the electric field;

3. Electronic polarization loses its response in the ultraviolet region around 10^{15} Hz.

In the frequency region above ultra-violet, permittivity approaches the constant ε_0 in every substance, where ε_0 is the permittivity of the free space. Because permittivity indicates the strength of the relation between an electric field and polarization, if a polarization process loses its response, permittivity decreases.

Dielectric Relaxation

Dielectric relaxation is the momentary delay (or lag) in the dielectric constant of a material. This is usually caused by the delay in molecular polarization with respect to a changing electric field in a dielectric medium (*e.g.*, inside capacitors or between two large conducting surfaces). Dielectric relaxation in changing electric fields could be considered analogous to hysteresis in changing magnetic fields (for inductors or transformers). Relaxation in general is a delay or lag in the response of a linear system, and therefore dielectric relaxation is measured relative to the expected linear steady state (equilibrium) dielectric values. The time lag between electrical field and polarization implies an irreversible degradation of free energy (G).

In physics, **dielectric relaxation** refers to the relaxation response of a dielectric medium to an external electric field of microwave frequencies. This relaxation is often described in terms of permittivity as a function of frequency, which can, for ideal systems, be described by the Debye equation. On the other hand, the distortion related to ionic and electronic polarization shows behaviour of the resonance or oscillator type. The character of the distortion process depends on the structure, composition, and surroundings of the sample.

The number of possible wavelengths of emitted radiation due to dielectric relaxation can be equated using Hemmings' first law (named after Mark Hemmings)

$$n = \frac{l^2 - l}{2}$$

where :

n is the number of different possible wavelengths of emitted radiation

l is the number of energy levels (including ground level).

Debye Relaxation

Debye relaxation is the dielectric relaxation response of an ideal, non-interacting population of dipoles to an alternating external electric field. It is usually expressed in the complex permittivity ε of a medium as a function of the field's frequency ω :

$$\hat{\varepsilon}(\omega) \; = \; \varepsilon_\infty + \frac{\Delta\varepsilon}{1 + i\omega\tau},$$

where ε_∞ is the permittivity at the high frequency limit, $\Delta\varepsilon = \varepsilon_s - \varepsilon_\infty$ where ε_s is the static, low frequency permittivity, and τ is the characteristic relaxation time of the medium.

This relaxation model was introduced by and named after the physicist Peter Debye (1913).

Variants of the Debye Equation

- Cole–Cole equation
- Cole–Davidson equation
- Havriliak–Negami relaxation
- Kohlrausch–Williams–Watts function (Fourier transform of stretched exponential function).

Para-electricity

Para-electricity is the ability of many materials (specifically ceramic crystals) to become polarized under an applied electric field. Unlike ferroelectricity, this can happen even if there is no permanent electric dipole that exists in the material, and removal of the fields results in the polarization in the material returning to zero. The mechanisms that cause **para-electric** behaviour are the distortion of individual ions (displacement of the electron cloud from the nucleus) and polarization of molecules or combinations of ions or defects.

Para-electricity occurs in crystal phases where electric dipoles are unaligned (*i.e.*, unordered domains that are electrically charged) and thus have the potential to align in an external electric field and strengthen it. In comparison to the ferroelectric phase, the domains are unordered and the internal field is weak.

The $LiNbO_3$ crystal is ferroelectric below 1430 K, and above this temperature it transforms into a disordered paraelectric phase. Similarly, other perovskites also exhibit para-electricity at high temperatures.

Para-electricity has been explored as a possible refrigeration mechanism; polarizing a paraelectric by applying an electric field under adiabatic process conditions raises the temperature, while removing the field lowers the temperature. A heat pump that polarizes the paraelectric, allows it to return to ambient temperature, then brings it into contact with the object to be cooled, and depolarizes it, would result in refrigeration.

Tunability

Tunable dielectrics are insulators whose ability to store electrical charge changes when a voltage is applied.

Generally, strontium titanate ($SrTiO_3$) is used for devices operating at low temperatures, while barium strontium titanate ($Ba1-xSrxTiO_3$) substitutes for room temperature devices. Other potential materials include microwave dielectrics and carbon nanotube (CNT) composites.

In 2013 multi-sheet layers of strontium titanate interleaved with single layers of strontium oxide produced a dielectric capable of operating at up to 125GHz. The material was created via molecular beam epitaxy. The two have mismatched crystal spacing that produces strain within the strontium titanate layer that makes it less stable and tunable.

Systems such as $Ba1-xSrxTiO_3$ have a paraelectric–ferroelectric transition just below ambient temperature, providing high tunability. Such films suffer significant losses arising from defects.Here we report the experimental realization of a highly tunable ground state arising from the emergence of a local ferroelectric instability in biaxially strained $Srn+1TinO_3n+1$ phases with $n \geq 3$ at frequencies up to 125 GHz. In contrast to traditional methods of modifying ferroelectrics — doping or strain — in this unique system an increase in the separation between the (SrO)2 planes, which can be achieved by changing n, bolsters the local ferroelectric instability. This new control parameter, n, can be exploited to achieve a figure of merit at room temperature that rivals all known tunable microwave dielectrics.

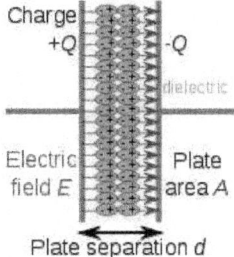

Fig. : Charge separation in a parallel-plate capacitor causes an internal electric field. A dielectric (orange) reduces the field and increases the capacitance.

Applications

Capacitors

Commercially manufactured capacitors typically use a solid dielectric material with high permittivity as the intervening medium between the stored positive and negative charges. This material is often referred to in technical contexts as the *capacitor dielectric*.

The most obvious advantage to using such a dielectric material is that it prevents the conducting plates the charges are stored on from coming into direct electrical contact. More significantly, however, a high permittivity allows a greater stored charge at a given voltage. This can be seen by treating the case of a linear dielectric with permittivity ε and thickness d between two conducting plates with uniform charge density σ_ε. In this case the charge density is given by

$$\sigma_\varepsilon = \varepsilon \frac{V}{d}$$

and the capacitance per unit area by

$$c = \frac{\sigma_\varepsilon}{V} = \frac{\varepsilon}{d}$$

From this, it can easily be seen that a larger ε leads to greater charge stored and thus greater capacitance.

Dielectric materials used for capacitors are also chosen such that they are resistant to ionization. This allows the capacitor to operate at higher voltages before the insulating dielectric ionizes and begins to allow undesirable current.

Dielectric Resonator

A *dielectric resonator oscillator* (DRO) is an electronic component that exhibits resonance for a narrow range of frequencies, generally in the microwave band. It consists of a "puck" of ceramic that has a large dielectric constant and a low dissipation factor. Such resonators are often used to provide a frequency reference in an oscillator circuit. An unshielded dielectric resonator can be used as a Dielectric Resonator Antenna (DRA).

Some Practical Dielectrics

Dielectric materials can be solids, liquids, or gases. In addition, a high vacuum can also be a useful, nearly lossless dielectric even though its relative dielectric constant is only unity.

Solid dielectrics are perhaps the most commonly used dielectrics in electrical engineering, and many solids are very good insulators. Some examples include porcelain, glass and most plastics. Air, nitrogen and sulfur hexafluoride are the three most commonly used gaseous dielectrics.

- Industrial coatings such as parylene provide a dielectric barrier between the substrate and its environment.
- Mineral oil is used extensively inside electrical transformers as a fluid dielectric and to assist in cooling. Dielectric fluids with higher dielectric constants, such as electrical grade castor oil, are often used in high voltage capacitors to help prevent corona discharge and increase capacitance.
- Because dielectrics resist the flow of electricity, the surface of a dielectric may retain *stranded* excess electrical charges. This may occur accidentally when the dielectric is rubbed (the triboelectric effect). This can be useful, as in a Van de

Graaff generator or electrophorus, or it can be potentially destructive as in the case of electrostatic discharge.

- Specially processed dielectrics, called electrets (which should not be confused with ferroelectrics), may retain excess internal charge or "frozen in" polarization. Electrets have a semi-permanent external electric field, and are the electrostatic equivalent to magnets. Electrets have numerous practical applications in the home and industry.

- Some dielectrics can generate a potential difference when subjected to mechanical stress, or change physical shape if an external voltage is applied across the material. This property is called piezoelectricity. Piezoelectric materials are another class of very useful dielectrics.

- Some ionic crystals and polymer dielectrics exhibit a spontaneous dipole moment, which can be reversed by an externally applied electric field. This behaviour is called the ferro-electric effect. These materials are analogous to the way ferro-magnetic materials behave within an externally applied magnetic field. Ferro-electric materials often have very high dielectric constants, making them quite useful for capacitors.

Chapter 12

ELECTRONIC CIRCUIT

An **electronic circuit** is composed of individual electronic components, such as resistors, transistors, capacitors, inductors and diodes, connected by conductive wires or traces through which electric current can flow. The combination of components and wires allows various simple and complex operations to be performed : signals can be amplified, computations can be performed, and data can be moved from one place to another. Circuits can be constructed of discrete components connected by individual pieces of wire, but today it is much more common to create interconnections by photolithographic techniques on a laminated substrate (a printed circuit board or PCB) and solder the components to these interconnections to create a finished circuit. In an integrated circuit or IC, the components and interconnections are formed on the same substrate, typically a semi-conductor such as silicon or (less commonly) gallium arsenide.

Breadboards, perfboards, and stripboards are common for testing new designs. They allow the designer to make quick changes to the circuit during development.

An electronic circuit can usually be categorized as an analog circuit, a digital circuit, or a mixed-signal circuit (a combination of analog circuits and digital circuits).

ANALOGUE ELECTRONICS

Analogue electronics (or **analog** in American English) are electronic systems with a continuously variable signal, in contrast to digital electronics where signals usually take only two different levels. The term "analogue" describes the proportional relationship between a signal and a voltage or current that represents the signal. The word analogue is derived from the Greek word ανάλογος (analogos) meaning "proportional".

Analogue Signals

An analogue signal uses some attribute of the medium to convey the signal's information. For example, an aneroid barometer uses the angular position of a needle as the signal to convey the information of changes in atmospheric pressure. Electrical signals may represent information by changing their voltage, current, frequency, or total charge. Information is converted from some other physical form (such as sound, light, temperature, pressure, position) to an electrical signal by a transducer which converts one type of energy into another (*e.g.* a microphone).

The signals take any value from a given range, and each unique signal value represents different information. Any change in the signal is meaningful, and each level of the signal represents a different level of the phenomenon that it represents. For example, suppose the signal is being used to represent temperature, with one volt representing one degree Celsius. In such a system 10 volts would represent 10 degrees, and 10.1 volts would represent 10.1 degrees.

Another method of conveying an analogue signal is to use modulation. In this, some base carrier signal has one of its properties altered : amplitude modulation (AM) involves altering the amplitude of a sinusoidal voltage waveform by the source information, frequency modulation (FM) changes the frequency. Other techniques, such as phase modulation or changing the phase of the carrier signal, are also used.

In an analogue sound recording, the variation in pressure of a sound striking a microphone creates a corresponding variation in the current passing through it or voltage across it. An increase in the volume of the sound causes the fluctuation of the current or voltage to increase proportionally while keeping the same waveform or shape.

Mechanical, pneumatic, hydraulic and other systems may also use analogue signals.

Inherent Noise

Analogue systems invariably include noise; that is, random disturbances or variations, some caused by the random thermal vibrations of atomic particles. Since all variations of an analogue signal are significant, any disturbance is equivalent to a change in the original signal and so appears as noise. As the signal is copied and re-copied, or transmitted over long distances, these random variations become more significant and lead to signal degradation. Other sources of noise may include external electrical signals or poorly designed components. These disturbances are reduced by shielding, and using low-noise amplifiers (LNA).

Analogue vs. Digital Electronics

Since the information is encoded differently in analogue and digital electronics, the way they process a signal is consequently different. All operations that can be performed on an analogue signal such as amplification, filtering, limiting and others, can also be duplicated in the digital domain. Every digital circuit is also an analogue circuit, in that the behaviour of any digital circuit can be explained using the rules of analogue circuits.

The first electronic devices invented and mass-produced were analogue. The use of microelectronics has made digital devices cheap and widely available.

Noise

Because of the way information is encoded in analogue circuits, they are much more susceptible to noise than digital circuits, since a small change in the signal can represent a significant change in the information present in the signal and can cause the information present to be lost. Since digital signals take on one of only two different values, a disturbance would have to be about one-half the magnitude of the digital signal to cause an error; this property of digital circuits can be exploited to make signal processing noise-resistant. In digital electronics, because the information is quantized, as long as the signal stays inside a range of values, it represents the same information. Digital circuits use this principle to regenerate the signal at each logic gate, lessening or removing noise.

Precision

A number of factors affect how precise a signal is, mainly the noise present in the original signal and the noise added by processing. Fundamental physical limits such as the shot noise in components limits the resolution of analogue signals. In digital electronics additional precision is obtained by using additional digits to represent the signal; the practical limit in the number of digits is determined by the performance of the analogue-to-digital converter (ADC), since digital operations can usually be performed without loss of precision. The ADC takes an analogue signal and changes into a series of binary numbers. The ADC may be used in simple digital display devices e. g. thermometers, light meters but it may also be used in digital sound recording and in data acquisition. However, a digital-to-analogue converter (DAC) is used to change a digital signal to an analogue signal. A DAC takes a series of binary numbers and converts it to an analogue signal. It is common to find a DAC in the gain-control system of an op-amp which in turn may be used to control digital amplifiers and filters.

Design Difficulty

Analogue circuits are harder to design, requiring more skill, than comparable digital systems. This is one of the main reasons why digital systems have become more common than analogue devices. An analogue circuit must be designed by hand, and the process is much less automated than for digital systems. However, if a digital electronic device is to interact with the real world, it will always need an analogue interface. For example, every digital radio receiver has an analogue preamplifier as the first stage in the receive chain.

DIGITAL ELECTRONICS

Three Digital Circuits

Digital electronics, or **digital (electronic) circuits**, represent signals by discrete bands of analog levels, rather than by a continuous range. All levels within a band represent the same signal state. Relatively small changes to the analog signal levels due to manufacturing tolerance, signal attenuation or parasitic noise do not leave the discrete envelope, and as a result are ignored by signal state sensing circuitry.

In most cases the number of these states is two, and they are represented by two voltage bands : one near a reference value (typically termed as "ground" or zero volts) and a value near the supply voltage, corresponding to the "false" ("0") and "true" ("1") values of the Boolean domain respectively.

Digital techniques are useful because it is easier to get an electronic device to switch into one of a number of known states than to accurately reproduce a continuous range of values.

Digital electronic circuits are usually made from large assemblies of logic gates, simple electronic representations of Boolean logic functions.

Advantages

An advantage of digital circuits when compared to analog circuits is that signals represented digitally can be transmitted without degradation due to noise. For example, a continuous audio signal transmitted as a sequence of 1s and 0s, can be reconstructed without error, provided the noise picked up in transmission is not enough to prevent identification of the 1s and 0s. An hour of music can be stored on a compact disc using about 6 billion binary digits.

In a digital system, a more precise representation of a signal can be obtained by using more binary digits to represent it. While this requires more digital circuits to process the signals, each digit is handled by the same kind of hardware. In an analog system, additional resolution requires fundamental improvements in the linearity and noise characteristics of each step of the signal chain.

Computer-controlled digital systems can be controlled by software, allowing new functions to be added without changing hardware. Often this can be done outside of the factory by updating the product's software. So, the product's design errors can be corrected after the product is in a customer's hands.

Information storage can be easier in digital systems than in analog ones. The noise-immunity of digital systems permits data to be stored and retrieved without degradation. In an analog system, noise from aging and wear degrade the information stored. In a digital system, as long as the total noise is below a certain level, the information can be recovered perfectly.

Disadvantages

In some cases, digital circuits use more energy than analog circuits to accomplish the same tasks, thus producing more heat which increases the complexity of the circuits such as the inclusion of heat sinks. In portable or battery-powered systems this can limit use of digital systems.

For example, battery-powered cellular telephones often use a low-power analog front-end to amplify and tune in the radio signals from the base station. However, a base station has grid power and can use power-hungry, but very flexible software radios. Such base stations can be easily reprogrammed to process the signals used in new cellular standards.

Digital circuits are sometimes more expensive, especially in small quantities.

Most useful digital systems must translate from continuous analog signals to discrete digital signals. This causes quantization errors. Quantization error can be reduced if the system stores enough digital data to represent the signal to the desired degree of fidelity. The Nyquist-Shannon sampling theorem provides an important guideline as to how much digital data is needed to accurately portray a given analog signal.

In some systems, if a single piece of digital data is lost or mis-interpreted, the meaning of large blocks of related data can completely change. Because of the cliff effect, it can be difficult for users to tell if a particular system is right on the edge of failure, or if it can tolerate much more noise before failing.

Digital fragility can be reduced by designing a digital system for robustness. For example, a parity bit or other error management method can be inserted into the signal path. These schemes help the system detect errors, and then either correct the errors, or at least ask for a new copy of the data. In a state-machine, the state transition logic can be designed to catch unused states and trigger a reset sequence or other error recovery routine.

Digital memory and transmission systems can use techniques such as error detection and correction to use additional data to correct any errors in transmission and storage.

On the other hand, some techniques used in digital systems make those systems more vulnerable to single-bit errors. These techniques are acceptable when the underlying bits are reliable enough that such errors are highly unlikely.

A single-bit error in audio data stored directly as linear pulse code modulation (such as on a CD-ROM) causes, at worst, a single click. Instead, many people use audio compression to save storage space and download time, even though a single-bit error may corrupt the entire song.

Design Issues in Digital Circuits

Digital circuits are made from analog components. The design must assure that the analog nature of the components doesn't dominate the desired digital behaviour.

Digital systems must manage noise and timing margins, parasitic inductances and capacitances, and filter power connections.

Bad designs have intermittent problems such as "glitches", vanishingly fast pulses that may trigger some logic but not others, "runt pulses" that do not reach valid "threshold" voltages, or unexpected ("undecoded") combinations of logic states.(Y. K. Chan and S. Y. Lim, Faculty of Engineering & Technology, Multimedia University, Malaysia.)

Additionally, where clocked digital systems interface to analog systems or systems that are driven from a different clock, the digital system can be subject to metastability where a change to the input violates the set-up time for a digital input latch. This situation will self-resolve, but will take a random time, and while it persists can result in invalid signals being propagated within the digital system for a short time.

Since digital circuits are made from analog components, digital circuits calculate more slowly than low-precision analog circuits that use a similar amount of space and power. However, the digital circuit will calculate more repeatably, because of its high noise immunity. On the other hand, in the high-precision domain (for example, where 14 or more bits of precision are needed), analog circuits require much more power and area than digital equivalents.

Construction

A digital circuit is often constructed from small electronic circuits called logic gates that can be used to create combinational logic. Each logic gate represents a function of boolean logic. A logic gate is an arrangement of electrically controlled switches, better known as transistors.

Each logic symbol is represented by a different shape. The actual set of shapes was introduced in 1984 under IEEE\ANSI standard 91-1984. "The logic symbol given under this standard are being increasingly used now and have even started appearing in the literature published by manufacturers of digital integrated circuits."

The output of a logic gate is an electrical flow or voltage, that can, in turn, control more logic gates.

Logic gates often use the fewest number of transistors in order to reduce their size, power consumption and cost, and increase their reliability.

Integrated circuits are the least expensive way to make logic gates in large volumes. Integrated circuits are usually designed by engineers using electronic design automation software.

Another form of digital circuit is constructed from lookup tables, (many sold as "programmable logic devices", though other kinds of PLDs exist). Lookup tables can perform the same functions as machines based on logic gates, but can be easily reprogrammed without changing the wiring. This means that a designer can often repair design errors without changing the arrangement of wires. Therefore, in small volume products, programmable logic devices are often the preferred solution. They are usually designed by engineers using electronic design automation software.

When the volumes are medium to large, and the logic can be slow, or involves complex algorithms or sequences, often a small micro-controller is programmed to make an embedded system. These are usually programmed by software engineers.

When only one digital circuit is needed, and its design is totally customized, as for a factory production line controller, the conventional solution is a programmable logic controller, or PLC. These are usually programmed by electricians, using ladder logic.

Structure of Digital Systems

Engineers use many methods to minimize logic functions, in order to reduce the circuit's complexity. When the complexity is less, the circuit also has fewer errors and less electronics, and is therefore less expensive.

The most widely used simplification is a minimization algorithm like the Espresso heuristic logic minimizer within a CAD system, although historically, binary decision diagrams, an automated Quine–McCluskey algorithm, truth tables, Karnaugh maps, and Boolean algebra have been used.

Representations are crucial to an engineer's design of digital circuits. Some analysis methods only work with particular representations.

The classical way to represent a digital circuit is with an equivalent set of logic gates. Another way, often with the least electronics, is to construct an equivalent system of electronic switches (usually transistors). One of the easiest ways is to simply have a memory containing a truth table. The inputs are fed into the address of the memory, and the data outputs of the memory become the outputs.

For automated analysis, these representations have digital file formats that can be processed by computer programs. Most digital engineers are very careful to select computer programs ("tools") with compatible file formats.

To choose representations, engineers consider types of digital systems. Most digital systems divide into "combinational systems" and "sequential systems." A combinational system always presents the same output when given the same inputs. It is basically a representation of a set of logic functions, as already discussed.

A sequential system is a combinational system with some of the outputs fed back as inputs. This makes the digital machine perform a "sequence" of operations. The simplest sequential system is probably a flip flop, a mechanism that represents a binary digit or "bit".

Sequential systems are often designed as state machines. In this way, engineers can design a system's gross behaviour, and even test it in a simulation, without considering all the details of the logic functions.

Sequential systems divide into two further sub-categories. "Synchronous" sequential systems change state all at once, when a "clock" signal changes state. "Asynchronous" sequential systems propagate changes whenever inputs change. Synchronous sequential systems are made of well-characterized asynchronous circuits such as flip-flops, that change only when the clock changes, and which have carefully designed timing margins.

The usual way to implement a synchronous sequential state machine is to divide it into a piece of combinational logic and a set of flip flops called a "state register." Each time a clock signal ticks, the state register captures the feedback generated from the previous state of the combinational logic, and feeds it back as an unchanging input to the combinational part of the state machine. The fastest rate of the clock is set by the most time-consuming logic calculation in the combinational logic.

The state register is just a representation of a binary number. If the states in the state machine are numbered (easy to arrange), the logic function is some combinational logic that produces the number of the next state.

In comparison, asynchronous systems are very hard to design because all possible states, in all possible timings must be considered. The usual method is to construct a table of the minimum and maximum time that each such state can exist, and then adjust the circuit to minimize the number of such states, and force the circuit to periodically wait for all of its parts to enter a compatible state (this is called "self-resynchronization"). Without such careful design, it is easy to accidentally produce asynchronous logic that is "unstable", that is, real electronics will have unpredictable results because of the cumulative delays caused by small variations in the values of the electronic components. Certain circuits (such as the synchronizer flip-flops, switch debouncers, arbiters, and the like which allow external unsynchronized signals to enter synchronous logic circuits) are inherently asynchronous in their design and must be analyzed as such.

As of 2010, almost all digital machines are synchronous designs because it is much easier to create and verify a synchronous design—the software currently used to simulate digital machines does not yet handle asynchronous designs. However, asynchronous logic is thought to be superior, if it can be made to work, because its speed is not constrained by an arbitrary clock; instead, it runs at the maximum speed of its logic gates. Building an asynchronous circuit using faster parts makes the circuit faster.

Many digital systems are data flow machines. These are usually designed using synchronous register transfer logic, using hardware description languages such as VHDL or Verilog.

In register transfer logic, binary numbers are stored in groups of flip flops called registers. The outputs of each register are a bundle of wires called a "bus" that carries that number to other calculations. A calculation is simply a piece of combinational logic. Each calculation also has an output bus, and these may be connected to the inputs of several registers. Sometimes a register will have a multiplexer on its input, so that it can store a number from any one of several buses. Alternatively, the outputs of several items may be connected to a bus through buffers that can turn off the output of all of the devices except one. A sequential state machine controls when each register accepts new data from its input.

In the 1980s, some researchers discovered that almost all synchronous register-transfer machines could be converted to asynchronous designs by using first-in-first-out synchronization logic. In this scheme, the digital machine is characterized as a set of data flows. In each step of the flow, an asynchronous "synchronization circuit" determines when the outputs of that step are valid, and presents a signal that says, "grab the data" to the stages that use that stage's inputs. It turns out that just a few relatively simple synchronization circuits are needed.

The most general-purpose register-transfer logic machine is a computer. This is basically an automatic binary abacus. The control unit of a computer is usually designed as a micro-program run by a micro-sequencer. A micro-program is much like a player-piano roll. Each table entry or "word" of the micro-program commands the state of every bit that controls the computer. The sequencer then counts, and the count addresses the memory or combinational logic machine that contains the micro-program. The bits from the micro-program control the arithmetic logic unit, memory and other parts of the computer, including the micro-sequencer itself.

In this way, the complex task of designing the controls of a computer is reduced to a simpler task of programming a collection of much simpler logic machines.

Computer architecture is a specialized engineering activity that tries to arrange the registers, calculation logic, buses and other parts of the computer in the best way for some purpose. Computer architects have applied large amounts of ingenuity to computer design to reduce the cost and increase the speed and immunity to programming errors of computers. An increasingly common goal is to reduce the power used in a battery-powered computer system, such as a cell-phone. Many computer architects serve an extended apprenticeship as microprogrammers.

"Specialized computers" are usually a conventional computer with a special-purpose micro-program.

Automated Design Tools

To save costly engineering effort, much of the effort of designing large logic machines has been automated. The computer programs are called "electronic design automation tools" or just "EDA."

Simple truth table-style descriptions of logic are often optimized with EDA that automatically produces reduced systems of logic gates or smaller lookup tables that still produce the desired outputs. The most common example of this kind of software is the Espresso heuristic logic minimizer.

Most practical algorithms for optimizing large logic systems use algebraic manipulations or binary decision diagrams, and there are promising experiments with genetic algorithms and annealing optimizations.

To automate costly engineering processes, some EDA can take state tables that describe state machines and automatically produce a truth table or a function table for the combinational logic of a state machine. The state table is a piece of text that lists each state, together with the conditions controlling the transitions between them and the belonging output signals.

It is common for the function tables of such computer-generated state-machines to be optimized with logic-minimization software such as Minilog.

Often, real logic systems are designed as a series of sub-projects, which are combined using a "tool flow." The tool flow is usually a "script," a simplified computer language that can invoke the software design tools in the right order.

Tool flows for large logic systems such as micro-processors can be thousands of commands long, and combine the work of hundreds of engineers.

Writing and debugging tool flows is an established engineering specialty in companies that produce digital designs. The tool flow usually terminates in a detailed computer file or set of files that describe how to physically construct the logic. Often it consists of instructions to draw the transistors and wires on an integrated circuit or a printed circuit board.

Parts of tool flows are "debugged" by verifying the outputs of simulated logic against expected inputs. The test tools take computer files with sets of inputs and outputs, and highlight discrepancies between the simulated behaviour and the expected behaviour.

Once the input data is believed correct, the design itself must still be verified for correctness. Some tool flows verify designs by first producing a design, and then scanning the design to produce compatible input data for the tool flow. If the scanned data matches the input data, then the tool flow has probably not introduced errors.

The functional verification data are usually called "test vectors." The functional test vectors may be preserved and used in the factory to test that newly constructed logic works correctly. However, functional test patterns don't discover common fabrication faults. Production tests are often designed by software tools called "test pattern generators". These generate test vectors by examining the structure of the logic and systematically generating tests for particular faults. This way the fault coverage can closely approach 100%, provided the design is properly made testable.

Once a design exists, and is verified and testable, it often needs to be processed to be manufacturable as well. Modern integrated circuits have features smaller than the wavelength of the light used to expose the photoresist. Manufacturability software adds interference patterns to the exposure masks to eliminate open-circuits, and enhance the masks' contrast.

Design for Testability

There are several reasons for testing a logic circuit. When the circuit is first developed, it is necessary to verify that the design circuit meets the required functional and timing specifications. When multiple copies of a correctly designed circuit are being manufactured, it is essential to test each copy to ensure that the manufacturing process has not introduced any flaws.

A large logic machine (say, with more than a hundred logical variables) can have an astronomical number of possible states. Obviously, in the factory, testing every state is impractical if testing each state takes a microsecond, and there are more states than the number of microseconds since the universe began. Unfortunately, this ridiculous-sounding case is typical.

Fortunately, large logic machines are almost always designed as assemblies of smaller logic machines. To save time, the smaller sub-machines are isolated by permanently installed "design for test" circuitry, and are tested independently.

One common test scheme known as "scan design" moves test bits serially (one after another) from external test equipment through one or more serial shift registers known as "scan chains". Serial scans have only one or two wires to carry the data, and minimize the physical size and expense of the infrequently used test logic.

After all the test data bits are in place, the design is reconfigured to be in "normal mode" and one or more clock pulses are applied, to test for faults (*e.g.* stuck-at low or stuck-at high) and capture the test result into flip-flops and/ or latches in the scan shift register(s). Finally, the result of the test is shifted out to the block boundary and compared against the predicted "good machine" result.

In a board-test environment, serial to parallel testing has been formalized with a standard called "JTAG" (named after the "Joint Test Action Group" that proposed it).

Another common testing scheme provides a test mode that forces some part of the logic machine to enter a "test cycle." The test cycle usually exercises large independent parts of the machine.

Trade-offs

Several numbers determine the practicality of a system of digital logic : cost, reliability, fanout and speed. Engineers explored numerous electronic devices to get an ideal combination of these traits.

Cost

The cost of a logic gate is crucial. In the 1930s, the earliest digital logic systems were constructed from telephone relays because these were inexpensive and relatively reliable. After that, engineers always used the cheapest available electronic switches that could still fulfill the requirements.

The earliest integrated circuits were a happy accident. They were constructed not to save money, but to save weight, and permit the Apollo Guidance Computer to control an inertial guidance system for a spacecraft. The first integrated circuit logic gates cost nearly $50 (in 1960 dollars, when an engineer earned $10,000/year). To everyone's surprise, by the time the circuits were mass-produced, they had become the least-expensive method of constructing digital logic. Improvements in this technology have driven all subsequent improvements in cost.

With the rise of integrated circuits, reducing the absolute number of chips used represented another way to save costs. The goal of a designer is not just to make the simplest circuit, but to keep the component count down. Sometimes this results in slightly more complicated designs with respect to the underlying digital logic but nevertheless reduces the number of components, board size, and even power consumption.

For example, in some logic families, NAND gates are the simplest digital gate to build. All other logical operations can be implemented by NAND gates. If a circuit already required a single NAND gate, and a single chip normally carried four NAND gates, then the remaining gates could be used to implement other logical operations like logical and. This could eliminate the need for a separate chip containing those different types of gates.

Reliability

The "reliability" of a logic gate describes its mean time between failure (MTBF). Digital machines often have millions of logic gates. Also, most digital machines are "optimized" to reduce their cost. The result is that often, the failure of a single logic gate will cause a digital machine to stop working.

Digital machines first became useful when the MTBF for a switch got above a few hundred hours. Even so, many of these machines had complex, well-rehearsed repair procedures, and would be non-functional for hours because a tube burned-out, or a moth got stuck in a relay. Modern transistorized integrated circuit logic gates have MTBFs greater than 82 billion hours (8.2×10^{10}) hours, and need them because they have so many logic gates.

Fanout

Fanout describes how many logic inputs can be controlled by a single logic output without exceeding the current ratings of the gate. The minimum practical fanout is about five. Modern electronic logic using CMOS transistors for switches have fanouts near fifty, and can sometimes go much higher.

Speed

The "switching speed" describes how many times per second an inverter (an electronic representation of a "logical not" function) can change from true to false and back. Faster logic can accomplish more operations in less time. Digital logic first became useful when switching speeds got above fifty hertz, because that was faster than a team of humans operating mechanical calculators. Modern electronic digital logic routinely switches at five gigahertz (5×10^9 hertz), and some laboratory systems switch at more than a terahertz (1×10^{12} hertz).

Logic Families

Design started with relays. Relay logic was relatively inexpensive and reliable, but slow. Occasionally a mechanical failure would occur. Fanouts were typically about ten, limited by the resistance of the coils and arcing on the contacts from high voltages.

Later, vacuum tubes were used. These were very fast, but generated heat, and were unreliable because the filaments would burn out. Fanouts were typically five to seven, limited by the heating from the tubes' current. In the 1950s, special "computer tubes" were developed with filaments that omitted volatile elements like silicon. These ran for hundreds of thousands of hours.

The first semi-conductor logic family was resistor-transistor logic. This was a thousand times more reliable than tubes, ran cooler, and used less power, but had a very low fan-in of three. Diode-transistor logic improved the fanout up to about seven, and reduced the power. Some DTL designs used two power-supplies with alternating layers of NPN and PNP transistors to increase the fanout.

Transistor transistor logic (TTL) was a great improvement over these. In early devices, fanout improved to ten, and later variations reliably achieved twenty. TTL was also fast, with some variations achieving switching times as low as twenty nanoseconds. TTL is still used in some designs.

Emitter coupled logic is very fast but uses a lot of power. It was extensively used for high-performance computers made up of many medium-scale components (such as the Illiac IV).

By far, the most common digital integrated circuits built today use CMOS logic, which is fast, offers high circuit density and low-power per gate. This is used even in large, fast computers, such as the IBM System z.

Recent Developments

In 2009, researchers discovered that memristors can implement a boolean state storage (similar to a flip flop, implication and logical inversion, providing a complete logic family with very small amounts of space and power, using familiar CMOS semi-conductor processes.

The discovery of superconductivity has enabled the development of rapid single flux quantum (RSFQ) circuit technology, which uses Josephson junctions instead of transistors. Most recently, attempts are being made to construct purely optical computing systems capable of processing digital information using non-linear optical elements.

CIRCUIT BREAKER

Origins

An early form of circuit breaker was described by Thomas Edison in an 1879 patent application, although his commercial power distribution system used fuses. Its purpose was to protect lighting circuit wiring from accidental short-circuits and overloads. A modern miniature circuit breaker similar to the ones now in use was patented by Brown, Boveri & Cie in 1924. Hugo Stotz, an engineer who had sold his company, to BBC, was credited as the inventor on DRP (*Deutsches Reichspatent*) 458392. Stotz's invention was the forerunner of the modern thermal-magnetic breaker commonly used in household load centers to this day.

Interconnection of multiple generator sources into an electrical grid required development of circuit breakers with increasing voltage ratings and increased ability to safely interrupt the increasing short circuit currents produced by networks. Simple air-break manual switches produced hazardous arcs when interrupting high currents; these gave way to oil-enclosed contacts, and various forms using directed flow of pressurized air, or of pressurized oil, to cool and interrupt the arc. By 1935, the specially constructed circuit breakers used at the Boulder Dam project use eight series breaks and pressurized oil flow to interrupt faults of up to 2,500 MVA, in three cycles of the AC power frequency.

Operation

All circuit breakers have common features in their operation, although details vary substantially depending on the voltage class, current rating and type of the circuit breaker.

The circuit breaker must detect a fault condition; in low voltage circuit breakers this is usually done within the breaker enclosure. Circuit breakers for large currents or high voltages are usually arranged with protective relay pilot devices to sense a fault condition and to operate the trip opening mechanism. The trip solenoid that releases the latch is usually energized by a separate battery, although some high-voltage circuit breakers are self-contained with current transformers, protective relays and an internal control power source.

Once a fault is detected, contacts within the circuit breaker must open to interrupt the circuit; some mechanically-stored energy (using something such as springs or compressed air) contained within the breaker is used to separate the contacts, although some of the energy required may be obtained from the fault current itself. Small circuit breakers may be manually operated, larger units have solenoids to trip the mechanism, and electric motors to restore energy to the springs.

The circuit breaker contacts must carry the load current without excessive heating, and must also withstand the heat of the arc produced when interrupting (opening) the circuit. Contacts are made of copper or copper alloys, silver alloys and other highly conductive materials. Service life of the contacts is limited by the erosion of contact material due to arcing while interrupting the current. Miniature and molded-case circuit breakers are usually discarded when the contacts have worn, but power circuit breakers and high-voltage circuit breakers have replaceable contacts.

When a current is interrupted, an arc is generated. This arc must be contained, cooled and extinguished in a controlled way, so that the gap between the contacts can again withstand the voltage in the circuit. Different circuit breakers use vacuum, air, insulating gas or oil as the medium the arc forms in. Different techniques are used to extinguish the arc including :

- Lengthening / deflection of the arc
- Intensive cooling (in jet chambers)
- Division into partial arcs
- Zero point quenching (Contacts open at the zero current time crossing of the AC waveform, effectively breaking no load current at the time of opening. The zero crossing occurs at twice the line frequency, *i.e.* 100 times per second for 50 Hz and 120 times per second for 60 Hz AC)
- Connecting capacitors in parallel with contacts in DC circuits.

Finally, once the fault condition has been cleared, the contacts must again be closed to restore power to the interrupted circuit.

Arc Interruption

Low-voltage MCB uses air alone to extinguish the arc. Larger ratings have metal plates or non-metallic arc chutes to divide and cool the arc. Magnetic blowout coils or permanent magnets deflect the arc into the arc chute.

In larger ratings, oil circuit breakers rely upon vapourization of some of the oil to blast a jet of oil through the arc.

Gas (usually sulfur hexafluoride) circuit breakers sometimes stretch the arc using a magnetic field, and then rely upon the dielectric strength of the sulfur hexafluoride (SF_6) to quench the stretched arc.

Vacuum circuit breakers have minimal arcing (as there is nothing to ionize other than the contact material), so the arc quenches when it is stretched a very small amount (< 2–3 mm). Vacuum circuit breakers are frequently used in modern medium-voltage switchgear to 38,000 volts.

Air circuit breakers may use compressed air to blow out the arc, or alternatively, the contacts are rapidly swung into a small sealed chamber, the escaping of the displaced air thus blowing out the arc.

Circuit breakers are usually able to terminate all current very quickly : typically the arc is extinguished between 30 ms and 150 ms after the mechanism has been tripped, depending upon age and construction of the device.

Short-circuit Current

Circuit breakers are rated both by the normal current that they are expected to carry, and the maximum short-circuit current that they can safely interrupt.

Under short-circuit conditions, a current many times greater than normal can exist. When electrical contacts open to interrupt a large current, there is a tendency for an arc to form between the opened contacts, which would allow the current to continue. This condition can create conductive ionized gases and molten or vapourized metal, which can cause further continuation of the arc, or creation of additional short circuits, potentially resulting in the explosion of the circuit breaker and the equipment that it is installed in. Therefore, circuit breakers must incorporate various features to divide and extinguish the arc.

In air-insulated and miniature breakers an *arc chute* structure consisting (often) of metal plates or ceramic ridges cools the arc, and magnetic blowout coils deflect the arc into the arc chute. Larger circuit breakers such as those used in electrical power distribution may use vacuum, an inert gas such as sulfur hexafluoride or have contacts immersed in oil to suppress the arc.

The maximum short-circuit current that a breaker can interrupt is determined by testing. Application of a breaker in a circuit with a prospective short-circuit current higher than the breaker's interrupting capacity rating may result in failure of the breaker to safely interrupt a fault. In a worst-case scenario the breaker may successfully interrupt the fault, only to explode when reset.

MCB used to protect control circuits or small appliances may not have sufficient interrupting capacity to use at a panel board; these circuit breakers are called "supplemental circuit protectors" to distinguish them from distribution-type circuit breakers.

Standard Current Ratings

Circuit breakers are manufactured in standard sizes, using a system of preferred numbers to cover a range of ratings. Miniature circuit breakers have a fixed trip setting; changing the operating current value requires changing the whole circuit breaker. Larger circuit breakers can have adjustable trip settings, allowing standardized elements to be appplied but with a setting intended to improve protection. For example, a circuit breaker with a 400 ampere "frame size" might have its overcurrent detection set to operate at only 300 amperes, to protect a feeder cable.

International Standard − IEC 60898-1 and European Standard EN 60898-1 define the *rated current* I_n of a circuit breaker for low voltage distribution applications as the maximum current that the breaker is designed to carry continuously (at an ambient air temperature of 30 °C). The commonly-available preferred values for the rated current are 6 A, 10 A, 13 A, 16 A, 20 A, 25 A, 32 A, 40 A, 50 A, 63 A, 80 A, 100 A, and 125 A (Renard series, slightly modified to include current limit of British BS 1363 sockets). The circuit breaker is labelled with the rated current in amperes, but without the unit symbol "A". Instead, the ampere figure is preceded by a letter "B", "C" or "D", which indicates the *instantaneous tripping current* — that is, the minimum value of current that causes the circuit breaker to trip without intentional time delay (*i.e.*, in less than 100 ms), expressed in terms of I_n :

Type	Instantaneous tripping current
B	above 3 I_n up to and including 5 I_n
C	above 5 I_n up to and including 10 I_n
D	above 10 I_n up to and including 20 I_n
K	above 8 I_n up to and including 12 I_n For the protection of loads that cause frequent short duration (approximately 400 ms to 2 s) current peaks in normal operation.
Z	above 2 I_n up to and including 3 I_n for periods in the order of tens of seconds. For the protection of loads such as semi-conductor devices or measuring circuits using current transformers.

Circuit breakers are also rated by the maximum fault current that they can interrupt; this allows use of more economical devices on systems unlikely to develop the high short-circuit current found on, for example, a large commercial building distribution system.

In the United States, Underwriters Laboratories (UL) certifies equipment ratings, called Series Ratings (or "integrated equipment ratings"), using a two-tier rating. For example, a 22/10 rating. This rating means that the meter pack has a 22 kAIC tenant breaker, feeding a 10 kAIC loadcenter with 10 kAIC branches, where kAIC stands for "Thousand Ampere Interrupting Capacity." Common meter pack ratings are 22/10, 42/10 and 100/10.

Types of Circuit Breakers

Fig. : Front panel of a 1250 A air circuit breaker manufactured by ABB. This low voltage power circuit breaker can be withdrawn from its housing for servicing. Trip characteristics are configurable via DIP switches on the front panel.

Many different classifications of circuit breakers can be made, based on their features such as voltage class, construction type, interrupting type, and structural features.

Low-voltage Circuit Breakers

Low-voltage (less than 1,000 V_{AC}) types are common in domestic, commercial and industrial application, and include :

- MCB (Miniature Circuit Breaker) — rated current not more than 100 A. Trip characteristics normally not adjustable. Thermal or thermal-magnetic operation.

    There are three main types of MCBs : 1. Type B - trips between 3 and 5 times full load current; 2. Type C - trips between 5 and 10 times full load current; 3. Type D - trips between 10 and 20 times full load current. In the UK all MCBs *must* be selected in accordance with BS 7671.

- MCCB (Molded Case Circuit Breaker) — rated current up to 2,500 A. Thermal or thermal-magnetic operation. Trip current may be adjustable in larger ratings.

- Low-voltage power circuit breakers can be mounted in multi-tiers in low-voltage switchboards or switchgear cabinets.

    The characteristics of low-voltage circuit breakers are given by international standards such as IEC 947. These circuit breakers are often installed in draw-out enclosures that allow removal and interchange without dismantling the switchgear.

    Large low-voltage molded case and power circuit breakers may have electric motor operators so they can open and close under remote control. These may form part of an automatic transfer switch system for standby power.

    Low-voltage circuit breakers are also made for direct-current (DC) applications, such as DC for subway lines. Direct current requires special breakers because the arc is continuous — unlike an AC arc, which tends to go out on each half cycle. A direct current circuit breaker has blow-out coils that generate a magnetic field that rapidly stretches the arc. Small circuit breakers are either installed directly in equipment, or are arranged in a breaker panel.

    The DIN rail-mounted thermal-magnetic miniature circuit breaker is the most common style in modern domestic consumer units and commercial electrical distribution boards throughout Europe. The design includes the following components :

1. Actuator lever - used to manually trip and reset the circuit breaker. Also indicates the status of the circuit breaker (On or Off/tripped). Most breakers are designed so they can still trip even if the lever is held or locked in the "on" position. This is sometimes referred to as "free trip" or "positive trip" operation.

2. Actuator mechanism - forces the contacts together or apart.

3. Contacts - Allow current when touching and break the current when moved apart.
4. Terminals
5. Bimetallic strip.
6. Calibration screw - allows the manufacturer to precisely adjust the trip current of the device after assembly.
7. Solenoid
8. Arc divider/extinguisher.

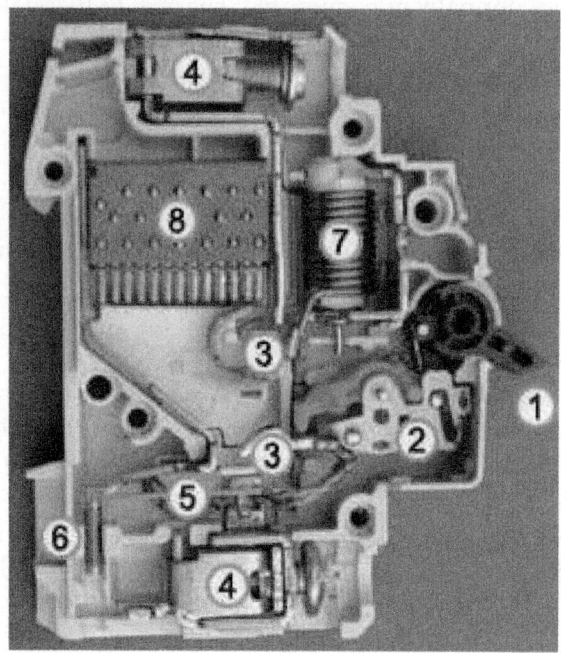

Fig. : Inside of a circuit breaker.

Magnetic Circuit Breakers

Magnetic circuit breakers use a solenoid (electromagnet) whose pulling force increases with the current. Certain designs utilize electromagnetic forces in addition to those of the solenoid. The circuit breaker contacts are held closed by a latch. As the current in the solenoid increases beyond the rating of the circuit breaker, the solenoid's pull releases the latch, which lets the contacts open by spring action. Some magnetic breakers incorporate a hydraulic time delay feature using a viscous fluid. A spring restrains the core until the current exceeds the breaker rating. During an overload, the speed of the solenoid motion is restricted by the fluid. The delay permits brief current surges beyond normal running current for motor starting, energizing equipment, etc. Short circuit currents provide sufficient solenoid force to release the latch regardless of core position thus bypassing the delay feature. Ambient temperature affects the time delay but does not affect the current rating of a magnetic breaker

Thermal Magnetic Circuit Breakers

Thermal magnetic circuit breakers, which are the type found in most distribution boards, incorporate both techniques with the electromagnet responding instantaneously to large surges in current (short circuits) and the bimetallic strip responding to less extreme but longer-term over-current conditions. The thermal portion of the circuit breaker provides an "inverse time" response feature, which trips the circuit breaker sooner for larger overcurrents.

Common Trip Breakers

When supplying a branch circuit with more than one live conductor, each live conductor must be protected by a breaker pole. To ensure that all live conductors are interrupted when any pole trips, a "common trip" breaker must be used. These may either contain two or three tripping mechanisms within one case, or for small breakers, may externally tie the poles together via their operating handles. Two-pole common trip breakers are common on 120/240-volt systems where 240 volt loads (including major appliances or further distribution boards) span the two live wires. Three-pole common trip breakers are typically used to supply three-phase electric power to large motors or further distribution boards.

Two- and four-pole breakers are used when there is a need to disconnect multiple phase AC, or to disconnect the neutral wire to ensure that no current flows through the neutral wire from other loads connected to the same network when workers may touch the wires during maintenance. Separate circuit breakers must never be used for live and neutral, because if the neutral is disconnected while the live conductor stays connected, a dangerous condition arises : the circuit appears de-energized (appliances don't work), but wires remain live and some RCDs may not trip if someone touches the live wire (because some RCDs need power to trip). This is why only common trip breakers must be used when neutral wire switching is needed.

Medium-voltage Circuit Breakers

Medium-voltage circuit breakers rated between 1 and 72 kV may be assembled into metal-enclosed switchgear line ups for indoor use, or may be individual components installed outdoors in a sub-station. Air-break circuit breakers replaced oil-filled units for indoor applications, but are now themselves being replaced by vacuum circuit breakers (up to about 40.5 kV). Like the high voltage circuit breakers described below, these are also operated by current sensing protective relays operated through current transformers. The characteristics of MV breakers are given by international standards such as IEC 62271. Medium-voltage circuit breakers nearly always use separate current sensors and protective relays, instead of relying on built-in thermal or magnetic overcurrent sensors.

Medium-voltage circuit breakers can be classified by the medium used to extinguish the arc :

- Vacuum circuit breakers—With rated current up to 6,300 A, and higher for generator circuit breakers. These breakers interrupt the current by creating and extinguishing the arc in a vacuum container - aka "bottle". Long life bellows are designed to travel the 6 to 10 mm the contacts must part. These are generally applied for voltages up to about 40,500 V, which corresponds roughly to the medium-voltage range of power systems. Vacuum circuit breakers tend to have longer life expectancies between overhaul than do air circuit breakers.

- Air circuit breakers—Rated current up to 6,300 A and higher for generator circuit breakers. Trip characteristics are often fully adjustable including configurable trip thresholds and delays. Usually electronically controlled, though some models are microprocessor controlled via an integral electronic trip unit. Often used for main power distribution in large industrial plant, where the breakers are arranged in draw-out enclosures for ease of maintenance.

- SF_6 circuit breakers extinguish the arc in a chamber filled with sulfur hexafluoride gas.

Medium-voltage circuit breakers may be connected into the circuit by bolted connections to bus bars or wires, especially in outdoor switchyards. Medium-voltage circuit breakers in switchgear line-ups are often built with draw-out construction, allowing breaker removal without disturbing power circuit connections, using a motor-operated or hand-cranked mechanism to separate the breaker from its enclosure. Some important manufacturer of VCB from 3.3 kV to 38 kV are Eaton, ABB, Siemens, HHI (Hyundai Heavy Industry), S&C Electric Company, Jyoti and BHEL.

High-voltage Circuit Breakers

Electrical power transmission networks are protected and controlled by high-voltage breakers. The definition of *high voltage* varies but in power transmission work is usually thought to be 72.5 kV or higher, according to a recent definition by the International Electrotechnical Commission (IEC). High-voltage breakers are nearly always solenoid-operated, with current sensing protective relays operated through current transformers. In substations the protective relay scheme can be complex, protecting equipment and buses from various types of overload or ground/earth fault.

High-voltage breakers are broadly classified by the medium used to extinguish the arc.

- Bulk oil
- Minimum oil
- Air blast
- Vacuum
- SF_6
- CO_2.

Some of the manufacturers are ABB, Alstom,General Electric, Hitachi, Hyundai Heavy Industry(HHI), Mitsubishi Electric, Pennsylvania Breaker, Siemens, Toshiba, Končar HVS, BHEL, CGL, and Becker/SMC (SMC Electrical Products).

Due to environmental and cost concerns over insulating oil spills, most new breakers use SF_6 gas to quench the arc.

Circuit breakers can be classified as *live tank,* where the enclosure that contains the breaking mechanism is at line potential, or *dead tank* with the enclosure at earth potential. High-voltage AC circuit breakers are routinely available with ratings up to 765 kV. 1,200 kV breakers were launched by Siemens in November 2011, followed by ABB in April the following year.

High-voltage circuit breakers used on transmission systems may be arranged to allow a single pole of a three-phase line to trip, instead of tripping all three poles; for some classes of faults this improves the system stability and availability.

A high-voltage direct current circuit breaker uses DC transmission lines rather than the AC transmission lines that dominate as of 2013. An HVDC circuit breaker can be used to connect DC transmission lines into a DC transmission grid (which is more efficient than an AC transmission grid), thereby making it possible to link renewable energy sources and even out local variations in wind and solar power.

Sulfur Hexafluoride (SF6) High-voltage Circuit Breakers

A sulfur hexafluoride circuit breaker uses contacts surrounded by sulfur hexafluoride gas to quench the arc. They are most often used for transmission-level voltages and may be incorporated into compact gas-insulated switchgear. In cold climates, supplemental heating or de-rating of the circuit breakers may be required due to liquefaction of the SF6 gas.

Disconnecting Circuit Breaker (DCB)

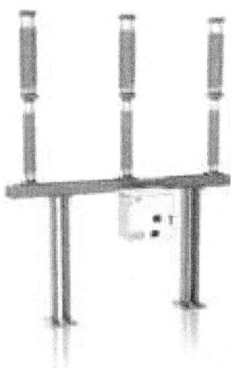

Fig. : 72.5 kV carbon dioxide high-voltage circuit breaker.

The disconnecting circuit breaker (DCB) was introduced in 2000 and is a high-voltage circuit breaker modelled after the SF_6-breaker. It presents a technical solution where the disconnecting function is integrated in the breaking chamber, eliminating the need for separate disconnectors. This increases the availability, since open-air disconnecting switch main contacts need maintenance every 2–6 years, while modern circuit breakers have maintenance intervals of 15 years. Implementing a DCB solution also reduces the space requirements within the sub-station, and increases the reliability, due to the lack of separate disconnectors.

In order to further reduce the required space of sub-station, as well as simplifying the design and engineering of the sub-station, a fiber optic current sensor (FOCS) can be integrated with the DCB. A 420 kV DCB with integrated FOCS can reduce a sub-station's footprint with over 50 % compared to a conventional solution of live tank breakers with disconnectors and current transformers, due to reduced material and no additional insulation medium.

Carbon Dioxide (CO_2) High-voltage Circuit Breakers

In 2012 ABB presented a 75 kV high-voltage breaker that uses carbon dioxide as the medium to extinguish the arc. The carbon dioxide breaker works on the same principles as an SF_6 breaker and can also be produced as a disconnecting circuit breaker. By switching from SF_6 to CO_2 it is possible to reduce the CO_2 emissions by 10 tons during the product's life cycle.

Other Breakers

The following types are described in separate articles :

*   Breakers for protections against earth faults too small to trip an over-current device :
    o   Residual-current device (RCD, formerly known as a *residual current circuit breaker*) — detects current imbalance, but does not provide over-current protection.
    o   Residual current breaker with over-current protection (RCBO) — combines the functions of an RCD and an MCB in one package. In the United States and Canada, panel-mounted devices that combine ground (earth) fault detection and over-current protection are called Ground Fault Interrupter (GFI) breakers; a wall mounted outlet device or separately enclosed plug-in device providing ground fault detection and interruption only (no overload protection) is called a Ground Fault Circuit Interrupter (GFCI).
    o   Earth leakage circuit breaker (ELCB) — This detects earth current directly rather than detecting imbalance. They are no longer seen in new installations for various reasons.
*   Recloser — A type of circuit breaker that closes automatically after a delay. These are used on overhead electric power distribution systems, to prevent short duration faults from causing sustained outages.

- Polyswitch (polyfuse) — A small device commonly described as an automatically resetting fuse rather than a circuit breaker.

OIL CIRCUIT BREAKER BULK AND MINIMUM OIL CIRCUIT BREAKER

Mineral oil has better insulating property than air. In **oil circuit breaker** the fixed contact and moving contact are immerged inside the insulating oil. Whenever there is a separation of electric current carrying contacts in the oil, the arc in circuit breaker is initialized at the moment of separation of contacts, and due to this arc the oil is vapourized and decomposed in mostly hydrogen gas and ultimately creates a hydrogen bubble around the arc. This highly compressed gas bubble around the arc prevents re-striking of the arc after current reaches zero crossing of the cycle. The **oil circuit breaker** is the one of the oldest type of circuit breakers.

Operation of Oil Circuit Breaker

The **operation of oil circuit breaker** is quite simple let's have a discussion. When the current carrying contacts in the oil are separated an arc is established in between the separated contacts. Actually, when separation of contacts has just started, distance between the current contacts is small as a result the voltage gradient between contacts becomes high. This high voltage gradient between the contacts ionized the oil and consequently initiates arcing between the contacts. This arc will produce a large amount of heat in surrounding oil and vapourizes the oil and decomposes the oil in mostly hydrogen and a small amount of methane, ethylene and acetylene. The hydrogen gas can not remain in molecular form and its is broken into its atomic form releasing lot of heat. The arc temperature may reach up to 5000_0K. Due to this high temperature the gas is liberated surround the arc very rapidly and forms an excessively fast growing gas bubble around the arc. It is found that the mixture of gases occupies a volume about one thousand times that of the oil decomposed. We can assume how fast the gas bubble around the arc will grow in size. If this growing gas bubble around the arc is compressed by any means then rate of de – ionization process of ionized gaseous media in between the contacts will accelerate which rapidly increase the dielectric strength between the contacts and consequently the arc will be quenched at zero crossing of the current cycle. This is the basic **operation of oil circuit breaker**. In addition to that cooling effect of hydrogen gas surround the arc path also helps, the quick arc quenching in oil circuit breaker.

Types of Oil Circuit Breakers

There are mainly two **types of oil circuit breakers** available

Bulk Oil Circuit Breaker or BOCB

Bulk oil circuit breaker or **BOCB** is such **types of circuit breakers** where oil is used as arc quenching media as well as insulating media between current carrying contacts and earthed parts of the breaker. The oil used here is same as transformer insulating oil.

Minimum Oil Circuit Breaker or MOCB

These types of circuit breakers utilize oil as the interrupting media. However, unlike **bulk oil circuit breaker**, a **minimum oil circuit breaker** places the interrupting unit in insulating chamber at live potential. The insulating oil is available only in interrupting chamber. The features of designing **MOCB** is to reduce requirement of oil, and hence these breaker are called **minimum oil circuit breaker**.

Bulk Oil Circuit Breaker

Construction of Bulk Oil Circuit Breaker

The basic construction of bulk oil circuit breaker is quite simple. Here all moving contacts and fixed contacts are immerged in oil inside closed iron vessel or iron tank. Whenever the current carrying contacts are being open within the oil the arc is produced in between the separated contacts. The large energy will be dissipated from the arc in oil which vapourizes the oil as well as decomposes it. Because of that a large gaseous pressure is developed inside the oil which tries to displace the liquid oil from surrounding of the contacts. The inner wall of the oil tank has to withstand this large pressure of the displaced oil. Thus the oil tank of bulk oil circuit breaker has to be sufficiently strong in construction. An air cushion is necessary between the oil surface and tank roof to accommodate the displaced oil when gas forms around the arc. That is why the oil tank is not totally filled up with oil it is filled up to certain level above which the air is tight in the tank. The breaker tank top cover should be securely bolted on the tank body and total breaker must be properly locked with foundation otherwise it may jump out during interruption of high fault current. In these type of equipment where expansible oil is enclosed in an air tight vessel (oil tank) there must be a gas vent fitted on the tank cover. Naturally some form of gas vent always is provided on the cover of bulk oil circuit breaker tank. This is very basic features for construction of bulk oil circuit breaker.

Arc Quenching in Bulk Oil Circuit Breaker

When the current carrying contacts in the oil are separat ed an arc is established in between the separated contacts. This arc will produce rapidly growing gas bubble around the arc. As the moving contact move away from fixed contact the length of arc is increased as a result the resistance of the arc increases. The increased resistance causes lowering the temperature and hence reducing the formation of gasses surround the arc. The arc quenching in bulk oil circuit breaker takes place when electric current passes through zero crossing. If we go through the arc quenching phenomenon more thoroughly we will find many other factors effects the arc quenching in bulk oil circuit breaker. As the gas bubble is enclosed by the oil inside the totally air tight vessel, the oil surround it will apply high pressure on the bubble, which results highly compressed gas around the arc. As the pressure is increased the de – ionization of gas increases which helps the arc quenching. The cooling effect of hydrogen gas also helps in arc quenching in oil circuit breaker.

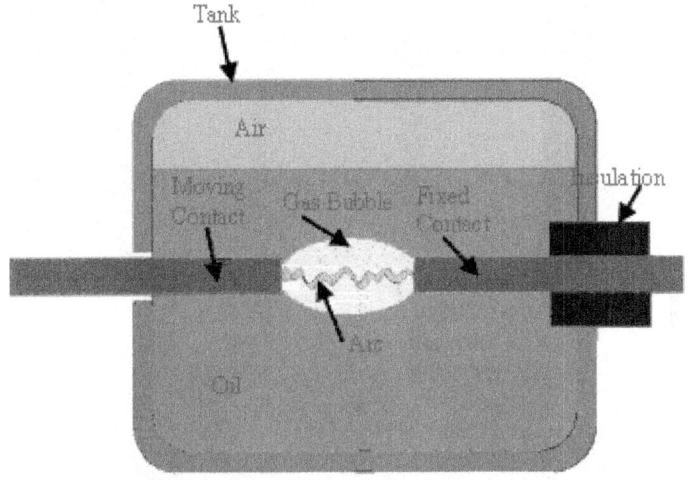

Fig. : Conceptual view of Bulk Oil Circuit Breaker.

Single Break Bulk Oil Circuit Breaker

In single break bulk oil circuit breaker there is one pair of current carrying contacts for each phase of power circuit. The each pair of current carrying contacts in this bulk oil circuit breaker consists of one fixed contact and one moving contact. Fixed contact is stationary contact and moving contact moves away from fixed contact during opening of the circuit breaker. As the moving contact is being moved away from fixed contact the arc is produced in between the contacts and it is extinguished during zero crossing of the fault current. As the days go on further research works have been done to improve better arc control in single break bulk oil circuit breaker. The main aim of development of bulk oil circuit breaker is to increase the pressure developed by the vapourization and dissociation of oil. Since in large gas pressure, the mean free paths of electrons and ions are reduced which results in effective de-ionization. So if the pressure can be increased, the rate of de-ionization is increased which helps to quick arc extinction. It has been found that if the opening of fixed and moving contacts is done inside a semi-closed insulated chamber then the gas bubble created around the arc will get less space of expansion, hence it becomes highly compressed. These semi-closed insulated arcing chamber in bulk oil circuit breaker is known as side vented explosion pot or cross jet pot. The principle of operation of cross jet pot is quite simple let's have a discussion. The pressure developed by the vapourization and dissociation of the oil is retained in the side vented explosive pot by withdrawing the moving contact through a stack of insulating plates having a minimum radial clearance around the contact. Thus there is practically no release of pressure until the moving contact uncovers one of the side vents. The compressed hydrogen gas can then escape across the arc path, thus exerting a powerful cooling action on the ionized column.

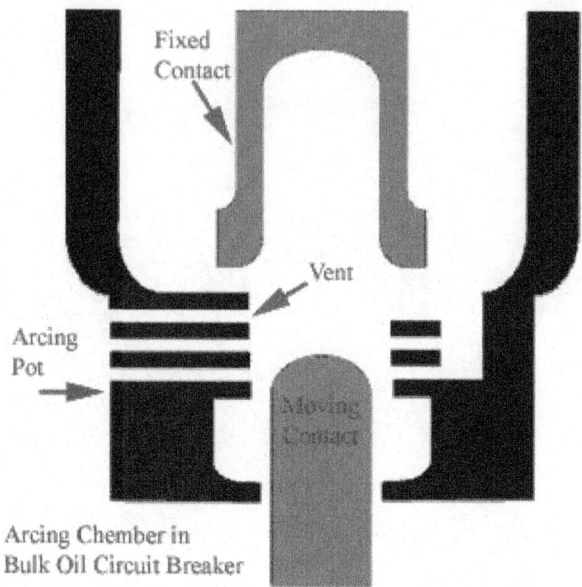

Arcing Chember in
Bulk Oil Circuit Breaker

When the current zero is reached, the post arc resistance increased rapidly due this cooling action. At higher breaking currents larger will be the pressure generated and a bulk oil circuit breaker gives its best performance at the highest current within its rating. These single break bulk oil circuit breaker may have problem during clearing low currents such as load current of the breaker.

Various improvement in the design of pressure chamber or side vented explosive chamber have been suggested to overcome the problem of low current interruption. One solution of this is providing a supplementary oil chamber below the side vents. This supplementary oil chamber is known as compensating chamber which provides fresh source of oil to be vapourised in order to feed more clean gas back across the arc path during clearing low current.

Double Break Bulk Oil Circuit Breaker

Various improvements in the design of bulk oil circuit breaker have been suggested to satisfactory and safe arc interruption especially at currents below the rated maximum. One solution to this problem is to use an intermediate contact between tow current carrying contacts. The arc is here split into two parts in series. The aim here is to extinguish the second arc quickly by using the gas pressure and oil momentum due to the first arc. In double break bulk oil circuit breaker, there are two fixed contact and are bridged by one moving contact. The moving contact is fitted with driving mechanism of the oil circuit breaker by means of an insulated rod. As the moving contact bridge moves downwards the contact gaps are created with fixed contacts at both end of the intermediate moving contact bridge. Hence arcs are produced at both contacts gap.

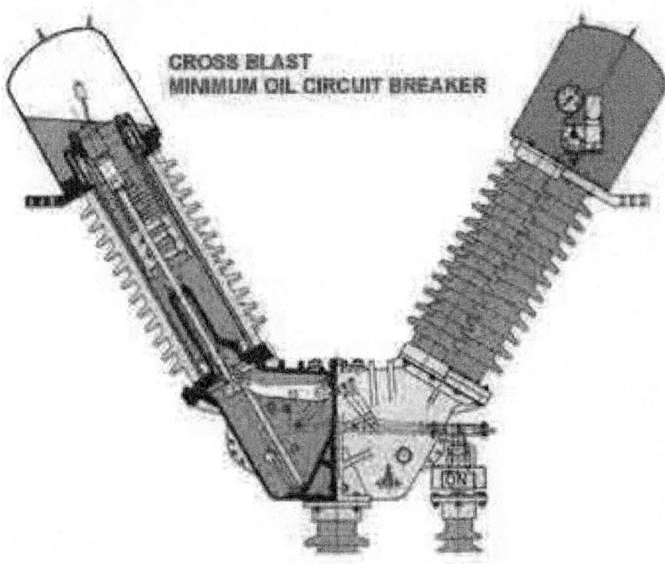

Minimum Oil Circuit Breaker

As the volume of the oil in bulk oil circuit breaker is huge, the chances of fire hazard in bulk oil system are more. For avoiding unwanted fire hazard in the system, one important development in the design of oil circuit breaker has been introduced where use of oil in the circuit breaker is much less than that of bulk oil circuit breaker. It has been decided that the oil in the circuit breaker should be used only as arc quenching media not as an insulating media. Then the concept of **minimum oil circuit breaker** comes. In this type of circuit breaker the arc interrupting device is enclosed in a tank of insulating material which as a whole is at live potential of system. This chamber is called arcing chamber or interrupting pot. The gas pressure developed in the arcing chamber depends upon the current to be interrupted. Higher the current to be interrupted causes larger the gas pressure developed in side the chamber, hence better the arc quenching. But this put a limit on the design of the arc chamber for mechanical stresses. With use of better insulating materials for the arcing chambers such as glass fiber, reinforced synthetic resin etc, the **minimum oil circuit breaker** are able to meet easily the increased fault levels of the system.

Working Principle or Arc Quenching in Minimum Oil Circuit Breaker

Working Principle of minimum oil circuit breaker or arc quenching in minimum oil circuit breaker is described below. In a **minimum oil circuit breaker**, the arc drawn across the current carrying contacts is contained inside the arcing chamber.

Hence the hydrogen bubble formed by the vapourized oil is trapped inside the chamber. As the contacts continue to move, after its certain travel an exit vent becomes available for exhausting the trapped hydrogen gas. There are two different types of arcing chamber is available in terms of venting are provided in the arcing chambers. One is axial venting and other is radial venting. In axial venting, gases (mostly Hydrogen), produced due to vapourization of oil and decomposition of oil during arc, will sweep the arc in axial or longitudinal direction.

Let's have a look on **working principle Minimum Oil Circuit Breaker** with axial venting arc chamber

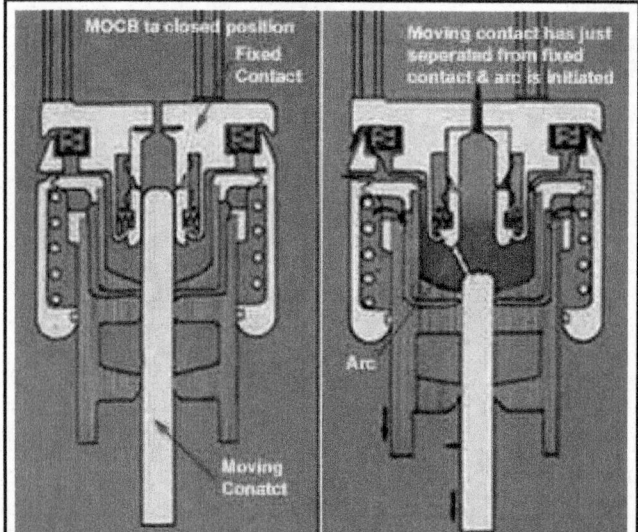	The moving contact has just been separated and arc is initiated in MOCB.
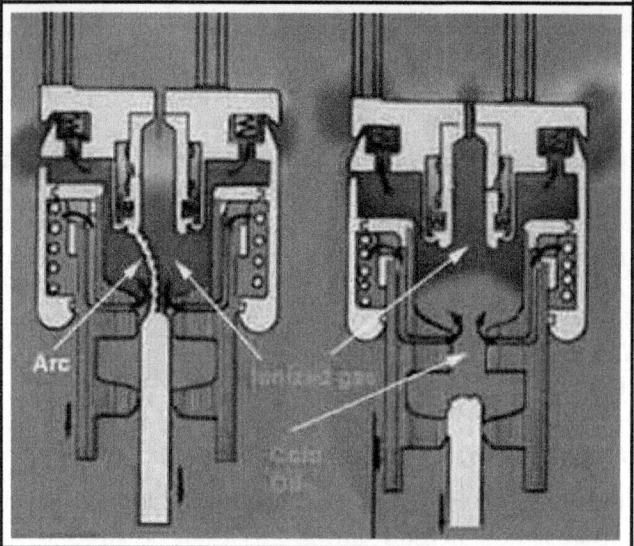	The ionized gas around the arc sweep away through upper vent and cold oil enters into the arcing chamber through the lower vent in axial direction as soon as the moving contact tip crosses the lower vent opening and final arc quenching in minimum oil circuit breaker occurs

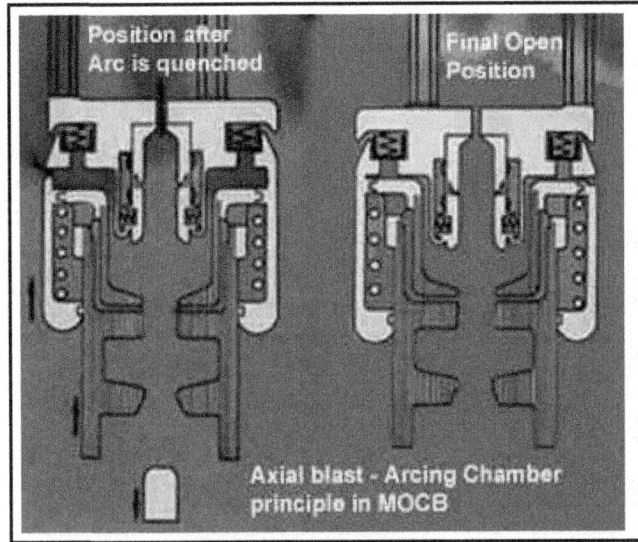

The cold oil occupies the gap between fixed contact and moving contact and the minimum oil circuit breaker finally comes into open position.

Whereas in case of radial venting or cross blast, the gases (mostly Hydrogen) sweep the arc in radial or transverse direction.

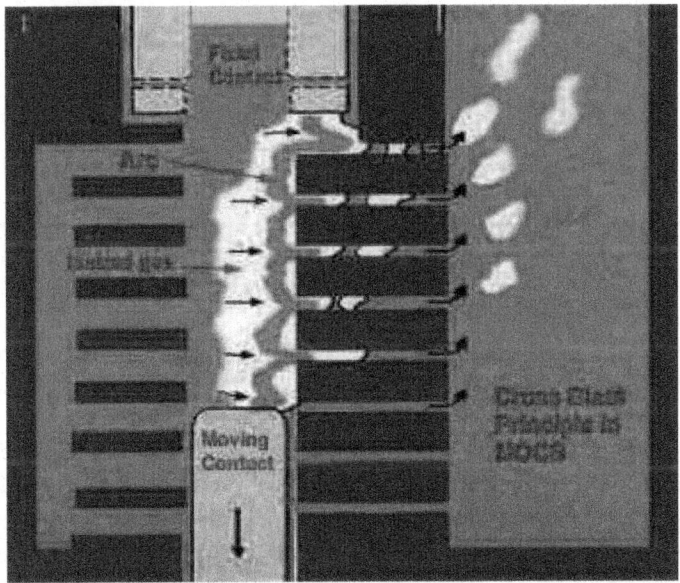

The axial venting generates high gas pressure and hence has high dielectric strength, so it is mainly used for interrupting low current at high voltage. On the other hand radial venting produces relatively low gas pressure and hence low dielectric strength so it can be used for low voltage and high current interruption. Many times the combination of both is used in minimum oil circuit breaker so that the chamber is equally efficient to interrupt low current as well as high current. These types of circuit breaker are available up to 8000 MVA at 245 KV.

Chapter 13

FIBER OPTIC SENSOR FOR ACOUSTIC DETECTION OF PARTIAL DISCHARGES IN OIL-PAPER INSULATED ELECTRICAL SYSTEMS

Julio Posada-Roman *, Jose A. Garcia-Souto and Jesus Rubio-Serrano

Department of Electronics Technology, Optoelectronics and Laser Technology Group, Universidad Carlos III de Madrid, Av. Universidad 30, E-28911 Leganés, Madrid, Spain; E-Mails: jsouto@ing.uc3m.es (J.A.G.-S.); jrserran@ing.uc3m.es (J.R.-S.)

* Author to whom correspondence should be addressed; E-Mail: jposada@ing.uc3m.es; Tel.: +34-91-624-5988; Fax: +34-91-624-9430.

ABSTRACT

A fiber optic interferometric sensor with an intrinsic transducer along a length of the fiber is presented for ultrasound measurements of the acoustic emission from partial discharges inside oil-filled power apparatus. The sensor is designed for high sensitivity measurements in a harsh electromagnetic field environment, with wide temperature changes and immersion in oil. It allows enough sensitivity for the application, for which the acoustic pressure is in the range of units of Pa at a frequency of 150 kHz. In addition, the accessibility to the sensing region is guaranteed by immune fiber-optic cables and the optical phase sensor output. The sensor design is a compact and rugged coil of fiber. In addition to a complete calibration, the *in-situ* results show that two types of partial discharges are measured through their acoustic emissions with the sensor immersed in oil.

Keywords

Optical fiber sensors; interferometry; acoustic emission; ultrasounds; partial discharges; transformers

1. INTRODUCTION

Oil-paper insulation systems are commonly used in power apparatus such as power transformers and high voltage cables, among others. In power transformers the insulation system is the most important part of the internal insulation and it largely determines its operational reliability. Thus, condition assessment of the insulation system in a timely manner, ensures reliable operation of the transformer, and maximizes equipment utilization [1].

Partial discharges (PD) are a cause and a symptom of the degradation of the insulation system and its activity monitoring is used as a tool for insulation condition assessment in power transformers [2]. PD are small electrical sparks present in an insulator as result of the electrical breakdown of a gas (for example air) contained within a void or in a highly non-uniform electric field [3]. The sudden release of energy caused when a PD occurs produces a number of effects like chemical and structural changes in the materials, electromagnetic signal generation and acoustic emissions (AE) [4]. These induced effects are used for its detection. Techniques such as dissolved gas analysis (DGA) electrical measurements of high frequency transients (HF-VHF), detection of electromagnetic signals generated in the UHF band and ultrasound AE detection are used for this propose. Among these techniques AE ultrasound detection offers great advantages such as the possibility of on-line testing and the ability to locate where PD activity is occurring, which is helpful in large test objects like power transformers.

The UHF technique has been applied to transformers for the detection and location of PD and show promising results [5,6]. However, the installation of the sensors is a drawback because they require an electromagnetic wave "view" into the tank, so the installation of a dielectric window that provides a mounting point for the sensors on the transformer tank is needed. Moreover, electromagnetic waves from PD are affected by reflections and refractions produced by obstacles such as the core, copper conductors, *etc.* producing multiple paths before arriving to the sensors, which complicates the localization of the source.

The detection and location of PD using AE techniques are commonly done with external piezoelectric (PZT) ultrasound sensors mounted on the tank wall, which have narrowband detection at 150 kHz. However, they suffer the same problem as UHF detection related to a multi-path signal and, in addition, to a low level detected signal as a consequence of the attenuation of the acoustic waves through the oil. Therefore, it is desirable to have sensors that can be placed inside the transformer, close to the PD, that overcome these problems.

Recently several fiber optic (FO) sensors of different types have been developed for PD detection within power transformers, such as those based on FO Fabry-Perot cavities with resonant response at around 150-kHz [7–9], or the ones based on Fiber Bragg Gratings [10]. However, the sensitivity of these sensors is moderate and they show a great dependence on the technology of integration.

Another PD sensing approach for internal installation is based on the interferometric measurement of AE with a single-mode optical fiber as the intrinsic transducer of ultrasounds into optical phase [11,12]. More sensitivity can be achieved with this technique by using long fibers in the sensing arm but the frequency response would be a drawback.

This work is focused on the development, characterization and testing of a fiber optic intrinsic sensor for acoustic detection of PD. It is organized as follows: Section 2 is devoted to the principle of sensing of the FO sensor and the description of the read-out system. The calibration of the sensor is presented in Section 3. Several tests under real conditions are presented in Section 4. These were carried out in a high voltage laboratory where the sensor was tested with PD generated in transformer oil. Finally, the conclusions are included in Section 5.

2. PRINCIPLE OF SENSING AND INTERFEROMETRIC READ-OUT SYSTEM

2.1. Sensing Principle

Acoustic emissions are pressure variations in an elastic medium. The principle of sensing of an intrinsic fiber optic acoustic sensor is based on the change in the optical path length produced by the strain induced by the acoustic pressure waves. For an interferometric approach, where the optical phase of the interfering light contains the information of the measured magnitude, the phase of the light (Φ) passing through a piece of optical fiber of longitude L is given by:

$$\phi = \beta L = \frac{2\pi n_{eff}}{\lambda} L \tag{1}$$

where β is the propagation constant, n_{eff} is the effective refractive index of the fiber and λ is the optical wavelength. The change in the phase is then:

$$\Delta\phi = \beta\,\Delta L + L\Delta\beta$$
$$= \Delta\phi_1 + \Delta\phi_2 \tag{2}$$

The first term in the last expression represents the phase shift due to the axial stretching of the fiber:

$$\Delta\phi_1 = -\frac{\beta L}{E}(1-2v)\Delta P \tag{3}$$

where v is the Poisson ratio, E is the Young modulus and ΔP is the acoustic pressure change.

The second term in Equation (2) is the change of the propagation constant which depends on the change of the refractive index (strain-optic effect) and the fiber diameter produced by the strain. However, the effect of the change on diameter is proved to be negligible [13] so $\Delta\Phi_2$ can be written as:

$$\Delta\phi_2 = \frac{\beta L n^2}{2E}(1-2v)(p_{11}+2p_{12})\Delta P \tag{4}$$

where p_{11} and p_{12} are elements of the strain-optic tensor. Substituting Equations (3) and (4) in Equation (2) it is obtained:

$$\frac{\Delta\phi}{\phi\Delta P} = \frac{(1-2\upsilon)}{E}\left[\frac{n^2}{2}(p_{11}+2p_{12})-1\right] \tag{4}$$

the expression $\Delta\Phi/(\Phi\Delta P)$ is known as the normalized acoustic phase responsivity (NR) and it is expressed in Pa^{-1}.

2.2. Design of the Sensor Probe

Some specific characteristics of the external ultrasound sensors that are commonly used for PD detection have been taken for the development of the proposed fiber optic sensor: sensitivity of 10 mV/Pa at the frequency of 150 kHz and bandwidth between 100 kHz and 300 kHz [14]. Moreover, PD acoustic emissions are expected in a typical range of 1 Pa to 10 kPa. This makes the interferometric measurement suitable for the application due to the high range-resolution that it is able to achieve. In this case, the most important parameter of design is the resolution (1 Pa). The proposed sensitivity is like that of an R15i sensor (10 mV/Pa) and corresponds to -180 dB re rad μPa^{-1} (1×10^{-9} rad/μPa). Under these conditions and assuming a quasi-linear output range of 20 V in the interferometer (stabilized for homodyne detection), the input dynamic range in the system is 2 kPa. These requirements should be appropriate and they are the initial requirements of design.

The sensor design is based on a fiber optic coil in a multilayer configuration which will be exposed to AE. Since the phase sensitivity is inversely proportional to λ, as can be seen in Equation (5), a short optical wavelength (633 nm) single-mode fiber is selected for the construction of the sensor in order to obtain higher phase sensitivity. The fiber used to build the sensor is the model: SCSM-633-HP1, which is a coated fiber with an operating wavelength range of 600-760 nm. A simple calculation of the fiber length that is needed to obtain the desired sensitivity is done using the NR for a typical coated fiber, which is ≈ -330 dB re μPa^{-1} [15], and the optical wavelength of 633 nm aforementioned for the interrogation of the sensor. With these parameters the total optical phase (Equation (1)) in 1 m of fiber corresponds to 143 dB re rad, therefore the sensitivity of the sensor will be -187 dB re rad μPa^{-1} m^{-1}. It means that 2 m of fiber is needed to obtain the desired sensitivity.

Since the value of NR that is used for the calculation of the sensor fiber length is an approximated value, an experimental measurement was carried out in order to obtain the real value of the responsivity for this fiber. The set up for the test is shown in Figure 1.

In this experiment a FO segment of 5-cm is immersed in water and a calibrated hydrophone (B&K 8103) is placed at the same distance to the source. This is observed in Figure 1(b): both signals start at the same time. Both are exposed to the same AE of 150-kHz emitted with a transducer placed at 5-cm. Figure 1(a) — the reference hydrophone and the FO segment are separated for clarity. The segment of fiber is interrogated by a Mach-Zhender interferometer with an output range of 12-V. The reference hydrophone has a sensitivity of 200 µV/Pa@150 kHz.

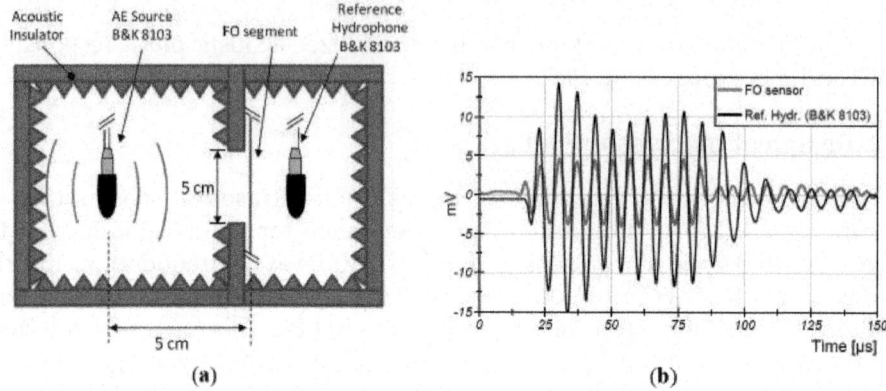

(a)   (b)

Figure 1. (a) Experimental set-up for the measurement of fiber sensitivity; (b) The same AE detected with the FO segment and the calibrated hydrophone.

The results of the measured sensitivity are shown in Figure 1b. The AE of 100-Pa is observed in the fiber segment as 1.45×10^{-3}-rad. This result shows a responsivity of -190 dB re rad-μPa^{-1}·m^{-1}. Therefore, the length of fiber needed for the sensor is 3.2-m.

Before the construction of the sensor, a design problem related with the acoustic frequency of interest (150 kHz) and the dimensions of the sensor has to be analyzed. When an acoustic wave of 150 kHz is propagating in the transformer oil (velocity of 1,400 m/s) the acoustic wavelength is ~10 mm. The minimum diameter achievable in the coil of conventional fiber is around 30-mm in order to avoid excessive optical losses. Therefore, since the dimensions of the sensor are greater than the desired wavelength, a phase difference of the acoustic field is present across the coil and this causes a reduction of the average pressure onto the sensor. This effect reduces the sensitive zone of the sensor that brings a net contribution free of the averaging effect and it is about 1/4 of the total fiber length of the coil (Figure 2). As a consequence, a fiber length of at least 13 m is needed for the building of the sensor.

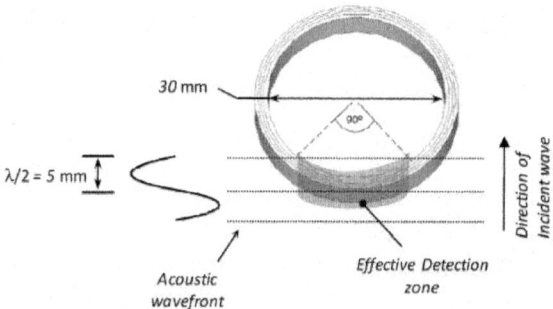

Figure 2. Detail of the effective sensitive zone of the sensor at the desired frequency.

The sensor was made by winding multiple layers of fiber around a former of 30 mm in diameter. The fiber in the coil was 17-m long. This amount of fiber is arranged in five layers of 5 mm of width.

2.3. Interferometric Read-Out System and Electronic Conditioning

In order to prove the sensor, a scheme based on an all-fiber Mach-Zehnder interferometer with homodyne demodulation was used. The block diagram of the implemented optoelectronic instrumentation system is shown in Figure 3.

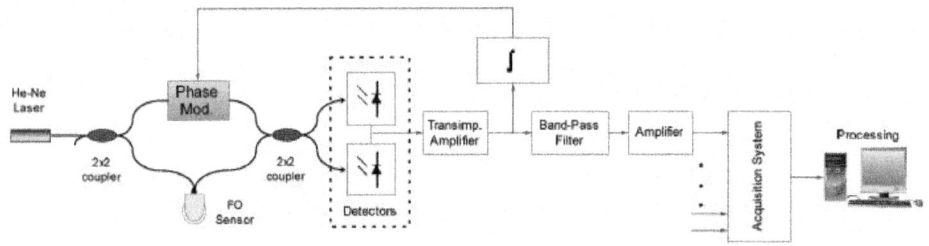

Figure 3. Diagram of the optoelectronic instrumentation system.

In this scheme a He-Ne (633 nm) laser coherent light source is used to interrogate the sensor. The detection of the optical phase encoded in the intensity output is done through two balanced photodetectors. Differential transimpedance amplification is used to compensate the effect of the mean optical power onto each photodetector in order to improve the performance with a DC null configuration. A homodyne demodulation is used to set the interferometer operating point at the maximum sensitivity and in the middle of the quasi-linear output range ($-\pi/4$ rad to $\pi/4$ rad). It is done by a feedback loop that integrates the error signal and actuates through a phase modulator which is connected to the fiber reference arm to compensate the temperature drift and other low frequency disturbances on the interferometer. The phase modulator implemented here is able to compensate up to 50π rad at frequencies below 200 Hz. Moreover, a control of the optic polarization has been included in order to avoid the interference signal fading.

Once the optical phase is converted into a voltage signal, it is conditioned through a band-pass filtering stage with a resonance frequency of 150 kHz in order to adjust the bandwidth of the FO sensor to the acoustic PD emissions. It equalizes the frequency response obtained with the sensing probe, as well as it attenuates the disturbances that are produced by other acoustic sources in the transformers (such as the Barkhausen noise [14]).

The stabilized interferometer has a voltage output proportional to the optical phase change given by:

$$Vs = 2I_0 \Delta\phi R\eta G_T G_F \tag{6}$$

where I_0 is the mean optical power at each photo-detector, R is the responsivity of the photo-detectors, η is a factor between 0 and 1 that is determined by the contrast of the interference (in this case 0.6). G_T and G_F are the transimpedance gain and the band-pass filter gain respectively.

3. CALIBRATION OF THE SENSOR

The calibration of the sensor was carried out in an acoustic test bench in water (Figure 4). In this scenario the AE produced by PD can be reproduced, but without high voltage elements present in the set-up. This provides a controlled and safety environment for the calibration. The test bench includes an immersed calibrated hydrophone (Brüel & Kjaër 8103) and an external sensor (model: R15i-AST), which is commonly used in acoustic PD detection, mounted on the wall of the tank. The generation of the acoustic emission is done by using a transducer identical to the reference hydrophone.

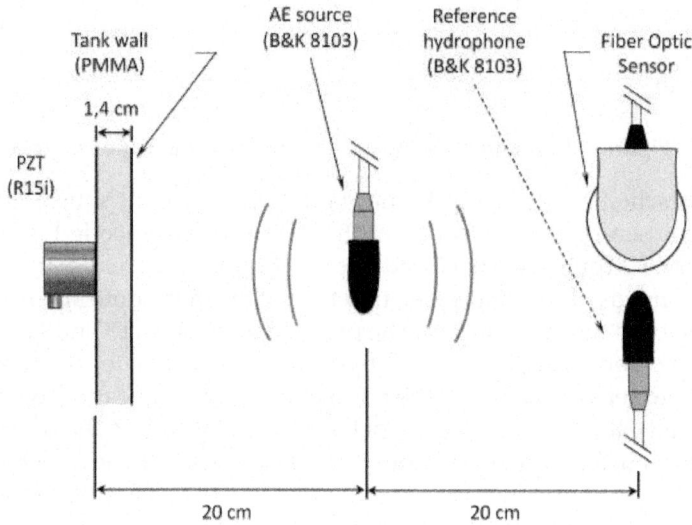

Figure 4. Experimental set-up for the calibration of the sensor in water.

3.1. Frequency Response of the Sensor and Sensitivity at 150 kHz

A comparative analysis of the frequency response between the internal FO sensor and the external R15i sensor was done in order to evaluate if the bandwidth of the fiber optic sensor is suitable for the application. In this test, a frequency sweep was applied from 50 kHz to 200 kHz: in the range of frequencies in which the hydrophone is calibrated. The results are shown in Figure 5(a).

The sensitivity obtained for the FO sensor at 150-kHz was 1.1-mrad/Pa (11 mV/Pa for 20-V output range in the interferometer). Moreover it has broader bandwidth (100 kHz–300 kHz were observed with a flat response). This is useful for PD identification because part of the emitted acoustic energy is found in that portion of the spectrum. The resultant noise equivalent bandwidth of the sensor is a drawback in this case.

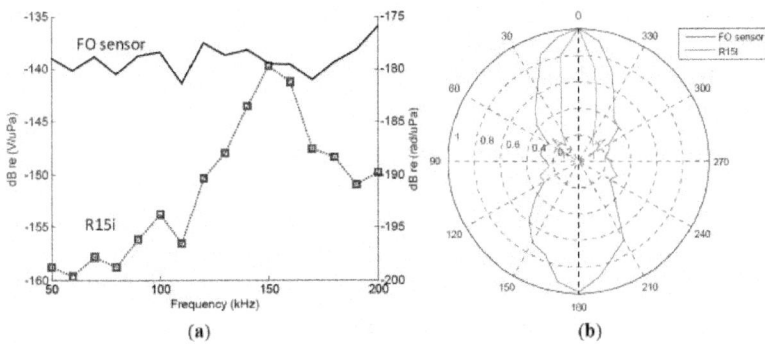

Figure 5. (a) Frequency response of the FO sensor; (b) Comparative analysis of the directivity of the external sensor and the fiber optic internal sensor at 150-kHz.

3.2. Directionality

The directional characteristics were characterized at 150-kHz and the results are shown in Figure 5(b). A wider detection field was found in the FO sensor. The directivity span is approximately ±30° compared to ±15° of the external sensor R15i. This result shows an advantage for PD detection with FO sensors over conventional external detection because wide zones should be inspected in large transformers.

4. TESTING THE FIBER OPTIC SENSOR WITH PD GENERATED IN TRANSFORMER OIL

A four layer sensor was manufactured and tested with real PD generated in transformer oil in order to test its performance in real conditions. In this test bench PD are generated with high voltage electrodes immersed in oil. Two different types were generated in order to produce representative PD: plane-plane electrodes for internal PD and needle-plane electrodes for surface PD (Figure 6(a)).

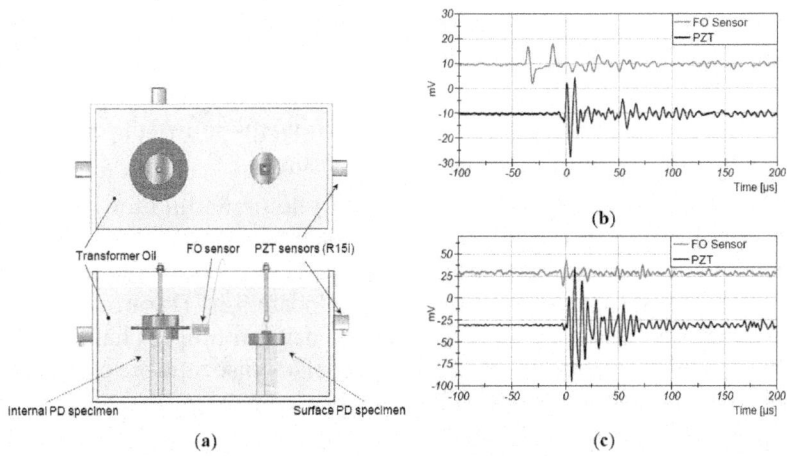

Figure 6. (a) High-voltage test bench of PD generation in oil; (b) Average of eight signals detected for internal PD; (c) Signals detected for surface PD.

The set-up was configured to detect with two sensors, internal (FO sensor) and external (R15i), at the same time. The FO sensor was located at 9-cm and 15-cm from the internal PD source and the surface PD source, respectively. The external sensor was located at 15 cm from the source in both cases.

The results of the tests are shown in Figure 6(b,c). The FO sensor shows suitable sensitivity to detect the acoustic emission of PD in both cases. The amplitude of the detected acoustic emissions is about ~1.3 Pa. Different waveforms are detected by the two sensors because their different frequency response. The broader bandwidth of the FO sensor produces shorter transient response than the PZT.

In addition, two peaks are detected by the FO sensor, which are delayed 23 μs. In this experiment, the acoustic absorber is removed from the middle of the coil. Therefore, the AE is detected by the first sensitive face and it travels to the second sensitive face where is also detected with a delay.

Based on previous works, if bare fiber is exposed for long time periods to the conditions inside transformers, it becomes tinted with the color of the oil. However, optical fibers withstand harsh environments and do not show permanent changes of the optical transmission and of the elasto-optic properties [16–18]. Hard coatings such as Teflon give better acoustic sensitivity, flat frequency response and thermal drift desensitization. Soft coatings such as acrylate based elastomers have narrower band-pass sensitivity to ultrasounds, which is desirable in transformers.

5. CONCLUSIONS

A fiber optic sensor for PD detection in transformer oil has been developed. It has been characterized in an acoustic test bench in water and also tested with real PD generated in oil. In the characterization a sensitivity of 1.1 m rad/Pa at 150-kHz (11 mV/Pa with 20 V of range at interferometer output) and a wider detection field (±30°) compared with an external sensor (±15°) has been shown. The test of the sensor with real PD shows that it has suitable sensitivity and enough resolution to detect acoustic pressure as low as 1.3 Pa.

It is a cost-effective solution compared with other fiber optic sensors which can also be immersed in oil. It offers the possibility of a multichannel configuration for an easy deployment of several sensors to locate the source of partial discharges through the time of flight of the acoustic emissions.

The optoelectronic scheme with homodyne demodulation that is used for the characterization of the FO sensor is practical for one channel application. However, a new optical interrogation scheme is needed for the implementation of multiple internal sensors (at least four) to locate the PD source. A heterodyne interferometer is proposed, which is able to drive multiple channels. With this technique, measurements with improved signal to noise ratio are expected because the measuring signal modulates a high frequency carrier.

Further work will be focused on the stability of the optical fiber sensor head and the AE sensitivity drift in controlled conditions of oil immersion and accelerated aging that emulates the exposition to the transformer environment for long time periods.

ACKNOWLEDGMENTS

This work was supported by the Spanish National Ministry of Science and Innovation, under the R&D project No. DPI 2009-14628-C03-01. PD tests have been made in collaboration with the LINEALT of University Carlos III de Madrid.

REFERENCES

1. Wang, Y.; Gong, S.; Grzybowski, S. Reliability evaluation method for oil-paper insulation in power transformers. *Energies* **2011**, *4*, 1362-1375.

2. Wang, M.; Vandermaar, A.J.; Srivastava, K.D. Review of condition assessment of power transformers in service. *IEEE Electr. Insul. Mag.* **2002**, *18*, 12-25.

3. Stone, G.C. Partial discharge diagnostics and electrical equipment insulation condition assessment. *IEEE Trans. Dielectr. Electr. Insul.* **2005**, *12*, 891-903.

4. Lundgaard, L.E. Partial discharge-part XIII: Acoustic partial discharge detection — Fundamental considerations. *IEEE Electr. Insul. Mag.* **1992**, *8*, 25-31.

5. Judd, M.D. Experience with UHF partial discharge detection and location in power transformers. In *Proceedings of the Electrical Insulation Conference (EIC)*, Annapolis, MD, USA, 5-8 June 2011; pp. 201-205.

6. Judd, M.D.; Hunter, I.B.B. Partial discharge monitoring for power transformer using UHF sensors. Part 2: Field experience. *IEEE Electr. Insul. Mag.* **2005**, *21*, 5-13.

7. Yu, B.; Kim, D.W.; Deng, J.; Xiao, H.; Wang, A. Fiber fabry-perot sensors for detection of partial discharges in power transformers. *Appl. Opt.* **2003**, *42*, 3241-3250.

8. Dong, B.; Han, M.; Sun, L.; Wang, J.; Wang, Y.; Wang, A. Sulfur hexafluoride-filled extrinsic Fabry-Pérot interferometric fiber-optic sensors for partial discharge detection in transformers. *IEEE Photonics Technol. Lett.* **2008**, *20*, 1566-1568.

9. Wang, X.; Li, B.; Xiao, Z.; Lee, S. H.; Roman, H.; Russo, O. L.; Chin, K. K.; Farmer, K. R. An ultra-sensitive optical MEMS sensor for partial discharge detection. *J. Micromech. Microengineering* **2005**, *15*, 521-527.

10. Lima, S.E.U.; Frazão, O.; Farias, R. G.; Araújo, F. M.; Ferreira, L. A.; Santos, J. L.; Miranda, V. Mandrel-based fiber-optic sensors for acoustic detection of partial discharges — A proof of concept. *IEEE Trans. Power Deliv.* **2010**, *25*, 2526-2534.

11. Zhiqiang, Z.; Macalpine, M.; Demokan, M.S. The directionality of an optical fiber high-frequency acoustic sensor for partial discharge detection and location. *J. Lightwave Technol.* **2000**, *18*, 795-806.

12. Macià, C.; Lamela Rivera, H.; Garcia-souto, J.A. Detection and wavelet analysis of partial discharges using an optical fibre interferometric sensor for high-power transformers. *J. Opt.*

13. Wild, G.; Hinckley, S. Acousto-ultrasonic optical fiber sensors: Overview and state-of-the-art. *IEEE Sens. J.* **2008**, *8*, 1184-1193.

14. IEEE Guide for the Detection and Location of Acoustic Emissions from Partial Discharges in Oil-Immersed Power Transformers and Reactors. IEEE Std C57.127™-2007, 31 August 2007.

15. Kirkendall, C.K.; Dandridge, A. Overview of high performance fibre-optic sensing. *J. Phys. D*

16. Lamela Rivera, H.; Garcia-Souto, J.A.; Sanz, J. Measurements of mechanical vibrations at magnetic cores of power transformers with fiber-optic interferometric intrinsic sensor. *IEEE J. Sel. Topics Quantum Electron.* **2000**, *6*, 788-797.

17. Biswas, D.R. Aging behavior of polyimide/acrylate-coated optical fibers in harsh environments. *Opt. Eng.* **1997**, *36*, 2169–2170.

18. Tae-Young, K.; Jong-Eun, K.; Kwang, S. On-line monitoring of transformer oil degradation based on fiber optic sensors. *Sens. Mater.* **2008**, *20*, 201–209.